POTATO ECOLOGY

AND MODELLING OF CROPS UNDER CONDITIONS LIMITING GROWTH

The C.T. de Wit Graduate School

PRODUCTION ECOLOGY

Current Issues in Production Ecology

VOLUME 3

The titles published in this series are listed at the end of this volume.

Potato Ecology

And modelling of crops under conditions limiting growth

Proceedings of the Second International Potato Modeling Conference, held in Wageningen 17–19 May, 1994

Edited by

A.J. HAVERKORT

DLO-Research Institute for Agrobiology and Soil Fertility, Wageningen, The Netherlands

and

D.K.L. MacKERRON

Scottish Crop Research Institute, Dundee, UK

Kluwer Academic Publishers

Dordrecht / Boston / London

Library of Congress Cataloging-in-Publication Data

Potato Ecology and Modelling of Crops under Conditions Limiting Growth
/ edited by A.J. Haverkort and D.K.L. MacKerron.
p. cm. -- (Current issues in production ecology ; v. 3)
Papers from a conference held in Wageningen, the Netherlands, May 1994.
ISBN 0-7923-3412-4 (hb : acid free paper)
1. Potatoes--Growth--Computer simulation--Congresses.
2. Potatoes--Physiology--Computer simulation--Congresses.
I. Haverkort, A. J. II. MacKerron, D. K. L. (Donald K. L.)
III. Series.
SB211.P8E338 1995
635'.21'011--dc20 95-7370

ISBN 0-7923-3412-4

Published by Kluwer Academic Publishers,
P.O. Box 17, 3300 AA Dordrecht, The Netherlands.

Kluwer Academic Publishers incorporates
the publishing programmes of Martinus Nijhoff,
Dr W. Junk, D. Reidel, and MTP Press.

Sold and distributed in the U.S.A. and Canada
by Kluwer Academic Publishers,
101 Philip Drive, Norwell, MA 02061, U.S.A.

In all other countries, sold and distributed
by Kluwer Academic Publishers Group,
P.O. Box 322, 3300 AH Dordrecht, The Netherlands.

Printed on acid-free paper

Contents

Preface

Potato is the fourth major staple food in the world and is still rapidly gaining importance, especially in the tropics. In May, 1994 the second international potato modelling conference was held in Wageningen, the Netherlands, as a summerschool of the C.T. de Wit Graduate School. The conference was sponsored by DLO, SCRI, SSCR, WAU and the LEB-Fund. Over 80 scientists participated, coming from 16 countries. Of each crop physiological and modelling subject, a leading scientist was requested to write a review of the most recent developments in his or her field. The reviews, with highlights from the authors' own work, are such that the physiological work described is of interest to the modeller and the modelling work to the crop physiologist. Applications of the quantitative approach are also reviewed in the concluding chapters that deal with decision support systems, breeding and agro-ecological zoning.

An outstanding point of this book is that both the crop ecology and the modelling of a broad range of biotic and abiotic factors are treated by scientists representing groups which are specialized in the subject. The two related disciplines met during the conference and thus wrote the chapters with each other's interest in mind. The book highlights the limitations for potato growth and development from the viewpoints of both the crop physiologist and the crop-systems analyst. The abiotic factors influencing potato growth include environmental factors such as temperature, solar radiation and carbon dioxide which increasingly affect potato crops in tropical climates and as a result from the global change. Air-borne diseases and pests of which the ecological effects on crop production are treated, are late blight and aphids and soil-borne organisms include potato wilting and potato cyst nematodes.

Although the potato crop is used as an example, this book highlights recent advances in knowledge of both the crop ecology and modelling of a field grown crop and as such is of interest to any scientist dealing with field grown arable and vegetable crops.

We hope that the book is relevant to those who apply quantitative approaches in crop physiology, crop ecology, breeding, decision support systems, modelling sustainability and agro-ecological zoning in agricultural production.

Anton J. Haverkort
AB-DLO, Wageningen, the Netherlands

Donald K.L. MacKerron
SCRI, Dundee, United Kingdom

1. Models in research and education, planning and practice

F.W.T. PENNING DE VRIES[1] and R. RABBINGE[2]

[1] *C.T. De Wit Graduate School Production Ecology and DLO-Research Institute for Agrobiology and Soil Fertility, Wageningen*

[2] *C.T. De Wit Graduate School Production Ecology and Department of Theoretical Production Ecology, Agricultural University, Wageningen*

Abstract. Models are tools in which knowledge about agricultural systems is integrated. The process of model building and the application of completed models are important. The models are used for the analysis of agricultural, biological and ecological systems, for prediction of future developments and for exploration of possible behaviour of such systems.

Phases of development of explanatory models are distinguished: preliminary models, comprehensive, summary models. The values of these phases for practical applications, for education and for science are markedly different. Models for exploration and prediction are different in nature than those for disciplinary research and analysis.

modelling has gained wide acceptance as a tool in research. It is also used education programs, and is a widely accepted instrument for policy making. Examples are given of models in research (for priority setting, development of science), in education, and of use for explorative studies (such as quantitative land evaluation).

While development of crop simulation models *per se* continues, their association with other quantitative techniques is already emerging. There is much potential for the combined use of models with Geographic Information Systems (GIS) and Linear Programming (LP) techniques for strategic planning, but current limitations of these tools are generally underplayed. User-friendly 'shells' are beginning to bring a wealth of possibilities for simulation and for analysis to many users. The number of applications could therefore grow exponentially.

As bottlenecks for applying models at a wide scale are recognised: scarcity of basic data, lack of data for evaluation of the models, hardware, and the number of trained scientists. Which of these factors limits progress most differs among countries. Scarcity of basic data and lack of trained staff often count most, particularly in developing countries.

Introduction

Building models is a way to integrate knowledge and to make it available for various purposes. Both process and product are important because they

- help to define and categorise the state of knowledge of the subject;
- help to set priorities for research, by helping to locate gaps in knowledge and to link scientists across disciplines, levels of aggregation and from fundamental and applied sciences;
- provide a means for disseminating knowledge;
- provide a tool to make integrated knowledge operational for policy making and for resource management.

A.J. Haverkort and D.K.L. MacKerron (eds.), Potato Ecology and Modelling of Crops under Conditions Limiting Growth, 1–18.

This introductory paper recalls briefly relevant definitions and concepts, and categorises models in relation to their uses. In this way, it provides a framework for the following chapters.

It is no coincidence that the various subjects in this volume are regarded from the experimental and also from the modelling side. Modelling has gained wide acceptance as a tool in research. It is also no longer restricted to the research environment, but has entered education programs, and is an instrument for policy makers. We will give some typical examples of use of models in research for priority setting and for development of scientific knowledge, of models in education, and of models used for quantitative land evaluation.

Finally, we address briefly new opportunities for linking crop models with Geographic Information Systems (GIS), Linear Programming (LP) and remote sensing (RS). Limitations for wide scale application of models are discussed, among which are prominent the lack of basic data and the shortage of scientists that are familiar with modelling approach and with uses of the available tools.

Concepts in Modelling

Systems

'One way of considering the real world is to divide it into systems. A functional description of a system is: a part of reality with strongly interacting elements, but with little influence on its environment. What part of the real world is singled out as a system depends first of all upon objectives. However, some elements of a system interact more than others, co-determining the shape of a system. Upon delimiting a system, one should thus take into account such natural contours, and consider all essential parts.' (Penning de Vries 1983). The boundaries of a system can move to expand or shrink the system under investigation in relation to the objective of the study. Ideally, boundaries are chosen such that the environment influences processes of the system, but the system itself does not influence its environment.

A theoretical delimitation of crop systems of growing crops was proposed by De Wit (De Wit and Penning de Vries 1982). He distinguished four levels of production, and corresponding systems (for their relational diagrams, see Penning de Vries et al. 1989):

- *Production level 1*: growth in conditions with ample soil water and nutrients. The crop growth rate is limited by weather conditions and amounts to 100–350 kg $ha^{-1}d^{-1}$. This situation is sometimes approached on the best farms and in glasshouses.
- *Production level 2*: growth is limited by shortage of water for part of the growing season. This occurs on well fertilized soils in semi-arid regions and in temperate climates, but is not common in natural ecosystems.
- *Production level 3*: production is limited by shortage of nutrients and

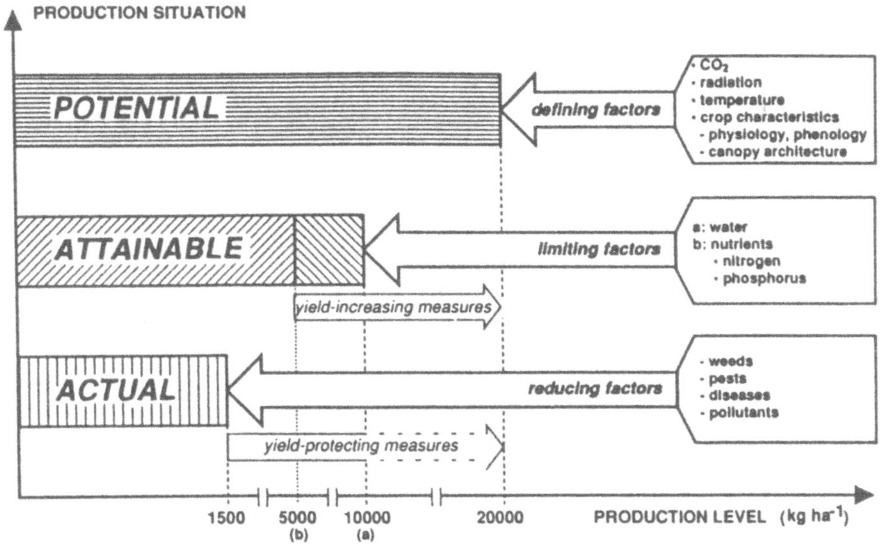

Figure 1. The relationship among potential:, attainable and actual yield and growth defining:, growth limiting and growth reducting factors (Rabbinge 1993).

sometimes by water. This is a common situation in agricultural systems using little fertilizer and in natuiral ecosystems.

- *Production level 4*: production is limited by availability of phospohorus or other minerals. Growth rates are 10–50 kg ha^{-1} d^{-1} during a short growing season. This situation occurs in heavily exploited areas where no fertilizer is used.

In practice, crop growth can be limited by several factors during the growing season, while pests may occur any time. It may also not be economical to attempt to reach potential yields. A practical, action oriented distinction has been proposed by Rabbinge (1993). He distinguished (Figure 1):

- *potential yield*, with the same definition as De Wit's Production level 1. Weather variables, including CO_2-level, and species characteristics are the yield defining factors. Modelling potential yield requires particularly the understanding of crop physiology and agrometeorology.
- *attainable yield*, roughly 50–20% below the potential yield; the reduction is caused by limiting factors such as water, nitrogen or phosphorus. Such limitations can largely be avoided by yield increasing measures (fertilizer, irrigation). Best farms attain this yield level. Soil physics and soil chemistry are important disciplines in the study of attainable yield, in additions to crop physiology and agrometeorology. 'Attainable yield' includes De Wit's production levels 2, 3 and 4.
- *actual yield*, roughly 50–0% below the attainable yield due to reducing factors, such as weeds, pests and pollutants. Yield a reduction can be avoided by crop protection measures such as integrated pest management. This situation is very common in many of the worlds agricultural crops.

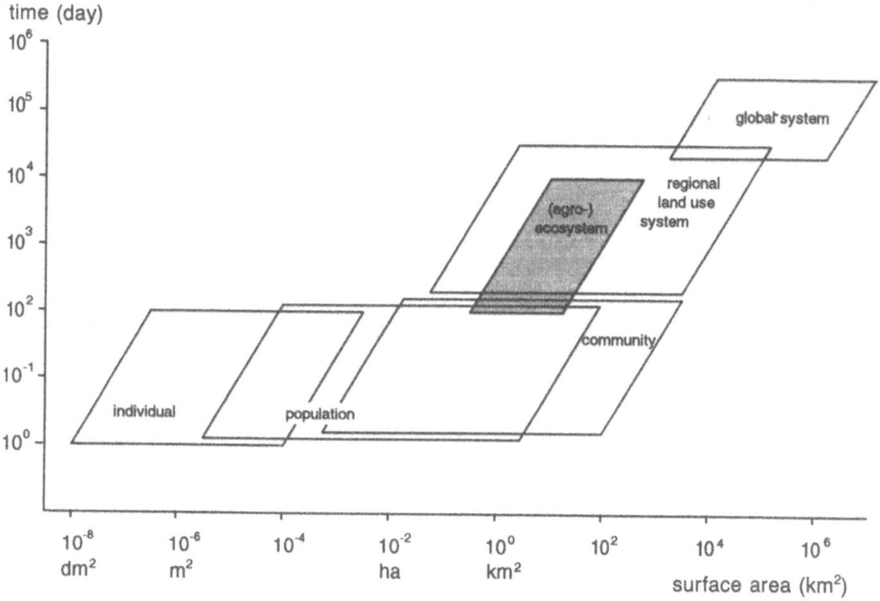

Figure 2. Temporal and spatial scales of units of analysis in production ecology. Source: Graduate School Production Ecology, Wageningen.

Systems of crop growth may also be categorised by hierarchical level. Typically at higher levels, the physical dimension ('size'), the time scale at which key processes react and progress (expressed as time coefficient), and the complexity of the system (expressed as the number of disciplines involved) increase. Figure 2 shows the scales commonly addressed in agricultural research. Important hierarchical levels are:

- the *plot level*: the system covers several square meters in surface area; processes have characteristic time coefficients of seconds to hours; many simulation studies address basic processes such as transport in soil or plants, interception, absorption of light and leaf photosynthesis; temporal changes of pests, etc.;
- the *field level*: the system covers around one hectare in surface area; processes have characteristic time coefficients of hours to days; studies address integration of basic processes to the field level; such as canopy photosynthesis, field evaporation, nutrient percolation, etc;
- the *farm level*: the system comprises several fields in one management unit, has a time horizon of several years and characteristic time coefficients of days to weeks; in addition to the processes at the field level, socio-economics is very important, as is management of machinery, stocks, personnel and cash;
- the *regional level*: the system comprises all land in a relatively uniform region of thousands of hectares, with time coefficients of months to years; socio-economics and limiting regional natural resources (e.g. water) are considered, possibilities and limitations are studies and mechanisms of change investigated.

Models

A *model* is a schematic representation of the conception of a system. In the real world, the appearance of biological systems changes, and their models are often called 'dynamic'. The most obvious change in a crop production system is growth. Analysis of dynamic systems is based on the assumption that the state of the system at any particular time can be expressed quantitatively, and that changes in the system can be described in mathematical terms (De Wit and Goudriaan 1974). This leads to formulation of state-determined models in which state variables, driving variables, rate variables and auxiliary variables are distinguished. State variables characterise and quantify the current state of the system, such as the amount of biomass, the leaf surface area, the N content of a part of the system etc. Their values change according to rules based on the underlying physiological, physical and biochemical processes. Driving variables, such as meteorological variables, characterise the influence of the environment. The formulation of the relations between all variables is called the 'structure' of a model. The 'behaviour' of a model is the total of the changes in the numerical values of variables during simulation. For a more extensive discussion of systems analysis and modelling, see Penning de Vries et al. (1989), Leffelaar (1993), and Goudriaan and Van Laar (1994).

Simulation is the scientific activity of building and utilising *explanatory models*. Models are called 'explanatory' when their behaviour is based on process knowledge at a lower hierarchical level. This contrasts with *demonstrative models*, whose behaviour resembles that of the real system, but the rules that make it behave in that way are quite different. For example, images of a film may look like the real world, but are only a demonstrative model of it. Explanatory and demonstrative models come in many forms: dynamic mathematical models, scale models, graphical models, electric analogy models. Each form has particular advantages and disadvantages. We focus at explanatory dynamic mathematical models: such models are flexible, have most potential for application and for further development.

In the process of *model development*, one passes from a conceptual phase, through a model construction phase to evaluation and application. Skills, data requirements, and techniques used are different in these phases. Ten different steps can be distinguished in the process of model development (Table 1). Typically, modelers go many times back and forth between these steps, and not always even reach the end. Indeed, in a long process, completion of each step may be seen as a small milestone. The first phase, developing a conceptual model, is often the most difficult one.

Uses of Models

Automobiles can be very useful when the type of the car matches the task: busses for moving people, trucks for bulk transport, and Formula 1's for racing. The

Table 1. Steps in systems research and model building. Source: Rabbinge and De Wit (1989)

Conceptual phase
1. formulation of objectives
2. definition of system boundaries
3. conceptualization of the system

Comprehensive modelling phase
4. quantification of input relations
5. model construction
6. model verification

Evaluation and application phase
7. model validation
8. sensitivity analysis
9. simplification: summary models, decision rules
10. feasibility and scenario studies

same applies to models: they are useful only when the proper type is selected for a specific task.

Models can be useful for development of science, for prediction and for instruction, but not all at the same time. Scientifically interesting models are often too detailed for application, while models for predictive or management purposes are often too trivial or too crude to challenge scientific interest. Three 'values' can be attributed to models (Table 2):

Table 2. The relative values of certain uses of models in different phases of development. Source: Penning de Vries (1983)

	Predictive value	Scientific value	Instructive value	Simplicity
preliminary model	+	+++	++	++
comprehensive model	++	+++	+	+
summary model	+++	+	+++	++

- the *scientific* value of a model expresses the extend to which it helps us to understand the real world, to integrate the relevant processes of the system, to bridge disciplines and aggregation levels, to evaluate alternative hypotheses, and to suggest experiments to falsify them;
- the *predictive value* of a model expresses the degree in which it simulates accurately the behaviour of a system. It measures the usefulness of the model as an instrument for application of knowledge in practice and for planning, and for explorative feasibility studies. The less detailed the desired results are, the simpler the predictive model can be. Simulation for different scenario's of the environment, such as for climate change, is a popular form of investigating possible consequences;
- the *instructive value* of a model refers to its use for disseminating knowledge

to graduate and post-graduate student, and to non-scientist end users such as extension services, farmers, policy makers. The model should convey the crucial behavioural aspects of an entire system in a transparent manner.

Evaluation

Evaluation is a broad term to describe the action of judging the value of a model. It comprises checking internal consistency and dimensions in computer programs, closed balances, comparison of model output with real world data, and judgement of practical utility. It is common experience that the behaviour of the model 'agrees' with that of experiments except for one of two aspects, which is 'corrected' by optimising the value of one or two parameters. This adjustment process, called *calibration*, should be avoided whenever possible, but is commonly used to tie simulation results to benchmark observations (e.g. Stol et al. 1992).

Evaluation of models remains often limited in depth as a result of too small a data base. Some models are only 'evaluated' by establishing a good correspondence between 'predicted' and 'observed' results, while the same observed results were used to derive constants in the model. This risks to reduce the explanatory nature of models. Strong experimentation is indispensable in parallel with modelling: experimentation at the explainable level for evaluation, and at the explanatory level for further improvement.

A source of concern is the brevity or inaccessibility of descriptions of many models and their programs. Too often, one is asked to believe an author without being able to check it. Some of the advantages of modelling are then lost. If documentation of a model is not rigorous, one must fear that also its evaluation was not thorough. There is a need for a standard of model quality.

Examples of simulation

Models in research

Explanatory simulation models are part and parcel of much of the quantitative research in agriculture and biology. Both preliminary and comprehensive models are common. The objective is to obtain a tool that explains the behaviour of the system. Generic models, such as SUCROS87 (Spitters et al. 1989), are often a starting point for their development. To account for specific processes (e.g. tillering in rice crops) or to target the models to specific objectives (e.g. understanding tuber formation dynamics in potato), research models are tailored to specific needs and circumstances. The procedure laid out in Table 1 applies very well to development of research models. Examples of such models are WHEAT for spring wheat (Van Keulen and Seligman 1987), and INTERCOM for crop-weed interactions (Kropff and Van Laar 1993).

Table 3. Categories of pest damage. Sources: Rabbinge (1983), Boote et al. (1983)

Category	Process	Example
assimilate sapper	tap assimilates	aphid
tissue consumer	removal tissue	insects
stand reducer	removal stem	stembore
photosynthetic rate reducer	damage chloroplast	diseases
leaf senecence accelerator	premature abcission	*Cercospora*
light stealer	shading	weeds
turgor reducer	modify water transfer	nematodes

Integration of disciplines leads to unifying principles. An example is the distinction of a limited number of categories of damage that the numerous species of insects and diseases can cause to crops, which is rooted in physiological mechanisms (Table 3).

Another example is the use of modelling for setting research priorities and for structuring of research activities. Simple models can already help to do so for broad conclusions, comprehensive models help to do so for specific hypotheses or knowledge gaps. For example, modelling identified the need for better knowledge of morphological development of crops and of maintenance respiration in order to make better predictive models (Spitters and Van Keulen 1990). However, these gaps have been found difficult to fill. In another case, use of models was more successful. Modelling studies hinted at the possibility that a sub-optimal N-content of leaves after flowering was a cause of sub optimal production in rice (Penning de Vries et al. 1991; Kropff et al. 1993). Yet, this was unexpected, since exhaustive experimentation two decades earlier provided the bases of the common fertilisation regime. Improving the post flowering N-status indeed raised the yield level significantly (Kropff et al. 1993). The cause was traced to a modified soil environment, in which the originally optimal fertilising regime had become sub-optimal. Stimulated by this result, several modelling studies are now ongoing at the International Rice Research Institute (Penning de Vries et al. 1991), and much research is organised in programs in which modelling provides the backbone (Kropff et al. 1994).

A third example is the analysis of causes of variability and risk in winterwheat due to aphids (Rossing et al. 1994). Careful mathematical and statistical analysis of field and laboratory data showed that uncertainty about the initial aphid population contributed most to uncertainty about the threshold level for spraying. They conclude that better field observations are needed more than better models or other types of data.

Models for instruction

Simple models are needed for instruction, and summary models are better than preliminary models. In addition to the contents, it is desirable that the models are packaged in an attractive, user-friendly form. Few products are in this stage, probably because for scientists there is little stimulus to derive summary models and for programmers it is a major job of restricted commercial value.

Simulation models in the simulation language CSMP, with clear textbooks and exercises, are a good form of instructive models (e.g. De Wit and Goudriaan 1974; Penning de Vries et al. 1989). They have lead to a wealth of models for different subjects and objectives. CSMP has been abandoned by its author (IBM), but a compiler has been developed (Van Kraalingen et al. 1994) by which CSMP-style models can be used in a flexible FORTRAN Simulation Environment (Van Kraalingen 1993). A package for computer aided instruction in crop simulation is now commercially available.

Models for prediction

'Prediction' comes in many forms, including straightforward and real time prediction of final crop yield during a growing season (e.g. De Koning et al. 1993), exploration of consequences of 'what if' questions on management (e.g. Keating et al. 1993), exploration of the impact of climate change (e.g. Rosenzweig 1993), and evaluation of new 'designs' of crop ideotypes (e.g. Dingkuhn et al. 1993). Special attention should be given to the matching of model and objective, to availability of data, and to evaluation of some results.

An example of use of models for prediction is in quantitative land use evaluation. Such models need much geographical referenced data on soils, climate, and sometimes on land use. Typically, output (crop production) and input (water, chemicals) are computed per soil-climate unit for one or more crops and weather patterns; outputs are expressed on maps, as averages or probabilities. A good example is a simulation study for India (Figure 3). In this vast country, consistent guidelines are sought as to where production of wheat should be stimulated, about optimum sowing dates, and what quantity of irrigation water would be needed (Aggarwal 1993). In a follow up of that study, consequences of inaccuracies in basic data (or of their absence) were quantified (Aggarwal 1994), thereby identifying the soil parameters (and to a smaller extend the crop parameters) of which field measurement would add most to accuracy of production redictions. In this way, the study helps agricultural planners and guides research services.

Since variability is more difficult to measure than to simulate, models are indispensable to quantify variability at the crop, cropping system, and regional level (Penning de Vries 1994).

For deriving solid advice to decision makers, crop models still have a number of weaknesses. It is the responsibility of scientists to cure them. Weak spots are:

- *poor quality control*. There is insufficient quality control of models, particularly when applied in practice. As a result, results may be wrong or irreproducible, and models are sometimes difficult to transfer to others. There is a strong need for guidelines for description (including range of validity), evaluation, documentation (scientific, technical), maintenance and distribution.

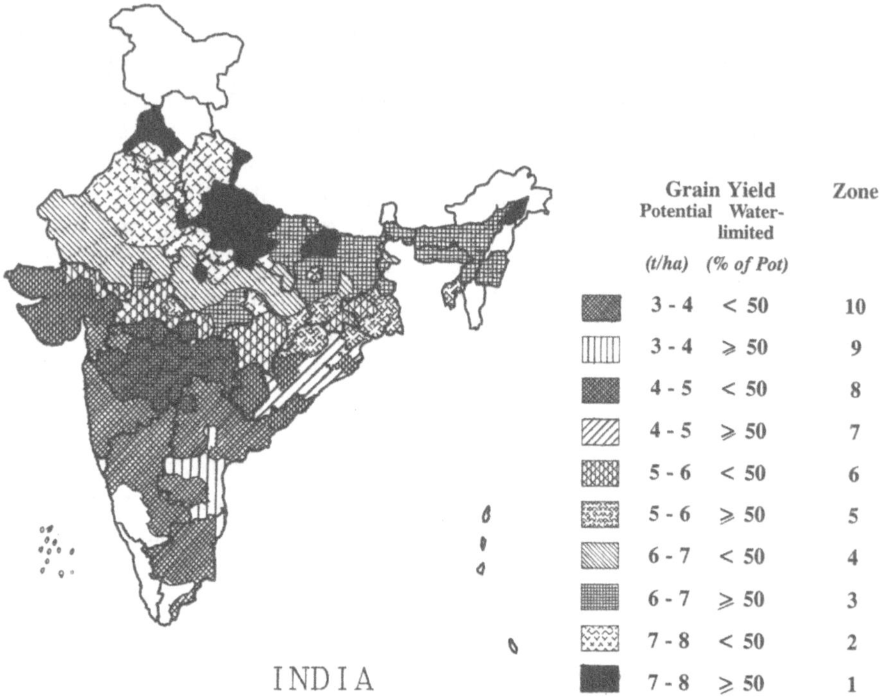

Figure 3. Iso-yield wheat zones of India based on potential and water limted productivity. Blank areas were not simulated. Source: Aggarwal (1993).

- *unknown accuracy*. Consequences of inaccuracies in basic inputs and uncertainties in model structure are not always determined, so that the precision of simulation results is unknown.
- *oversimplification*. Crop plants, pest populations and soils are not homogeneously distributed in space. Yet, this is rarely taken into account at plot, field or regional level. The consequences need investigation. Oversimplification of weather data (by averaging) was shown to cause significant deviations in simulated yields (Nonhebel 1993).

Future crop modelling

Geographic Information Systems

There is a large demand for predictions with models in a spatial context: for agro-ecological zoning (Aggarwal 1994), to quantify impacts of climate change, for regional yield prediction (De Koning et al. 1993) and land use feasibility and scenario studies (WRR 1992). The combination of GIS, that stores spatially distributed soil characteristics, data bases with weather data, and crop models is very powerful. Depending on the purpose, the basic soil, crop and weather data are observed values when explanation is sought, or hypothetical ones when new opportunities are sought. GIS-model combinations are ideal for exploring possibilities for land use, and for planning.

Yet, a warning is necessary:

- *basic data* in right formats are generally in short supply. Large scale studies generally must work with many approximate data. For weather data, large historic data sets of daily weather are scarce, and sometimes only commercially available. For soils data, international standardisation is an issue, and only few pedotransfer functions have been made (Bouma et al. 1993). A major quantitative and world-wide soil data base (SOTER: Oldeman 1994) will therefore not soon be available. For crop species and varieties, small data bases exist, but are scattered. The largest collections are for crop models of the Decision Support System for Agrotechnology Transfer (Uehara and Tsuji 1993) and for WOFOST (Boons et al. 1993).
- *aggregation* of simulation results at a plot or field level to the regional level has still unresolved questions, like how to deal with temporally and spatially varying inputs (e.g. sowing date across a region) and with uncertainty (Bouman et al. 1993), and how to deal with feed back between fields (e.g. pest populations, water storage).
- *evaluation* of simulated results is generally not possible in the traditional way (comparing simulated results with experimental data) simply because such experiments cannot be done and observations may be inadequate. E.g. in a recent study to predict regional crop yields for the Netherlands, we planned to compare simulated national crop yields with official statistics. Yet, we learned that official statistics are based on visual estimates, and have a relative uncertainty as large as that of the best simulations (around 5%, De Koning et al. 1993). In addition to testing as many model components as possible, new ways of evaluation ecoregional models must be developed. This is a necessity, because simulation without evaluation goes astray.

Multiple goal linear programming

Farms produce food and income to farmers, but can also cause air and ground water pollution, and provide employment. These intended and unintended 'outputs' can be seen as goal variables. The degree in which crop production contributes to these goals can be quantified in *technical coefficients*. Alternative cropping systems and farm management practices realise these goals to different degrees; technical coefficients link quantitatively every practice considered to every goal variable.

The actors involved (farmers, regional boards, governments) want to maximise/minimise these goals, but as these are often in conflict, a compromise must be sought. Modelling can provide such an instrument 'to help in negotiation' (R. Brinkman, FAO, pers. comm.). In recent years, a technique (called multiple goal linear programming) has been developed (De Wit et al. 1988) in which the production system can be optimised and the exchanges values

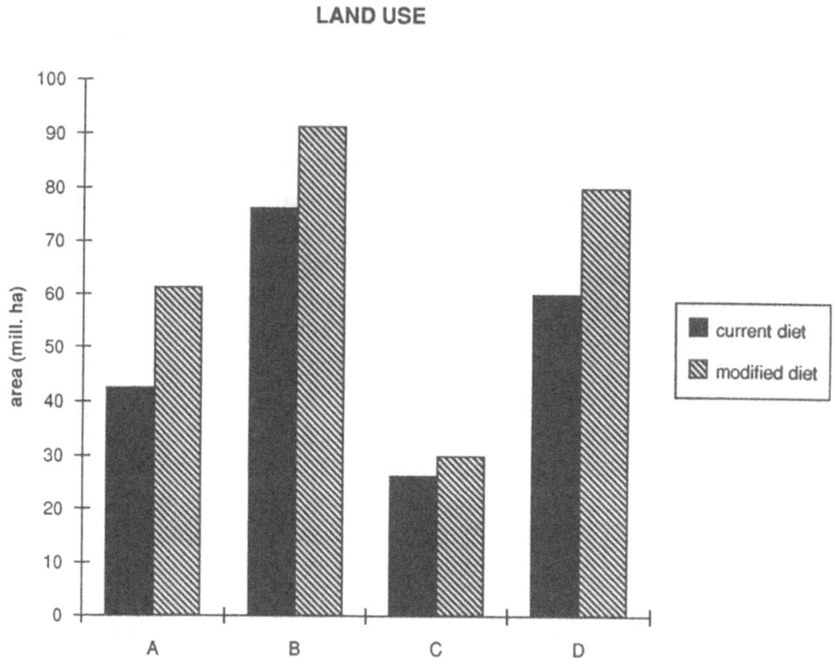

Figure 4. Differences in land use between four scenarios: A: free trade, B: regional development, C: nature and landscape, D: environmental protection. Source: WRR (1992).

can be determined between goal variables. With this technique, one can select from very many possible cropping systems and management practices the one that produces the best mix of a high income, low ground water pollution and little effort. Very important is that this technique necessitates collaboration with socio-economic sciences. In a large study at a regional scale, options for future agriculture and land use were quantified for many regions of the 12 EC-countries (WRR 1992). Optimisation towards main objectives of key actors in European agriculture showed how large the window is for policy making (Figure 4). Such modelling studies can help policy makers in tactical and strategic decisions.

This optimisation technique typically requires a large number of technical coefficients. Crop simulation will often be used to provide such data, as trials are expensive and time consuming. For instance Veeneklaas et al. (1990) investigated possibilities for future land use in a poor region in central Mali. Simulation models provided yields of millet, sorghum and range lands for areas that were inaccessible and for situations that could not be tested in trials (e.g. fertilisation). Expert judgement proved the behaviour of these models acceptable. Clearly, this land use planning study could not have occurred without crop models.

In another attempt, however, it appeared impossible to provide technical

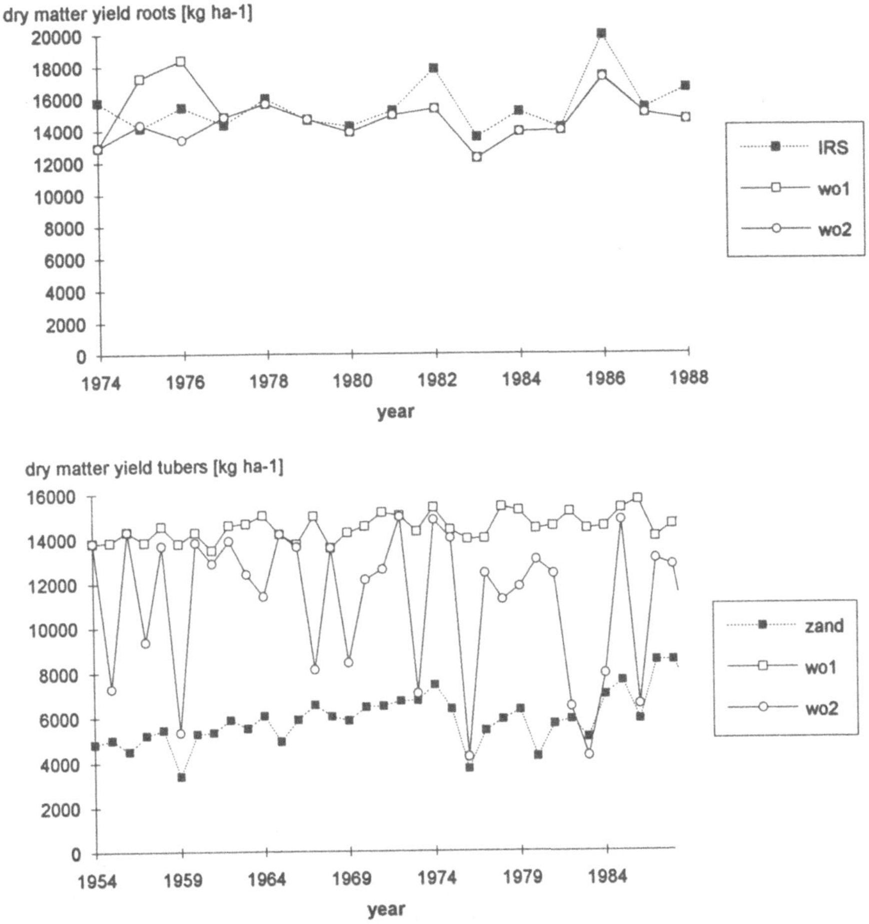

Figure 5. Experimental yields (black), simulated potential production (open squares) and simulated water limited production (open circles) of sugar beet roots (a) and potato tuber (b). Source: Habekotté (1994).

coefficients by modelling. In this case, a study for farms in the Netherlands, a coefficient was required for yield variability as function of crop and management. The generic model used did produce reasonable values for yield variability of some crops but not for others (Figure 5). This may be due to inaccuracy in the model (as not all crop and soil parameters were measured on the spot), but also because 'observed yields' have a significant inaccuracy. However, as variability was already approximately known and simulation could not improve its quality or provide extra data, this optimisation study is carried out with coefficients from field trials and expert judgement only (Schans and Habekotté 1995).

Though a very potent technique, we should not overlook the limitations. These are partially the same as we saw previously: lack of basic data, lack of data for evaluation of optimisation results. A specific limitation is that there is no procedure yet to translate uncertainty in basic data into uncertainty in end results. Particularly when crop models, GIS and LP are used all together, a number of the mentioned limitation apply simultaneously, and margins of error might well exceed 100%.

Remote Sensing

Yet another development is that crop models are used to interpret data collected by Remote Sensing. Certain ratios of specific wave lengths, per pixel, can be interpreted in terms of leaf area and canopy structure. Such values can serve to monitor the status of the crop in a region, or as inputs for a crop model to predict the end of season yield (Bouman 1992). In the near future, Remote Sensing will help to detect what crops grow where, provide data on the area under specific crops, and provide estimates of 'initial biomass' for growth simulation (Vossen, ISPRA, pers. comm.).

Userfriendliness

Over the past decade, some models have grown much in size, most grew in complexity. This can make them almost impossible to transport to other computer systems and other users. To ease such problems, 'shells' are introduced that reduce the need for technical expertise of the user and enable transportability. The IBSNAT project (Uehara and Tsuji 1993) has been among the first to develop a shell around its crop simulation models, called a Decision Support System for Agrotechnology Transfer (DSSAT), of which version 3.0 has been released. It provides a user-friendly environment to select models, to select sets of soil, crop and weather data, to specify crop management techniques, and to carry out simulations. DSSAT is widely used (Tsuji and Balas 1993). The Agricultural Production system Simulator (APSIM, McCown et al. 1995) is a modern shell, nearing completion. It contains several crop and soil models, whose structure is geared to cropping systems research. The models are accessible to users. For a group of models for rice based cropping systems, a first version of the 'SARP-shell' has been released (by the project 'Simulation and systems Analysis for Rice Production', Ten Berge 1993). In addition to easy selection of alternative crop and soil modules and corresponding data, users are free to modify all data and models, switching is possible between simulation of state variables or enforcing their values, and uncertainty analysis has been made easy.

Shells provide also interfaces with an increasing number of tools for inputs and outputs: weather data generation, GIS, statistical tools, uncertainty analysis, add-ons for economic analyses, etc. Much expansion is expected.

Training

One of the bottlenecks for applying models is the low number of skilled modelers. This is not solved by more user-friendly programs. For running a model, a limited amount of technical skill is needed, and this even decreases with the advent of more user-friendly models, data bases, and other software. However, effective use of these tools requires scientists trained to select the model best suited for the objectives, to recognise when sets of basic data are appropriate and complete, and to interpret often complex results. We conclude that while user-friendly tools bring modelling into the hands of more scientists, also more skilled modelers are needed to guide and help users. The Graduate School Production Ecology is one of the few locations where such training is provided. Another location is the Asian Institute of Technology. Particularly developing countries have to too few training centres. SARP (Penning de Vries et al. 1988) and IBSNAT (Uehara and Tsuji 1993) have been effective in training, but their courses still need institutionalisation (Singh et al. 1994). That is why concerted actions among various groups are being initiated.

Most development in modelling occurs in multidisciplinary groups of scientists of 4–6 scientists (cf. Penning de Vries 1983). Indeed, successful building of crop production models requires a 'critical mass' of at least 4–6 scientists from different but related disciplines, who collaborate closely for a long period. This concept was confirmed in a recent training and research project in Asia, where more than 18 modelling teams were formed and proved to be productive (Ten Berge 1993).

References

Aggarwal P K (1993) Agroecological zoning using crop growth simulation models: characterization of wheat environments of India. Pages 97–109 in Penning de Vries F W T, Teng P S, Metselaar K (Eds.) Systems Approaches for Sustainable Agriculture. Kluwer Academic Publishers, Dordrecht, Netherlands.

Aggarwal P K (1994) Uncertainties in plant, soil and weather inputs used in crop growth models: implications for simulated outputs and their applications. *Agricultural Systems* (in press).

Bastiaans L (1993) Understanding Yield Reduction in Rice Due to Leaf Blast. Ph.D. Thesis, Wageningen Agricultural University, Wageningen, The Netherlands.

Boons E R, De Koning G H J, Van Diepen C A, Penning de Vries F W T (1993) Crop-Specific Simulation Parameters for Yield Forecasting Across the European Community. Simulation Report CABO-TT 32, CABO, Wageningen, The Netherlands.

Boote, K J, Jones J W, Mishoe J W, Berger R D (1983) Coupling pests to crop growth simulators to predict yield reductions. Phytopathology 73:1581–1587.

Bouma, J, Wopereis M C S, Wösten J H M, Stein A (1993) Soil data for crop-soil models. Pages 207--220 in Penning de Vries F W T, Teng P S, Metselaar K (Eds.) Systems Approaches for Sustainable Agriculture. Kluwer Academic Publishers, Dordrecht, The Netherlands.

Bouman, B A M (1992) Linking physical remote sensing models with crop growth simulation models, applied for sugar beet. Int. J. Remote Sensing 13(14):2565–2581.

Bouman B A M, Jansen M J W (1993) RIGAUS. Random Input Generator for the Analysis of

Uncertainty in Simulation. Simulation Reports CABO-TT 34. CABO-DLO, Wageningen, The Netherlands.

De Wit C T, Goudriaan J (1974) Simulation of Ecological Processes. Simulation Monographs. PUDOC, Wageningen, The Netherlands.

De Wit C T, Penning de Vries F W T (1982) Pages 23–27 in La Productivité des Pâturages Sahéliens. Agricultural Research Report 918. PUDOC, Wageningen, The Netherlands.

De Wit C T, Van Keulen H, Seligman N G, Spharim I (1988) Application of interactive multiple goal programming techniques for analysis and planning of regional agricultural development. Agricultural Systems 26(3):211–230.

Dingkuhn M, Penning de Vries F W T, Miezan K M (1993) Improvement of rice plant type concepts: systems research enables interaction of physiology and breedings. Pages 19–36 in Penning de Vries F W T, Teng P S, Metselaar K (Eds.) Systems Approaches for Sustainable Agriculture. Kluwer Academic Publishers, Dordrecht, The Netherlands.

Goudriaan, J, Van Laar H H (1994) Modelling Potential Crop Growth Processes. Textbook with Exercises. Current Issues in Production Ecology, Vol. 2. Kluwer Academic Publishers, Dordrecht, The Netherlands.

Habekotté B (1994) Evaluatie van een Gewasgroeimodel voor Opbrengstberekening van Verschillende Gewassen. Simulation Reports CABO-TT 36 (24 pp). AB-DLO, P.O. Box 14, 6700 AA Wageningen, The Netherlands.

Keating B A, McCown R L, Wafula B M (1993) Adjustment of nitrogen inputs in response to a seasonal forecats in a region of high climatic risk. Pages 233–252 in Penning de Vries F W T, Teng P S, Metselaar K (Eds.) Systems Approaches for Sustainable Agriculture. Kluwer Academic Publishers, Dordrecht, The Netherlands.

Koning G H J, Jansen M J W, Boons-Prins E R, Van Diepen C A, Penning de Vries F W T (1993) Crop Growth Simulation and Statistical Validation for Regional Yield Forecasting Across the European Community. Simulation Report CABO-TT 31. CABO-DLO, Wageningen, The Netherlands.

Kropff M J, Van Laar H H (Eds.) (1993) Modelling Crop-Weed Interactions. CAB-international, Wallingford, UK.

Kropff M J, Cassman K G, Penning de Vries F W T, Van Laar H H (1993) Increasing the yield plateau in rice and the role of global climate change. J. Agric. Met. 48(5):795–798.

Kropff M J, Penning de Vries F W T, Teng P S (1994) Capacity building and human resource development for applying systems analysis in rice research. Pages 323–340 in Goldsworthy P, Penning de Vries F W T (Eds.) Opportunities, Use and Transer of Systems Research Methods in Agriculture to Developing Countries. Kluwer Academic Publishers, Dordrecht, The Netherlands.

Leffelaar P A (Ed.) (1993) On Systems Analysis and Simulation of Ecological Processes. Current Issues in Production Ecology, Vol. 1. Kluwer Academic Publishers, Dordrecht, The Netherlands.

McCown, R L, Hammer G L, Hargreaves J N G, Holzworth D P, Freebairn D M (1995) APSIM, a novel software system for model development, model testing, and simulation in agricultural systems research. Agric. Systems (in press).

Nonhebel S (1993) The Importance of Weather Data in Crop Growth Simulation Models and Assessment of Climate Change Effects. Ph.D. Thesis, Agricultural University, Wageningen, The Netherlands.

Oldeman L R (1994) Issues of development and management of databases. Pages 167–178 in Goldsworthy P, Penning de Vries F W T (Eds.) Oportunities, Use and Transer of Systems Research Methods in Agriculture to Developing Countries. Kluwer Academic Publishers, Dordrecht, The Netherlands.

Penning de Vries F W T (1983) Modelling of growth and production. Pages 118–150 in Lange O L, Nobel P S, Osmond C B, Ziegler H (Eds.) New Encyclopedia of Plant Physiology, New Series. Springer-Verlag, Berlin, Germany.

Penning de Vries F W T, Rabbinge R, Jansen D M, Bakema A (1988) Transfer of systems analysis and simulation in agriculture to developing countries. Agricultural Administration and Extension 29:85–96.

Penning de Vries F W T, Jansen D M, Ten Berge H F M, Bakema A H (1989) Ecophysiological Processes in Several Annual Crops. Simulation Monograph 29, PUDOC, Wageningen, The Netherlands and IRRI, Los Banos, Philippines.

Penning de Vries F W T, Kropff M J, Teng P S, Kirk G J D (Eds.) (1991) Systems Simulation at IRRI. IRRI Research Paper Series, 151, Los Banos, Philippines.

Penning de Vries F W T (1994) Computers, climate and risk. Pages 27–32 in Fifth International Congress for Computer Technology in Agriculture. Enabling Technologies for Land Use and Resource Management. Royal Agricultural Society of England, Warwickshire, UK.

Rabbinge R (1983) How to use combination models in crop protection? Pages 31–45 in Jeger M J (Ed.) The Development of Models for Practical Crop Protection. Bulletin IOBL/WPRS New Series VI, 2.

Rabbinge R (1993) The ecological background of food production. Pages 2–29 in Crop Protection and Sustainable Agriculture. CIBA Foundation Symposium, August 22–25. (Ed.) Crop Protection and Sustainable Agriculture. John Wiley & Sons, New York, USA.

Rabbinge R, De Wit C T (1989) Systems, models and simulation. Pages 3–15 in Rabbinge R, Ward S A, Van Laar H H (Eds.) Simulation and Systems Management in Crop Protection. Pudoc, Wageningen, The Netherlands.

Rosenzweig C (1993) Recent global assessments of crop responses to climate change. International Crop Science 1:265–272.

Rossing W A H, Daamen R A, Jansen M J W (1994) Undertainty analysis applied to supervised control of aphids and brown rust in winter wheat. II. Relative importance of different components of uncertainty. Agricultural Systems, 44:449–460.

Schans J, Habekotté B (1995) Optimization of cropping systems at the farm level (in prep.).

Singh G, Pathak B K, Penning de Vries F W T (1994) Requirements for systems research in agricultural and environmental sciences. Pages 255–266 in Goldsworthy P, Penning de Vries F W T (Eds.) Opportunities, Use and Transer of Systems Research Methods in Agriculture to Developing Countries. Kluwer Academic Publishers, Dordrecht, The Netherlands.

Spitters C J T, Van Keulen H, Van Kraalingen D W G (1989) Pages 147–177 in Rabbinge R, Ward S A, Van Laar H H (Eds.) Simulation and Systems Management in Crop Protection. PUDOC, Wageningen, The Netherlands.

Spitters C J T, Van Keulen H (1990) Toepassing van Simulatie en Systeemanalyse: Stand van Zaken en Knelpunten. CABO-Verslag 128, Centrum Agrobiologisch Onderzoek, Wageningen, The Netherlands.

Stol W, Rouse D I, Van Kraalingen D W G, Klepper O (1992) FSEOPT, a FORTRAN Program for Calibration and Uncertainty Analysis of Simulation Models. Simulation Report CABO-TT 24. CABO-DLO, Wageningen, The Netherlands.

Ten Berge H F M (1993) Building capacity for systems research at national agricultural research centres: SARP's experience. Pages 515–538 in Penning de Vries F W T, Teng P S, Metselaar K (Eds.) Systems Approaches for Sustainable Agriculture. Kluwer Academic Publishers, Dordrecht, The Netherlands.

Tsuji G Y, Balas S (1993) The IBSNAT Decade. Dept. Agronomy and Soil Science, College of Tropical Agriculture and Human Resources, University Hawaii, Honolulu, USA.

Uehara G, Tsuji G Y (1993) The IBSNAT project. Pages 505–514 in Penning de Vries F W T, Teng P S, Metselaar K (Eds.) Systems Approaches for Sustainable Agriculture. Kluwer Academic Publishers, Dordrecht, The Netherlands.

Van Keulen H, Seligman N G (1987) Simulation of Water Use, Nitrogen Nutrition and Growth of a Spring Wheat Crop. Simulation Monographs, PUDOC, Wageningen, The Netherlands.

Van Kraalingen, D W G (1993) The FSE System for Crop Simulation, Version 2.0. CABO-TPE Simulation Report 23. CABO-DLO, Wageningen, The Netherlands.

Van Kraalingen D W G, Rappoldt C, Van Laar H H (1994) The FORTRAN Simulation Translator (FST), a simulation language. In Goudriaan J, Van Laar H H (Eds.) Simulation of Crop Growth Processes. Kluwer Academic Publishers, Dordrecht, The Netherlands (in press).

Veeneklaas F R, Cissé S, Gosseye P A, Van Duivenbooden N, Van Keulen H (1990) Competition

for Limited Resources (Vol. 4). Centre for Agrobiological Research, P.O. Box 14, Wageningen, The Netherlands.
WRR (1992) Ground for Choices, four perspectives for the rural areas in the European Community. Netherlands Scientific Council for Government Policy, SDU-Publisher, The Hague, The Netherlands.

2. Crop physiology of potato (*Solanum tuberosum*): responses to photoperiod and temperature relevant to crop modelling

P.C. STRUIK[1] and E.E. EWING[2]

[1] *Department of Agronomy, Wageningen Agricultural University, Haarweg 333, 6709 RZ Wageningen, The Netherlands*

[2] *Department of Fruit and Vegetable Science, Cornell University, Ithaca, New York 14853–5908, USA*

Abstract. This chapter describes the complex effects of photoperiod and temperature on the behaviour of potato crops. Short days and low temperatures reduce branching and the number of leaves per stem, but increase the size of individual leaves. Short days also increase specific leaf area and photosynthesis per unit of leaf weight. High temperatures increase specific leaf area, but reduce photosynthesis and may shorten longevity of individual leaves.

Long days and high temperatures delay stolon and tuber initiation, delay and reduce partitioning of dry matter to the tubers and also reduce absolute maximum tuber bulking rates. These effects result in low harvest indices. A delay of tuber formation, however, may stimulate final yield when the growing season is sufficiently long to profit from the increased duration of ground cover.

Effects of photoperiod extension with full light are usually smaller than those with dim light. Short or partial exposure to low or high temperature may induce specific responses of the plant.

Introduction

Crop models for potato (*Solanum tuberosum* L.), like models for other crops, range from simple regression models that make no attempt to model morphology and physiology to highly detailed physiological models. Whether the model is simple or complex, any person developing a model for the potato crop should be aware of the effects of photoperiod and temperature on the growth and development of this plant. Photoperiod is a reliable, predictable climatic factor in any region, sensed by the leaves. Temperature is much more variable and unpredictable; it is sensed by each plant part. Both photoperiod and temperature affect the rate of accumulation of biomass; partitioning of biomass to leaves, stems, roots, stolons and tubers; specific leaf area; number and size of tubers; and rates of canopy development and crop senescence (hence per cent ground cover over time). These are variables that modelers must take into account, whether directly or indirectly.

Modelers would like to have precise relationships that would tell them the exact changes to be expected in the above variables in response to changes in photoperiod and temperature. This is of course impossible to provide, partly because photoperiod and temperature effects are strongly influenced by irradiance level, cultivar, N fertilization, and physiological condition of the seed

A.J. Haverkort and D.K.L. MacKerron (eds.), Potato Ecology and Modelling of Crops under Conditions Limiting Growth, 19–40.

tuber. Also, much information available has been obtained under highly controlled and thus artificial environments. The data from such experiments may provide insights into the qualitative nature of the changes to be expected, but information that can be used for quantitative relationships under field conditions typically is lacking.

A special problem is that physiologists have most often studied the effects of photoperiod by comparing short days with long days obtained through a dim light extension. The advantage of such a comparison is that the PAR (photosynthetically active radiation) is maintained constant among the photoperiod treatments; the disadvantage in terms of applying the information to crop modelling is that there often appears to be an exaggeration of the response to photoperiod when differences are obtained by dim light extension with an incandescent lamp (Wheeler and Tibbitts 1986).

This paper describes effects of photoperiod and temperature on organ development (number of stems, leaves, stolons and tubers), organ growth (size of stems, foliage, stolons and tubers), dry matter production and dry matter distribution. Effects of changes in photoperiod and temperature depend on their timing, duration and the localization of perception. We will, therefore, discuss:

a. effects of photoperiod and temperature during production and storage of seed tubers on behaviour of crops grown from these seed tubers;
b. effects of different levels of photoperiod and temperature throughout the growing period of a crop;
c. effects of short periods of changes in conditions;
d. effects of partial exposure to certain temperatures.

We will not go into details on the hormonal regulation of induction and initiation of stolons and tubers. For that topic we refer to recent reviews (Vreugdenhil and Struik 1989; Ewing 1990; Ewing and Struik 1992; Vreugdenhil and Helder 1992).

Before discussing the effects of photoperiod and temperature we will define plant parts and describe plant structure.

Definitions and plant structure

The usual method of propagation for the potato is to plant 'seed tubers', which produce stems and foliage and will produce the next generation of tubers. This next generation can be eaten, processed or used as seed tubers for a new, further generation. The growth vigour of the seed tubers and the physiological behaviour of the resulting plant strongly depend on the physiological age of the seed tubers (see e.g. Van der Zaag and Van Loon 1987).

There are many other methods of multiplication (Struik 1991), for example through true (botanical) potato seed, micro- or minitubers, *in vitro* plantlets and various sorts of cuttings or explants. Depending on the type of propagule used, the plant structure may vary. Number, shape and size of stems, leaves, stolons and tubers are all affected.

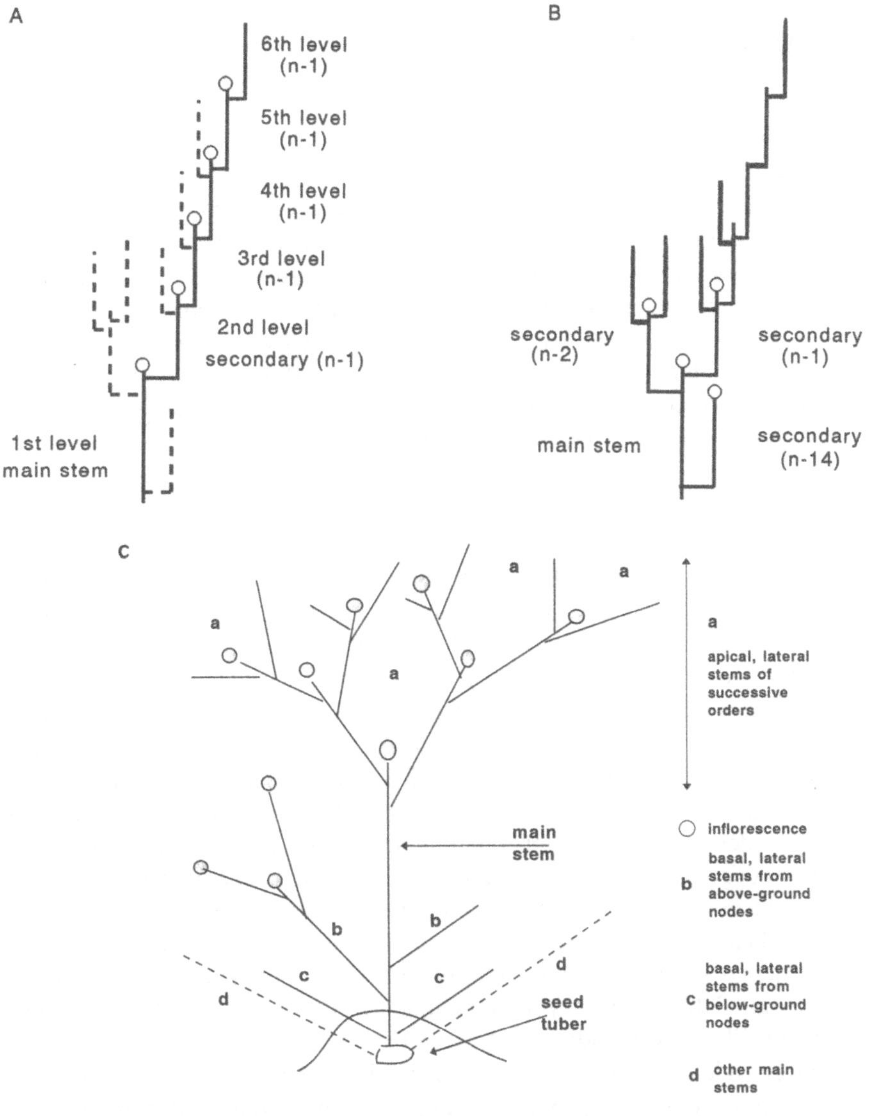

Figure 1. Structure of the above-ground parts of a potato plant. (A) A diagram of the sympodium of the potato plant, consisting of the main stem, and the lateral stems of different levels of growth, successively developing from the nodes n – 1. Each branch ends in an inflorescence depicted as a circle. After Almekinders and Struik (1994). (B) A simplified diagram of a shoot and its different types of secondary and higher order stems. Secondary stems are usually formed on nodes immediately below the inflorescence and on the lower nodes, whereas nodes in the middle of the main stem of the sympodium often do not form secondary stems. Tertiary and higher order stems are usually only formed on the secondary (n – 1) and (n – 2) stems, etc. After Almekinders and Struik (1994). The main stem may also branch on below-ground nodes. It is a matter of definition whether these below-ground branches should be called main stems or axillary branches. (C) A simplified diagram of the entire plant, including several types of shoots originating directly from the mother tuber or from a below-ground bud of another shoot. Revised after Vos (this volume).

This paper deals with the effects of photoperiod and temperature on the behaviour of plants grown from normal seed tubers with a physiological age allowing optimal growth. In some cases, however, we will refer to research involving structures other than whole plants from vigorous seed tubers.

Definitions of plant, stolon and tuber

A seed tuber may produce several (true) main stems and each of these may form a structure, consisting of several (types of) stems. When we refer to 'plant' we will mean the group of stems originating from one seed tuber.

Tubers are formed on stolons. Stolons are diageotropic stems with elongated internodes, usually hooked at the tip and bearing scale leaves (Cutter 1992). Usually they originate from the most basal stem nodes below soil level.

Tubers are greatly shortened and thickened stem parts, bearing scale leaves, each with a bud in its axil (Ewing and Struik 1992).

The shoot system

A potato plant is a cluster of true main stems (i.e. main stems directly originating from the seed tuber); each of these may develop into a complicated shoot structure. Each true main stem may develop diageotropic shoots (stolons) and below-ground branches (which may develop into normally functioning main stems) from buds that are below ground and above-ground branches from aerial buds. Above-ground shoots form a first level of vegetative development, terminating in a (primary) inflorescence. Secondary lateral stems develop from above-ground axillary buds on the main stem. They form a secondary level of vegetative development and terminate in secondary inflorescences. Continuation of shoot growth may lead to production of tertiary and higher levels of stems and inflorescences. After the main stem has formed an inflorescence, growth is usually taken over by the secondary stems developing from the nodes $n-1$ and $n-2$ (where n is the position of the last formed leaf) and, when secondary stems are terminated, by the tertiary stems from nodes $n-1$ and $n-2$ of the secondary stems, etc. Thus a sympodium is created (Figure 1A). Lateral stems may also develop from lower nodes of the main stem. The entire shoot thus consists of a constellation of the original main stem plus the apical and basal lateral stems and their branches (Figure 1B). The development of the shoot is strongly influenced by photoperiod and temperature (see below).

Since usually more than one true main stem arise from the seed tuber and since true main stems may form branches from below-ground nodes, each potato plant consists of a number of such constellations (Figure 1C).

The stolon system

Stolons are usually formed on the below-ground nodes of the stems. Stolonization starts at the nodes closest to the seed tuber and progresses

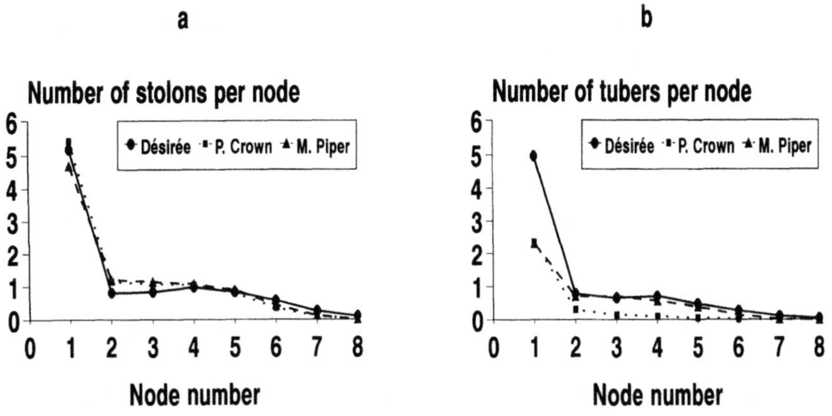

Figure 2. Relation between node number and number of stolons (2a) or tubers (2b) of cultivars Désirée, Pentland Crown and Maris Piper. Node 1 is the node closest to the seed tuber. Data are means of seven harvests. Based on data from Wurr (1977).

acropetally (Cutter 1992). The first stolons have a higher rate and longer duration of growth and branch more than those formed later (Struik and Van Voorst 1986). There tends to be more delay between stolon initiation and tuber initiation in the earlier formed stolons than in the later formed stolons (Vreugdenhil and Struik 1989). The number of stolons per node is (or seems) highest for the nodes closest to the seed tuber (see Figure 2a). The resulting structure of the stolon system, which is strongly influenced by photoperiod and temperature (Struik et al. 1990; 1991; see also below), determines the conditions the tubers encounter during their growth and the inter-tuber competition within one stem.

The tuber system

Developmental stages preceding and including tuberization may occur simultaneously upon the same stem (Vreugdenhil and Struik 1989). Typically most tubers that eventually grow to marketable sizes are initiated over a short time span.

An example of the distribution of tubers along the stem is given in Figure 2b. Position and number of tubers and possible interactions between them have been reported to depend on the structure of the stolon system, on other plant characteristics and on the micro-environment around each individual stolon tip or tuber (Struik et al. 1990; 1991).

Effects of photoperiod and temperature during production and storage of seed tubers on behaviour of crops grown from these seed tubers

Before seed tubers are planted, they are exposed to particular environmental conditions during their growth and storage. These conditions may have an effect on the behaviour of the seed tubers after planting or of the plants originating from them. Such effects require discussion.

Effects of conditions during growth of seed tubers

Van Ittersum (1992a) and Van Ittersum and Scholte (1992a) investigated the effects of conditions during tuber growth on subsequent physiological behaviour when seed tubers were stored under uniform conditions. Photoperiod during tuber growth only showed small, inconsistent effects on the behaviour of seed tubers during storage. The effects of very high day temperature were larger, but cultivar dependent. The number of sprouts shortly after breaking of dormancy was higher when temperature during tuber growth had been high. Such differences may soon disappear during prolonged storage.

Generally, the after-effects of environmental conditions during seed tuber growth on performance of the seed tuber when planted are absent or very small compared to the effects of conditions during storage and after planting; they are, therefore, not relevant to crop modelling.

Effects of conditions during storage of seed tubers

As soon as tubers become independent of an active shoot with green leaves, climatic factors may have a greater impact on their physiological development than during tuber bulking.

Information on the effects of photoperiod during storage on performance of seed tubers is scarce. Effects on dormancy are unlikely but, after dormancy is broken, photoperiod may have a significant effect. Both seed tubers and their sprouts are able to sense photoperiod when stored in light. Although exposure to diffuse light delays ageing of the seed tubers compared with storage in dark, Scholte (1989) showed that at 28 °C number and weight of sprouts were much lower at 12 h PAR plus 12 h dim light extension than at 12 h PAR. At 16 °C the effect of photoperiod was small.

Information on the effects of temperature during storage on the physiological ageing of seed tubers is abundant (for reviews see Van der Zaag and Van Loon 1987; Van Ittersum 1992b). Effects, however, are highly cultivar dependent. Dormancy can be shortened by warm storage, by a short heat treatment and in some cultivars also by a short cold treatment (Van Ittersum and Scholte 1992b). Such temperature treatments can also be used to advance growth vigour or increase stem number (Van Ittersum 1993). However, optimum temperatures for dormancy break and sprout growth differ. Moreover, growth vigour not only depends on the temperature sum during storage but also on the phasing of exposure to warm periods (Scholte 1987).

Storage conditions may also affect the number of leaves on the main stem before the inflorescence is produced and the number of tubers. Ageing sprouted tubers before planting resulted in more leaves before the first flower (Firman et al. 1991); and older seed tubers gave more tubers per stem (Van Loon 1987).

To the best of our knowledge no attempts have been made to include physiological age in potato growth models. Especially in areas with more than one growing season per year (and thus large differences in physiological age of the seed tubers used in different growing seasons) this could be highly relevant.

Effects of photoperiod and temperature after planting on haulm growth

Below 20 °C, lower soil temperatures delay and reduce sprout growth (Firman et al. 1992). Soil temperatures above 25 °C also delay emergence, and reduce plant survival and the number of main stems per plant (Sale 1979; Midmore 1984). After emergence, photoperiod and temperature together affect the development of the individual main stem by influencing the number of stems per shoot, the number of leaves per stem, the life span of the shoot and the leaf characteristics.

Effects on number of stems

The effects on number of lateral stems per sympodium and per entire shoot are illustrated in Table 1 for two contrasting cultivars. 'Atzimba' showed a much stronger response than 'Van Gogh', but there were some similarities. In 'Atzimba', long days stimulated the development of the sympodium, especially at the lower temperature, and the higher temperatures stimulated development at either daylength. In 'Van Gogh', the positive effect of long day was only significant at the lower temperature and the effect of higher temperature was positive in short days, but negative in long days. Results for the number of stems per entire shoot followed the same trends.

Effects on number of leaves and life span of the shoot

The number of leaves per stem is also affected by photoperiod and temperature. However, the effects are relatively small (Firman et al. 1991; Almekinders and Struik 1994) and in the case of photoperiod also inconsistent for the different levels of growth (Almekinders and Struik 1994), cultivars and experiments. Differences in pre-treatment of the seed tubers and in frequency of completion of a level of growth may cause these inconsistencies within and between plants. Figure 3 gives an example of the effects for an experiment with four temperature regimes and two photoperiods. In the first level of the main stems, the number of leaves was significantly higher at long days than at short days, but not in the secondary stems, except at the highest temperature. Effects of temperature were small in the lower temperature ranges, but were statistically significant on the first level of growth for 27 °C compared to lower temperatures. Higher order stems also showed hardly any photoperiod effect, but they showed a significant positive effect of temperature increase on number of leaves. Basal lateral stems responded similarly to the first level of the main stem.

The life span of the entire shoot is increased by long days. Effects of

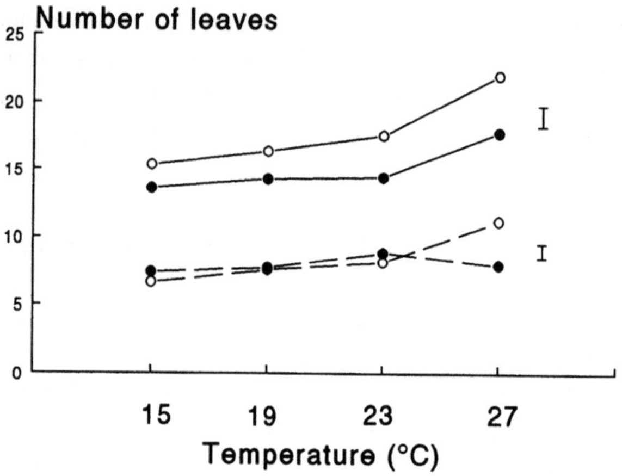

Figure 3. The effect of photoperiod (dim light extension) and temperature on the final number of leaves of the first level of the main stem (continuous lines) or of the secondary stem from node n – 1 (broken line). Closed circles: 12 h. Open circles: 18 h. Temperatures are averages of regimes with diurnal temperature fluctuations of 6 °C. Vertical bars indicate LSD (0.05). Data are averages for cultivars Spunta and Désirée. Only stems that completed their development were included in the analysis. After Almekinders and Struik (1994).

Table 1. Total number of stems (main stem plus lateral stems) with inflorescences per sympodium or per entire (single stem) shoot, as affected by photoperiod and temperature in cultivars Atzimba and Van Gogh. After Almekinders and Struik (1994).

Photoperiod (h)	12		16	
Temperature (°C)	20/10	30/20	20/10	30/20
Stems/sympodium				
'Atzimba'	3.2	4.6	3.8	4.8
'Van Gogh'	2.1	2.3	2.7	2.2
Stems/shoot				
'Atzimba'	14.0	27.1	27.0	33.1
'Van Gogh'	2.6	3.8	7.4	4.1

temperature on life span of the entire shoot are more complicated. Under long days high temperatures tend to increase the life span, but under short days high temperatures may decrease it. The development of the haulm is mainly affected through effects on the number of basal and apical lateral stems. Successful modelling of the effects of photoperiod and temperature must take account of these effects. However, the task is complicated by the large differences between cultivars.

Effects on leaf characteristics

Photoperiod hardly influences the rate of leaf appearance, but it affects individual leaf growth. It is generally accepted that short days stimulate pinna size (Edmundson 1941; Bodlaender 1963; Scaramella Petri 1963). Lorenzen and Ewing (1990) showed that leaf area increased rapidly under a 10-h (SD) or 18-h (LD) photoperiod of full light compared with a clear inhibition of short-term leaf growth at 10-h full light plus 8-h dim light (DE). Specific leaf area was highest for SD and lowest for LD, with DE being intermediate.

The effects of temperature are more complicated. The rate of leaf appearance shows a peak at a day-time temperature of 20 °C, irrespective of night temperature within a certain range (Borah and Milthorpe 1962). Final number of leaves per plant, however, may be highest at about 30 °C (Mosille 1985). Also, the optimum temperatures for rate and duration of individual leaf growth are low (25 °C according to Benoit et al. 1983), usually resulting in a reduced total leaf area at high temperature. Accurate data are lacking on the effects of temperature on the life span of individual leaves, but it is likely that optimum temperatures are low (see e.g. Burton 1972). Midmore and Prange (1991) showed for a large set of genotypes that the specific leaf area of plants grown under hot conditions was higher than under cool conditions.

Other effects

The optimum temperature for rate of stem elongation is above 25 °C (Borah and Milthorpe 1962), whereas the optimum temperatures for leaf characteristics are relatively low. Stem dry matter production has a higher optimum temperature than leaf dry matter production, resulting in a decline in the leaf/stem ratio at higher temperatures (Bodlaender 1963; Wheeler et al. 1986).

A special case of the effect of photoperiod on haulm growth has been reported by Cao and Tibbitts (1991). They reported stunting and chlorosis in some cultivars when grown under continuous light of high intensity. This effect could be controlled with daily temperature cycling (Tibbitts et al. 1990) and is therefore not relevant to crop modelling.

As a summary one can state that long photoperiods stimulate the development of haulm by increasing branching and number of leaves per stem, but reduce leaf size and specific leaf area. High temperature has an even greater positive effect on number of stems, number of leaves per stem and specific leaf area; but the optimum temperatures for size of individual leaves and perhaps also their longevity are relatively low.

Effects of photoperiod and temperature on photosynthesis

Short photoperiods increase the rate of dry matter production per unit of intercepted light (Lorenzen and Ewing 1990) and the rate of photosynthesis per unit leaf dry weight (Lorenzen and Ewing 1992). The latter effect was accompanied by more leaf starch accumulation during daytime (Lorenzen and Ewing 1992).

Temperature has a pronounced effect on photosynthesis in potato. Dwelle et al. (1981) found the optimum temperature for gross photosynthesis to be in the range 24–30 °C, whereas the optimum for net photosynthesis is not more than 25 °C (Winkler 1971; Ku et al. 1977; Dwelle et al. 1981). There is large variation in the photosynthetic response to heat among potato species and genotypes (Reynolds et al. 1990; Midmore and Prange 1991). Wolf et al. (1990) and Reynolds et al. (1990) showed that the photosynthetic system of potato has a high adaptive capacity to high temperatures, but Reynolds et al. (1990) concluded that low photosynthesis at very high temperatures was the result of accelerated senescence, chlorophyll loss, reduced stomatal conductance, and inhibition of dark reactions at high temperature.

Photosynthesis is believed to be higher following tuber initiation (Moll and Henninger 1978; Lorenzen and Ewing 1990), with the implication that photoperiod and temperature also influence photosynthesis through their effects on tuberization.

Effects of photoperiod and temperature on dry matter production

The effects of photoperiod and temperature on total dry matter yield result from their effects on haulm size, rate of net photosynthesis and leaf area duration, as previously described. These variables, however, interact with the physiology of tuber formation and dry matter partitioning.

The resultant effect of photoperiod and temperature strongly depends on the typical reaction of the *Solanum* species or cultivar in question (see the paper by Kooman and Haverkort, this volume).

In view of these effects it is clear that (although some induction is necessary to obtain the harvestable products) highest tuber yields are not found when tuber induction is strongest. This is confirmed by data of Wheeler et al. (1986), who showed that in the range 12–28 °C the harvest index increased with reducing temperatures both at 12- and 24-h PAR, whereas the optimal temperatures for both total dry matter yield and tuber dry matter yield were 20 °C at 12 h and 16 °C at 24 h.

Effects of photoperiod and temperature on dry matter partitioning

To obtain harvestable tubers the plants must be induced to tuberize, and must also be stimulated to partition part of the dry matter to the below-ground plant parts. Regulation of tuber initiation and of the shift in dry matter partitioning are partly independent. The shift in partitioning occurs even when the number of tubers is very low or when tuber growth is inhibited, for example by high soil temperature or exposure of the stolons to light. Excision of tubers or interference with translocation to the tubers does not stop the downward partitioning. In extreme cases aerial tubers or swollen stem parts and leaf petioles may be formed (Ewing and Struik 1992). Exposure to short days increased the rate of daytime assimilate transport within a few days of treatment initiation (J.H. Lorenzen and E.E. Ewing, unpublished). Ultimately, the harvest index is strongly reduced by long days (e.g. Bodlaender 1963).

Also, temperature has an immediate effect on assimilate partitioning. Sudden increases in temperature may even result in haulm growth being resumed after it had stopped. Basu and Minhas (1991) showed that in heat-sensitive cultivars high temperature resulted in more starch being accumulated in the leaves, and in a lower rate of assimilate transport from leaves, than in tolerant genotypes. They suggested that lack of transportable sucrose could be a major reason for low yields under high temperature and long days. High temperatures strongly reduce the harvest index (Marinus and Bodlaender 1975; Ben Khedher and Ewing 1985; Wheeler et al. 1986; Wolf et al. 1990).

We will use data from an experiment by Van Dam et al. (1995) to illustrate the effects of photoperiod and temperature on dry matter partitioning. In this experiment, cultivars Spunta and Désirée were grown in eight combinations of two photoperiods (12 and 18 h) and four temperatures (18/12, 22/16, 26/20 and 30/24 °C). During and shortly after tuber initiation, frequent harvests were carried out. Data on dry matter partitioning (five harvests of six plants per treatment) were used to fit the following curvilinear relationship:

$$f_{tuber} = f_{max}\{1 - \exp[-s(t - t_0)/f_{max}]\} \quad (1)$$

in which f_{tuber} = fraction of (newly produced) dry matter allocated to the tuber (g g^{-1}), f_{max} = maximum dry matter partitioning to tubers (g g^{-1}), s = relative partitioning rate (d^{-1}), t = number of days after planting (d), and t_0 = onset of tuber growth (d). It was shown that f_{max} was equal or very close to 1.0, which made it possible to simplify the equation. Table 2 shows the values of t_0 and s that could be assessed with adequate precision, based on early growth. Photoperiod extension delayed onset of tuber growth and reduced the relative partitioning of dry matter to the tubers. An increase in temperature also delayed tuber growth, especially in the upper range.

Wheeler et al. (1986) showed that as temperature increased, tubers contributed proportionally less to the total dry matter yield, both under 12-h and under 24-h PAR, while stems and leaves contributed more. However, they also showed that tuber yields are closely linked to the total quantity of photons obtained in each 24-h period, irrespective of the length of the light period. Long light periods encourage stem elongation and branching, but as long as sufficient photosynthates are available this effect does not reduce allocation of dry matter to the tubers, even though photoperiod is long.

Table 2. Onset of tuber growth t_0 (in d), relative partitioning rate s (in d^{-1}) and the fraction of variation explained (R^2; n = 30; in %) for 'Spunta' and 'Désirée' grown at six different photoperiod × temperature combinations. Calculations are based on Equation 1. Values for high temperatures and long days are not included because they could not be assessed reliably. Under these conditions tuber yields were very low, even after 102 days of culture. Photoperiod extension was by dim light. Temperatures are averages of regimes with diurnal temperature fluctuations of 6 °C. Data derived from Van Dam et al. (1995).

Cultivar	Photoperiod (h)	Temperature (°C)	t_0 (d)	s (d^{-1})	R^2 (%)
Spunta	12	15	36	0.090	89
		19	38	0.090	92
		23	42	0.055	84
		27	68	0.040	50
	18	15	44	0.086	94
		19	47	0.060	94
Désirée	12	15	33	0.085	86
		19	36	0.060	87
		23	40	0.045	98
		27	70	0.035	82
	18	15	40	0.051	91
		19	45	0.035	95

Attempts have been made to quantify partitioning for crop growth models on the basis of the cultivar-specific relation between temperature sum and partitioning values or harvest indices (Van Heemst 1986; Spitters 1990; Kooman and Haverkort, this volume), but it is questionable whether our current knowledge allows accurate quantification. Other factors, such as photoperiod, light intensity, physiological age of the seed tubers and nitrogen fertilization, play an important role. Moreover, cultivar differences in physiological behaviour and response to conditions during early phases of tuberization may be relevant.

Effects of photoperiod and temperature on initiation of stolons

The cultivars of *Solanum tuberosum* usually grown in Europe and other temperate regions do not require strong induction to form stolons. They may already form their first stolons before, or shortly after, plant emergence. During the early phases of plant development stolon initiation is stimulated by short days, but so is the switch from stolon initiation to tuberization. The resultant effect is that stolonization is stimulated when tuberization is delayed by long days. This is reflected in more stolons, longer stolons and more branches per stolon and, thus, in an increase in the number of potential tuber sites and changes in the competitive relations between potential tuber sites (Struik et al. 1988).

Ewing and Struik (1992), however, also indicated that when tuberization is inhibited by unfavourable conditions other than a long photoperiod, moderate shortening of the photoperiod may promote stolon production.

Initiation of stolons is strongly influenced by temperature. Midmore (1984) reported that stolon development was delayed but final stolon number was increased by higher temperature. Similar to the effects of long days, high temperatures (up to a certain optimum) stimulate the initiation of stolons, stolon growth and stolon branching, mainly by delaying tuberization (Bodlaender et al. 1964; Struik et al. 1989b). Such stimulation results in more stolons and more potential tuber sites per stolon (Struik et al. 1989a).

Effects of photoperiod and temperature on initiation of tubers

Initiation of tubers is favoured by days shorter than a critical photoperiod (Kopetz and Steineck 1954; Krug 1960a,b; Madec and Perennec 1962; Ewing 1978). This critical photoperiod is a heritable trait (Mendoza and Haynes 1977). At least ten gene loci on eight chromosomes have been mapped that control the trait in backcross populations obtained from hybridizing a diploid of *S. tuberosum* to a wild species, *S. berthaultii* (Van den Berg et al. 1993). After an adequate number of inductive cycles, the short-day signal is thought to trigger the production of a leaf factor that is one of the components of the stimulus to tuberize or it stimulates the transport of this factor to the below-ground plant parts (Struik et al. 1987; Ewing and Struik 1992; Helder et al. 1993). The critical photoperiod is shorter under warm conditions than under cool conditions (Snyder and Ewing 1989), i.e. shorter days are required for tuberization when the temperatures are high.

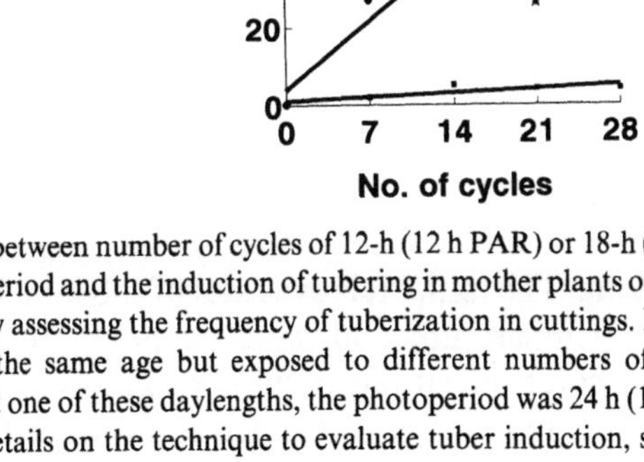

Figure 4. The relation between number of cycles of 12-h (12 h PAR) or 18-h (12 h PAR plus 6 h dim light extension) photoperiod and the induction of tubering in mother plants of cvs. Astarte and Rode Eersteling, measured by assessing the frequency of tuberization in cuttings. Leaf-bud cuttings were taken from plants of the same age but exposed to different numbers of cycles of 12 or 18 h photoperiod. When not one of these daylengths, the photoperiod was 24 h (12 h PAR plus 12 h dim light extension). For details on the technique to evaluate tuber induction, see Ewing (1985). Data from F. Havekes and P.C. Struik (unpublished).

When discussing the effects of photoperiod on tuber formation, distinction must be made between the effects of the length of each photoperiod in the cycle and the number of cycles to which the plant is exposed. Figure 4 illustrates this phenomenon in an experiment using cuttings of two cultivars differing in earliness. In both cultivars tuberization increased with an increase in the number of inductive cycles (12 h), whereas the response to an increase in the number of 18-h cycles was very weak.

Temperature modifies the response to photoperiod. High temperatures delay, impede or even inhibit tuber initiation. In particular, night temperature has a strong influence. Further, diurnal fluctuations are beneficial in some cultivars (Steward et al. 1981; Bennett et al. 1991).

Because of the effects of photoperiod and temperature on size and structure of the stolon system (and thereby on the number of potential tuber sites), number of tubers is usually increased by long days or high temperature although the growth rate of each individual tuber is low (Borah and Milthorpe 1962; Struik et al. 1988; 1989b). Other reports, however, indicate a low optimum temperature for number of tubers, especially under long days (Wheeler et al. 1986), or strong interactions between genotype and environment (e.g. Van Dam et al. 1995).

Long days and high temperatures also increase the within-plant variation in time of tuber initiation and in the rate and duration of tuber growth.

Effects of photoperiod and temperature on tuber growth

Thermal time functions for the onset and rates of tuber bulking have been developed both for temperate regions (Jefferies and MacKerron 1987; Spitters 1990) and for tropical conditions (Manrique and Hodges 1989). However, these relations are cultivar-specific and are affected by photoperiod (see also Kooman and Haverkort, this volume). To illustrate the effects of photoperiod and temperature on onset of tuber bulking, and on initial relative and linear tuber bulking rates we will use the data from Van Dam et al. (1995). Tuber yield data (5 harvests and 6 plants per harvest for each treatment) were analyzed using the formula:

$$Y = C_m/R_m \times \ln\{1 + \exp[R_m(t - t_b)]\} \qquad (2)$$

in which Y is tuber dry matter yield (g plant^{-1}), C_m is the maximum absolute rate of tuber bulking (g plant^{-1} d^{-1}), R_m is the initial maximum relative rate of tuber bulking (g g^{-1} d^{-1}), t is time (days), and t_b is time of onset of rapid tuber bulking (days).

Table 3 gives the values derived from literature or from the yield analysis for t_b, R_m and C_m, where possible. Short days and low temperatures advanced the onset of tuber bulking. The initial relative rates of tuber bulking were decreased by an increase in temperature and were supposed to be independent of photoperiod. The absolute linear rate of tuber bulking was reduced by very high temperatures and by long days if temperatures were high, although this interaction was difficult to quantify.

Table 3. Onset of rapid tuber bulking (days after planting; t_b), initial relative rate of tuber bulking ($g\ g^{-1}\ d^{-1}$; R_m; values derived from Ingram and McCloud 1984), and absolute maximum rate of tuber bulking in the linear phase ($g\ plant^{-1}\ d^{-1}$; C_m) for eight combinations of photoperiod and temperature and two cultivars. Calculations are based on Equation 2 with 30 data per treatment. Explained variation was between 89 and 98% in 9 out of 12 cases. Some values are missing, because they could not be assessed reliably. Photoperiod extension was by dim light. Based on Van Dam et al. (1995).

	t_b (d)		R_m ($g\ g^{-1}\ d^{-1}$)		C_m ($g\ plant^{-1}\ d^{-1}$)	
	Spunta	Désirée	Spunta	Désirée	Spunta	Désirée
12 h Photoperiod						
18/12 °C	58	55	0.38	0.38	4.7	3.6
22/16 °C	54	63	0.37	0.37	3.9	4.2
26/20 °C	65	68	0.34	0.34	3.7	3.4
30/24 °C	85	nr	0.32	0.32	1.8	nr
18 h Photoperiod						
18/12 °C	62	64	0.38	0.38	5.0	3.7
22/16 °C	67	nr	0.37	0.37	4.4	nr
26/22 °C	81	78	0.34	0.34	2.8	1.1
30/24 °C	nr	nr	0.32	0.32	nr	nr

nr = parameter could not be estimated reliably, because value approached the ending date of the experiment (t_b), or because growth rates were too low and yields were too variable among individual plants (R_m and C_m). In such cases the estimated t_b values were usually between 80 and 130, and actual growth rates were very small.

If tuber initiation took place under favourable conditions and photoperiod and temperature were only varied after some tuber growth had already taken place, the effects of long days were generally small and inconsistent (Van Ittersum 1992a), but effects of an increase in temperature were clearly negative over a large range of temperatures (Van Ittersum and Scholte 1992a).

Final tuber yield also depends on the effects of photoperiod and temperature on the duration of tuber bulking.

Diurnal fluctuations in temperatures provided better growth and tuber yield in some potato cultivars, but not in others (Bennett et al. 1991).

Effects of photoperiod and temperature on the physiology of the tuber

During tuber bulking photoperiod and especially temperature may influence processes in the tubers not directly reflected in changes in bulking rate. High temperatures stimulate synthesis of gibberellic acids, thus reducing the activity of ADP-glucosepyrophosphorylase and starch production. But there is also a direct effect: Mohabir and John (1988) found a very sharp optimum temperature of 21.5 °C for starch synthesis, whereas optimum temperatures for

ADP-glucosepyrophosphorylase and starch synthase were higher than 35 °C. The reasons for this low optimum temperature of starch synthesis are still unknown. The result, however, is that tubers grown under high temperatures will have a low starch content and an increased sugar content (Krauss and Marschner 1984). Associated with this phenomenon are the strong effects of heat on dry matter concentration (see e.g. Van den Berg et al. 1989). Haynes et al. (1988) showed that specific gravity (and thus dry matter content) was lower at higher temperatures but also observed a significant photoperiod × temperature interaction. Such effects are particularly relevant when fresh tuber yields are modeled.

Second growth and other physiological disorders such as heat sprouts, growth cracks, translucent ends, heat necrosis, etc. are enhanced by high temperatures (Lugt et al. 1964; Hiller et al. 1985; Van den Berg et al. 1989), especially when the tubers themselves are exposed to heat (Struik et al. 1989b). Second growth can also be obtained by lengthening the photoperiod after tuber induction in photoperiod sensitive cultivars (Madec and Perennec 1962).

Effects of partial exposure

Photoperiod is sensed by the leaves. Although the perception may be more intense in some leaves and partial exposure of the foliage may affect the response of the plant to photoperiod, this is not relevant to modelling. Temperature, however, is sensed by all plant organs and the effects of high or low temperatures on tuber yield may depend on the plant part(s) sensing them.

Tuber initiation is sensitive to temperature in different plant organs. Growing plants at high air temperatures gives poor tuberization in cuttings taken from these plants, independent of the soil temperature during plant growth (Reynolds and Ewing 1989). Exposing below-ground plant parts to high temperatures also inhibits tuber formation. When air temperatures are low, cuttings taken from these plants may tuberize, even when soil temperatures are high, but intact plants will not tuberize when stolons are exposed to heat, because the expression of the signal to tuberize can be blocked by high soil temperature (Reynolds and Ewing 1989). In fact, exposing plants to high soil temperatures causes about the same reduction in tuber yield as increasing air temperature (Menzel 1983).

Struik et al. (1989a,b,c) varied the temperatures in the zones of shoots, roots, and stolons plus tubers more or less independently of each other. Their results indicated that number of stolons may be stimulated by high root and stolon temperatures, the stolon growth may be enhanced by high shoot and stolon temperatures, and the diageotropic growth of stolons may be inhibited by high stolon and root temperatures. Earliness of induction to tuberize was particularly affected by shoot temperature, although the expression could be inhibited by a combination of high root and stolon temperatures. Numbers of tubers were much higher when stolon temperatures were high, but tuber yields were reduced by a temperature increase in any zone. Second growth was stimulated by any temperature increase, but an increase of the tuber temperature was especially effective. The dry matter partitioning to tubers was mainly influenced by air temperatures.

Soils can be cooled with mulches, resulting in earlier and higher tuber yields of better quality (Midmore 1984). The negative effects of high shoot temperatures may be reduced by selecting cultivars with leaf pubescence (Midmore and Mendoza 1984).

Exposing a single tuber to high or low temperatures affects its growth and the growth of the other tubers. Engels (1983) showed that cooling a single tuber reduced its growth rate, but increased the growth rates of other tubers. Krauss and Marschner (1984) indicated that subjecting single tubers to 30 °C for 6 days resulted in reduced growth and finally in a full stop of tuber growth but induced non-treated tubers (or tuber initials) (subjected to 20 °C) to increase their growth rates or to start bulking.

Effects of short periods of short days

Because plant size increases during early growth the timing of the number of inductive cycles is important. Time required for induction may be shorter when a plant has grown larger, because more leaf area is present to perceive the inductive cycles or to transmit their effects to the stolons (Kahn et al. 1983). Since short days induce a shift from shoot and stolon growth to tuber growth, a delay of induction may also result in more tubers and higher tuber yields. Such effects may be relevant when crops are grown in seasons with changes in daylength in the ranges in which significant responses may be expected.

Effects of short periods of high or low temperatures

Short periods of high temperatures can delay tuber formation or even temporarily arrest tuber growth. Unpublished data from P.C. Struik (Figure 5) showed that in Désirée, increasing temperature at around tuber initiation could result in a strong increase in number of tubers, although tuber yields were not affected. Consequently, the tuber size distribution was shifted towards the smaller sizes. In other experiments with similar treatments but other cultivars much smaller effects were obtained.

Tibbitts et al. (1994) found that warm temperatures during the first weeks of growth (i.e. before tuber initiation) followed by cooler temperatures increased tuber yield compared to maintaining cool temperatures during the entire growth period, probably because of the larger plant size. Delaying the warm period to later phases of growth reduced the tuber yield, whereas delaying a short cool period from early growth until after onset of tuber bulking greatly increased tuber yield.

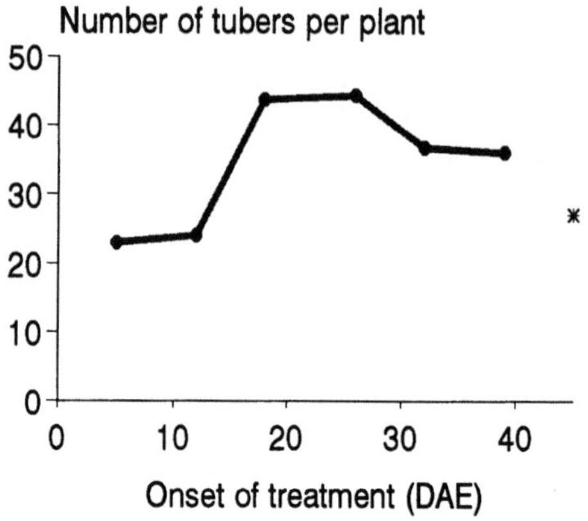

Figure 5. The relation between onset of a 7-day period of high temperature (32/27 °C instead of 20/15 °C) and number of tubers per plant in cv. Désirée. Data from P.C. Struik (unpublished). DAE = days after emergence. * = control (no warm treatment).

Burt (1964) showed that tuber initiation was advanced by 7-day periods of low temperatures in the range of 3 to 9 °C.

Conclusion

The qualitative effects of photoperiod and temperature on size of the haulm, its productivity, the timing of onset of tuber bulking, and rate and duration of tuber bulking are relatively well-known, although they are usually very dependent on cultivar. Quantitative information on the effects of photoperiod and temperature on the different aspects of tuber formation and their interactions is scarce, especially for conditions with high light intensity and for comparisons where photoperiod was prolonged with high rather than low irradiance levels. Because of limitations of many controlled environment facilities, even the full light intensity portions of the day have often been carried out under irradiance levels far lower than field conditions, especially in the tropics. Moreover, the potato plant is highly flexible in its physiological response to other environmental factors, which may interact with photoperiod and temperature. Insight is still lacking for more detailed models, that would enable us to predict or describe the effects of fluctuations in temperature or the effects of photoperiod and temperature on important characteristics such as number, average size and size distribution of tubers.

References

Almekinders C J M, Struik P C (1994) Photothermal response of sympodium development and flowering in potato (*Solanum tuberosum* L.) under controlled conditions. Netherlands Journal of Agricultural Science 42:311–329.

Basu P S, Minhas J S (1991) Heat tolerance and assimilate transport in different potato genotypes. Journal of Experimental Botany 42:861–866.

Ben Khedher M, Ewing E E (1985) Growth analyses of eleven potato cultivars grown in the greenhouse under long photoperiods with and without heat stress. American Potato Journal 62:537–554.

Bennett S M, Tibbitts T W, Cao W (1991) Diurnal temperature fluctuation effects on potatoes grown with 12 hr photoperiods. American Potato Journal 68:81–86.

Benoit G R, Stanley C D, Grant W J, Torrey D B (1983) Potato top growth as influenced by temperatures. American Potato Journal 60:489–501.

Bodlaender K B A (1963) Influence of temperature, radiation, and photoperiod on development and yield. Pages 199–210 in Ivins J D, Milthorpe F L (Eds.) Growth of the Potato. Butterworths, London, UK.

Bodlaender K B A, Lugt C, Marinus J (1964) The induction of second-growth in potato tubers. European Potato Journal 7:57–71.

Borah M N, Milthorpe F L (1962) Growth of the potato as influenced by temperature. Indian Journal of Plant Physiology 5:53–72.

Burt R L (1964) Influence of short periods of low temperature on tuber initiation in the potato. European Potato Journal 7:197–208.

Burton W G (1972) The response of the potato plant to temperature. Pages 217–233 in Rees A R, Kockahull K E, Hand D W, Hurd G R (Eds.) The Crop Processes in Controlled Environments. Academic Press, London, UK.

Cao W, Tibbitts T W (1991) Physiological response in potato plants under continuous irradiation. Journal of the American Society for Horticultural Science 116:525–527.

Cutter E G (1992) Structure and development of the potato plant. Pages 65–161 in Harris P M (Ed.) The Potato Crop. 2nd Edition. Chapman & Hall, London, UK.

Dwelle R B, Kleinkopf G E, Pavek J J (1981) Stomatal conductance and gross photosynthesis of potato (*Solanum tuberosum* L.) as influenced by irradiance, temperature and growth stage. Potato Research 24:49–59.

Edmundson W C (1941) Response of several varieties of potatoes to different photoperiods. American Potato Journal 18:100–112.

Engels Ch (1983) Wachstumsrate der Knollen von *Solanum tuberosum* Var. Ostara in Abhängigkeit von exogenen und endogenen Faktoren – Konkurrenz zwischen Einzelknollen um Assimilate –. Dissertation der Fakultät Pflanzenproduktion der Universität Hohenheim Stuttgart-Hohenheim, Germany.

Ewing E E (1978) Critical photoperiods for tuberization: a screening technique with potato cuttings. American Potato Journal 55:43–53.

Ewing E E (1985) Cuttings as simplified models of the potato plant. Pages 154–207 in Li P H (Ed.) Potato Physiology. Academic Press Inc., Orlando, USA.

Ewing E E (1990) Induction of tuberization in potato. Pages 25–41 in Vayda M E, Park W D (Eds.) The Molecular and Cellular Biology of the Potato. CAB International, Redwood Press, Melksham, UK.

Ewing E E, Struik P C (1992) Tuber formation in potato: induction, initiation, and growth. Horticultural Reviews 14:89–198.

Firman D M, O'Brien P J, Allen E J (1991) Leaf and flower initiation in potato (*Solanum tuberosum*) sprouts and stems in relation to number of nodes and tuber initiation. Journal of Agricultural Science, Cambridge 117:61–74.

Firman D M, O'Brien P J, Allen E J (1992) Predicting the emergence of potato sprouts. Journal of Agricultural Science, Cambridge 118:55–61.

Haynes K G, Haynes F L, Swallow W H (1988) Temperature and photoperiod effects on tuber production and specific gravity in diploid potatoes. HortScience 23:562–565.

Helder J, Miersch O, Vreugdenhil D, Sembdner G (1993) Occurrence of hydroxylated jasmonic acids in leaflets of *Solanum demissum* plants grown under long- and short-day conditions. Physiologia Plantarum 88:647–653.

Hiller L K, Koller D C, Thornton R E (1985) Physiological disorders of potato tubers. Pages 389–455 in Li P H (Ed.) Potato Physiology, Academic Press, Orlando, USA.

Ingram K T, McCloud D E (1984) Simulation of crop growth and development. Crop Science 24:21–27.

Jefferies R H, MacKerron D K L (1987) Thermal time as a non-destructive method of estimating tuber initiation in potatoes. Journal of Agricultural Science, Cambridge 108:249–252.

Kahn B A, Ewing E E, Senesac A H (1983) Effects of leaf age, leaf area, and other factors on tuberization of cuttings from induced potato (*Solanum tuberosum*) shoots. Canadian Journal of Botany 61:3193–3201.

Kooman P L, Haverkort A J (1995) Modelling development and growth of the potato crop influenced by temperature and daylength: LINTUL-POTATO. Pages 41–60 in Haverkort A J, MacKerron D K L (Eds.) Ecology and Modelling of Potato Crops under Conditions Limiting Growth. Kluwer Academic Publishers, Dordrecht, The Netherlands.

Kopetz L M, Steineck O (1954) Photoperiodische Untersuchungen an Kartoffelsämlingen. Züchter 24:69–77.

Krauss A, Marschner H (1984) Growth rate and carbohydrate metabolism of potato tubers exposed to high temperatures. Potato Research 27:297–303.

Krug H (1960a) Zum photoperiodischen Verhalten einiger Kartoffelsorten. I. Teil. European Potato Journal 3:47–79.

Krug H (1960b) Zum photoperiodischen Verhalten einiger Kartoffelsorten. II. Teil. European Potato Journal 3:107–136.

Ku S B, Edwards G E, Tanner C B (1977) Effects of light, carbon dioxide, and temperature on photosynthesis, oxygen inhibition of photosynthesis, and transpiration in *Solanum tuberosum*. Plant Physiology 59:868–872.

Lorenzen J H, Ewing E E (1990) Changes in tuberization and assimilate partitioning in potato (*Solanum tuberosum*) during the first 18 days of photoperiod treatment. Annals of Botany 66:457–464.

Lorenzen J H, Ewing E E (1992) Starch accumulation in potato (*Solanum tuberosum*) leaves during the first 18 days of photoperiod treatment. Annals of Botany 69:481–486.

Lugt C, Bodlaender K B A, Goodijk G (1964) Observations on the induction of second-growth in potato tubers. European Potato Journal 7:219–227.

Madec P, Perennec P (1962) Les relations entre l'induction de la tubérisation et la croissance chez la plante de pomme de terre (*Solanum tuberosum* L.). Annales de Physiologie Végétale 4:5–84.

Manrique L A, Hodges T (1989) Estimation of tuber initiation in potatoes grown in tropical environments based on different methods of computing thermal time. American Potato Journal 66:425–436.

Marinus J, Bodlaender K B A (1975) Response of some potato varieties to temperature. Potato Research 18:189–204.

Mendoza H A, Haynes F L (1977) Heritance of tuber initiation in tuber bearing *Solanum* as influenced by photoperiod. American Potato Journal 54:243–252.

Menzel C M (1983) Tuberization in potato (*Solanum tuberosum* cultivar Sebago) at high temperatures: interaction between shoot and root temperatures. Annals of Botany 52:65–70.

Midmore D J (1984) Potato (*Solanum* spp.) in the hot tropics. 1. Soil temperature effects on emergence, plant development and yield. Field Crops Research 8:255–271.

Midmore D J, Mendoza H A (1984) Improving adaptation of the potato (*Solanum* spp.) to hot climates – some physiological considerations. Pages 457–464 in Proc. 6th Symposium of the International Society of Tropical Root Crops, International Potato Center, Lima, Peru.

Midmore D J, Prange R K (1991) Sources of heat tolerance amongst potato cultivars, breeding lines, and *Solanum* species. Euphytica 55:235–245.

Mohabir G, John P (1988) Effect of temperature on starch synthesis in potato tuber tissue and in amyloplasts. Plant Physiology 88:1222–1228.

Moll A, Henninger W (1978) Genotypische Photosyntheserate von Kartoffeln und ihre mögliche Rolle für die Ertragsbildung. Photosynthetica 12:51–61.

Mosille E A W (1985) Der Einfluss verschiedener Kombinationen von Wurzelraumtemperaturen und Tageslängen auf Wachstum und Ertrag der Kartoffelpflanze (*Solanum tuberosum* L.). Thesis Institut für Nutzpflanzenforschung - Acker- und Pflanzenbau - der Technischen Universität Berlin, Berlin, Germany.

Reynolds M P, Ewing E E (1989) Effects of high air and soil temperature stress on growth and tuberization in *Solanum tuberosum*. Annals of Botany 64:241-247.

Reynolds M P, Ewing E E, Owens T G (1990) Photosynthesis at high temperature in tuber-bearing *Solanum* species. Plant Physiology 93:791-797.

Sale P M J (1979) Growth of potatoes (*Solanum tuberosum* L.) to the small tuber stage as related to soil temperature. Australian Journal of Agricultural Research 30:667-675.

Scaramella Petri P (1963) L'influence de la température sur la morphologie de la pomme de terre. European Potato Journal 6:242-257.

Scholte K (1987) Relation between storage T sum and vigour of seed potatoes. Pages 28-29 in EAPR Abstracts of Conference Papers and Posters, 10th Triennial Conference of the European Association for Potato Research, Aalborg, Denmark.

Scholte K (1989) Effect of daylength and temperature during storage in light on growth vigour of seed potatoes. Potato Research 32:214-215 (Abstr.).

Snyder R G, Ewing E E (1989) Interactive effects of temperature, photoperiod, and cultivar on tuberization of potato cuttings. HortScience 24:336-338.

Spitters C J T (1990) Crop growth models: their usefulness and limitations. In Dekker P H M, Booij R (Eds.) Proceedings of the 6th Symposium on the Timing of Field Production of Vegetables, Wageningen, August 1989. Acta Horticulturae 267: 349-368.

Steward F C, Moreno U, Roca W M (1981) Growth, form and composition of potato plants as affected by environment. Annals of Botany 48 (Supplement 2):1-45.

Struik P C (1991) Plant tissue culture techniques (with special reference to the potato plant). Pages 66-97 in Chapter 4 of Modern Biotechnology in Food Production. Open University & Thomas Polytechnic. Butterworth-Heinemann, Oxford, UK.

Struik P C, Boon E J, Vreugdenhil D (1987) Effects of extracellular extracts from leaves on the tuberization of cuttings of potato (*Solanum tuberosum* L.). Plant Physiology 84:214-217.

Struik P C, Geertsema J, Custers C H M G (1989a) Effects of shoot, root and stolon temperature on the development of the potato (*Solanum tuberosum* L.) plant. II. Development of stolons. Potato Research 32:143-150.

Struik P C, Geertsema J, Custers C H M G (1989b) Effects of shoot, root and stolon temperature on the development of the potato (*Solanum tuberosum* L.) plant. III. Development of tubers. Potato Research 32:151-158.

Struik P C, Geertsema J, Custers C H M G (1989c) Effects of shoot, root and stolon temperature on the development of the potato (*Solanum tuberosum* L.) plant. I. Development of the haulm. Potato Research 32:133-141.

Struik P C, Haverkort A J, Vreugdenhil D, Bus C B, Dankert R (1990) Manipulation of tuber-size distribution of a potato crop. Potato Research 33:417-432.

Struik P C, Van Heusden E, Burger-Meijer K (1988) Effects of short periods of long days on the development, yield and size distribution of tubers of *Solanum tuberosum* L. cv. Bintje. Netherlands Journal of Agricultural Science 36:11-22.

Struik P C, Van Voorst G (1986) Effects of drought on the initiation, yield and size distribution of tubers of *Solanum tuberosum* L. cv. Bintje. Potato Research 29:487-500.

Struik P C, Vreugdenhil D, Haverkort A J, Bus C B, Dankert R (1991) Possible mechanisms of hierarchy among tubers on one stem of a potato (*Solanum tuberosum* L.) plant. Potato Research 34:187-203.

Tibbitts T W, Bennett S M, Cao W (1990) Control of continuous irradiation injury on potatoes with daily temperature cycling. Plant Physiology 93:409-411.

Tibbitts T W, Cao W, Wheeler R M (1994) Growth of Potatoes for CELSS. NASA Contractor Report 177646. NASA, Ames Research Center, Moffett Field, CA, USA.

Van Dam J, Kooman P L, Struik P C (1995) Effects of temperature and photoperiod on early growth and final number of tubers in potato (*Solanum tuberosum* L.). (submitted).

Van den Berg J H, Bonierbale M W, Ewing E E, Plaisted R L, Tanksley S D (1993) Use of RFLPlinkage to study genetics and physiology of potato tuberization. Pages 278–279 in EAPR Abstracts, 12th Triennial Conference of the European Association for Potato Research, Paris, France.

Van den Berg J H, Struik P C, Ewing E E (1989) One-leaf cuttings as a model to study second growth in the potato (*Solanum tuberosum*) plant. Annals of Botany 66:273–280.

Van der Zaag D E, Van Loon C D (1987) Effects of physiological age on growth vigour of seed potatoes of two cultivars. 5. Review of literature and integration of some experimental results. Potato Research 30:451–472.

Van Heemst H J D (1986) The distribution of dry matter during growth of the potato crop. Potato Research 29:55–66.

Van Ittersum M K (1992a) Relation between growth conditions and dormancy of seed potatoes. 3. Effects of light. Potato Research 35:377–387.

Van Ittersum M K (1992b) Dormancy and Growth Vigour of Seed Potatoes. Ph.D. Thesis. Wageningen Agricultural University, Wageningen, The Netherlands.

Van Ittersum M K (1993) Advancing growth vigour of seed potatoes by storage temperature regimes. Netherlands Journal of Agricultural Science 41:23–36.

Van Ittersum M K, Scholte K (1992a) Relation between growth conditions and dormancy of seed potatoes. 2. Effects of temperature. Potato Research 35:365–375.

Van Ittersum M K, Scholte K (1992b) Shortening dormancy of seed potatoes by storage temperatures. Potato Research 35:389–401.

Van Loon C D (1987) Effect of physiological age on growth vigour of seed potatoes of two cultivars. 4. Influence of storage and storage temperature on growth and yield in the field. Potato Research 30:441–450.

Vos J (1995) Nitrogen and the growth of potato crops. Pages 115–128 in Haverkort A J, MacKerron D K L (Eds.) Potato Ecology and Modelling of Crops under Conditions Limiting Growth. Kluwer Academic Publishers, Dordrecht, The Netherlands.

Vreugdenhil D, Helder H (1992) Hormonal and metabolic control of tuber formation. Pages 393–400 in Karssen C M, Van Loon L C, Vreugdenhil D (Eds.) Progress in Plant Growth Regulation. Kluwer Academic Publishers, Dordrecht, The Netherlands.

Vreugdenhil D, Struik P C (1989) An integrated view of the hormonal regulation of tuber formation in potato (*Solanum tuberosum*). Physiologia Plantarum 75:525–531.

Wheeler R M, Steffen K L, Tibbitts T W, Palta J P (1986) Utilization of potatoes for life support systems. II. The effects of temperature under 24-h and 12-h photoperiods. American Potato Journal 63:639–647.

Wheeler R M, Tibbitts T W (1986) Growth and tuberization of potato (*Solanum tuberosum* L.) under continuous light. Plant Physiology 80:801–804.

Winkler E (1971) Kartoffelbau in Tirol. II. Photosynthesevermogen und Respiration von verschiedenen Kartoffelsorten. Potato Research 14:1–18.

Wolf S, Marani A, Rudich J (1990) Effects of temperature and photoperiod on assimilate partitioning in potato plants. Annals of Botany 66:515–520.

Wurr D C E (1977) Some observations on patterns of tuber formation and growth in the potato. Potato Research 20:63–75.

3. Modelling development and growth of the potato crop influenced by temperature and daylength: LINTUL-POTATO

P.L. KOOMAN[1,2] and A.J. HAVERKORT[1]

[1] *DLO-Research Institute for Agrobiology and Soil Fertility (AB-DLO), P.O. Box 14, 6700 AA Wageningen the Netherlands*

[2] *Department of Theoretical Production Ecology, Agricultural University, Bornsesteeg 65, 6708 PD Wageningen the Netherlands*

Abstract. A model 'LINTUL-POTATO' is described which bases total dry matter production on light use efficiency of intercepted light by a potato crop. The course of light interception by green leaves initially is temperature-sum dependent and is also steered by influences of daylength and temperature on assimilate partitioning. Dry matter allocation to the foliage, which determines the final length of the growing period, is determined by a daylength and temperature dependent period between emergence and tuber initiation and subsequent partitioning of dry matter to the tubers. Once the tuber daily growth rate equals total crop dry matter accumulation, no more foliage is formed and the crop senesces faster at higher temperatures. The aim of this paper is to explain effects of temperature and daylength on total tuber production from the effects of these environmental factors on the underlying processes. The relative effect of temperature on rates of emergence, light use efficiency, tuber initiation and tuber growth, and, the effect of daylength on development rate until tuber initiation are represented quantitatively. Potential tuber production at some temperature and daylength regimes is calculated exploratively and compared with results from the literature. The model can be used to simulate tuber dry matter production in specific environments and to select cultivars for such environments with suitable temperature and daylength reactions on tuber initiation and dry matter partitioning so as to maximise yields.

Introduction

The model LINTUL-POTATO is described in this paper: This model simulates the Light INTerception and Utilisation, temperature and daylength reactions of POTATO crops grown in different climates. The model is based on the light interception and utilisation model described by Spitters (1987) and Spitters and Schapendonk (1990) and adds temperature and daylength relations. A number of models simulating development and yield of potato have been described, which all perform relatively well under temperate long day conditions (Kabat et al. 1994). Such models, however, do not perform well in (sub)tropical conditions. In such climates, high temperatures during part of the year in lowland areas make the period or area unsuitable for potato growth. In addition to high temperatures, the prevailing short days, sometimes shorter than 11 h, also strongly affect the development of the crop, and often lead to inadequate yields (Haverkort 1990). Models developed under tropical conditions (e.g. Haverkort and Harris 1987) cannot be used under temperate conditions for

A.J. Haverkort and D.K.L. MacKerron (eds.), Potato Ecology and Modelling of Crops under Conditions Limiting Growth, 41–60.

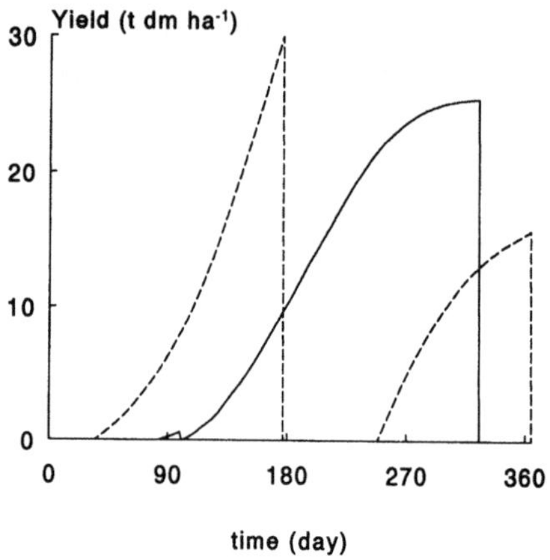

Figure 1. Potential total crop dry matter production in the Netherlands (——) and Tunisia (----).

similar reasons. There is a need for a model that includes crop reactions to temperature and daylength on those crucial processes that determine the length of the growth cycle and therefore the yield of the crop (Kooman 1995; Allen and Scott 1980). Exploration of the yielding ability of crops in different agro-ecological zones (Van Keulen and Stol, this volume) can then be carried out more accurately. Such a model may also offer breeders the scope to breed for cultivars which make the most effective use of a given environment described in terms of the yield-defining factors daylength, temperature and solar radiation which cannot be altered by the grower. Growers can affect yield limiting factors through irrigation and fertilisation and they can affect yield reducing factors through crop protection measures. The effect of yield limiting and yield reducing factors on crop growth and development are not included in LINTUL-POTATO.

Recently Kooman (1995) analysed the effect of temperature and daylength on growth and development of 8 cultivars in different environments in the Netherlands, Tunisia and Rwanda. Temperature and daylength affected crop production in different ways (see also Struik and Ewing, this volume). Temperature influences dry matter production directly through its influence on daily growth rate. Temperature and daylength influence production indirectly through their influence on development, thus limiting the duration of the growth cycle and yielding ability of the crop. This is illustrated in Figure 1 with long-term temperature and solar radiation data from Tunisia and the Netherlands. One single season in the Netherlands and two distinct growing seasons in Tunisia occur, assuming that no potato crop growth is possible below 2 °C and above 30 °C (Van Keulen and Stol, this volume). During winter, minimum daily temperatures in the Netherlands are constantly below 2 °C and below 28 °C in summer allowing a single, relatively long growing season of about six months. There is roughly one month period with temperatures that are too low and a three month period with mean daily maximum temperatures

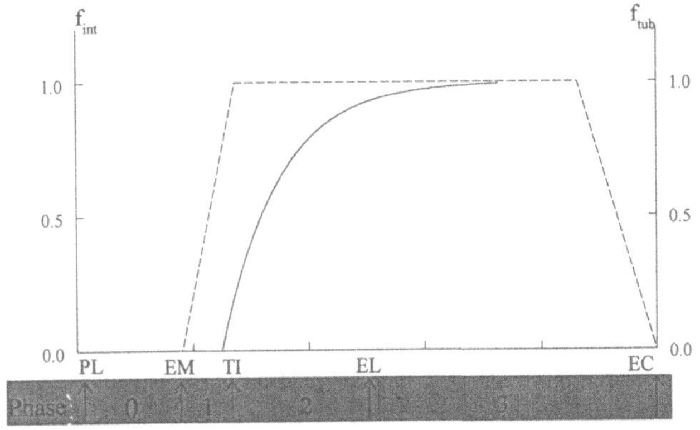

Figure 2. Schematic representation of the fraction of solar radiation intercepted by the crop (f_{int}) and the fraction of the daily assimilate that is partitioned to the tubers (f_{tub}) from planting (PL) to emergence (EM), the start of tuber growth (TI), the moment that f_{tub} reaches the value of 0.9 (EL) until the end of crop growth (EC).

above 28 °C, leading to two annual seasons suitable for potato production. With a potential light use efficiency of 2.5 g MJ^{-1} based on PAR. Figure 1 illustrates that potential dry matter production in the Tunisian spring is higher than in the temperate summer, that such a dry matter production can be achieved in much less time and that, in the Mediterranean area, potential yields in the autumn season are about half those in the spring season as was found by Fahem and Haverkort (1988). LINTUL-POTATO simulates potential dry matter production in different environments through the relative effect of temperature on rates of emergence, light use efficiency, tuber initiation and tuber growth. Moreover, the effect of daylength on development rate until tuber initiation is quantitatively represented and potential tuber dry matter production is calculated exploratively under some regimes of temperature and daylength.

An analysis of the system of tuber yield formation enables an approach that may be used in areas with widely varying temperature and daylength regimes. Total crop dry matter production is determined by the length of the growth cycle which is divided into four phases each starting and ending by a characteristic stage of development. Phase 0 starts at planting and ends at emergence. Phase 1, when only foliar growth takes place, is from emergence to tuber initiation. Phase 2 is from tuber initiation until the end of leaf growth (defined as the moment when 90% of the assimilates which are produced daily, are partitioned to the tubers). In this Phase 2 there is competition for assimilates between tuber and foliage. Phase 3, finally, is from the end of leaf growth until the end of crop growth. In Phase 3 all assimilates are partitioned to the tubers. The development of radiation interception and of the daily partitioning of dry matter to the tubers is represented schematically in Figure 2. The system is conveniently divided into the four phases, each with its own key processes described by variables and parameters with values depending on temperature and/or daylength:

- Phase 0 between planting and emergence (when 50% of the plants have emerged) the duration of which is determined by the sprout growth rate which depends on temperature;
- Phase 1 between emergence and tuber initiation (also defined as the start of tuber growth) the duration of which depends on the development rate until tuber initiation. In the Phases 1 through 3 the crop growth rate is calculated with the using foliar expansion rate, while the light use efficiency determines the total amount of dry matter produced;
- Phase 2 between tuber initiation and the moment when 90% of the daily acquired assimilates are partitioned to the tubers. This value is arbitrarily chosen as 100% partitioning is only reached asymptotically and, as such, cannot be determined precisely. The length of Phase 2 depends on the relative tuber growth rate which determines partitioning of dry matter between the tubers and the rest of the plant;
- Phase 3 lasts until the end of crop growth. The length of the period of Phase 3 is determined by the leaf senescence rate.

LINTUL-POTATO explains the effect of temperature and daylength on final tuber dry matter production through the quantification and integration of temperature and daylength effects on the major growth and development processes described in the Phases 0–3. The model is also aimed at simulating tuber dry matter production in specific environments and to select cultivars for such environments with suitable temperature and daylength reactions on tuber initiation and dry matter partitioning to maximise yields.

Temperature and daylength relations in LINTUL-POTATO

Phase 0

Rate of emergence: temperature relations. Under temperate conditions with low temperatures limiting potato growth in spring, the date of emergence may be calculated from planting depth, initial sprout length, soil temperature (often not available so in LINTUL-POTATO we use the mean daily air temperature) and the sprout growth rate of 1 mm per °Cd above a base temperature of 2 °C (MacKerron and Waister 1985). Daylength does not influence the sprout growth rate. At sub-optimal temperatures, whether low (Klemke and Moll 1990) or high (Manrique and Hodges 1989; Midmore 1984; 1988; Van der Zaag et al. 1986) sprout growth rate is reduced. Figure 3a shows the relative effect of temperature on the sprout growth rate compounded from the data collected by these authors: at a relative rate of 1.0, the sprout growth rate is 1 mm per °Cd.

Phase 1

Rate of foliar expansion: relation to daylength. Daylength is reported to influence above ground behaviour of potato plants and crops. Menzel (1985a),

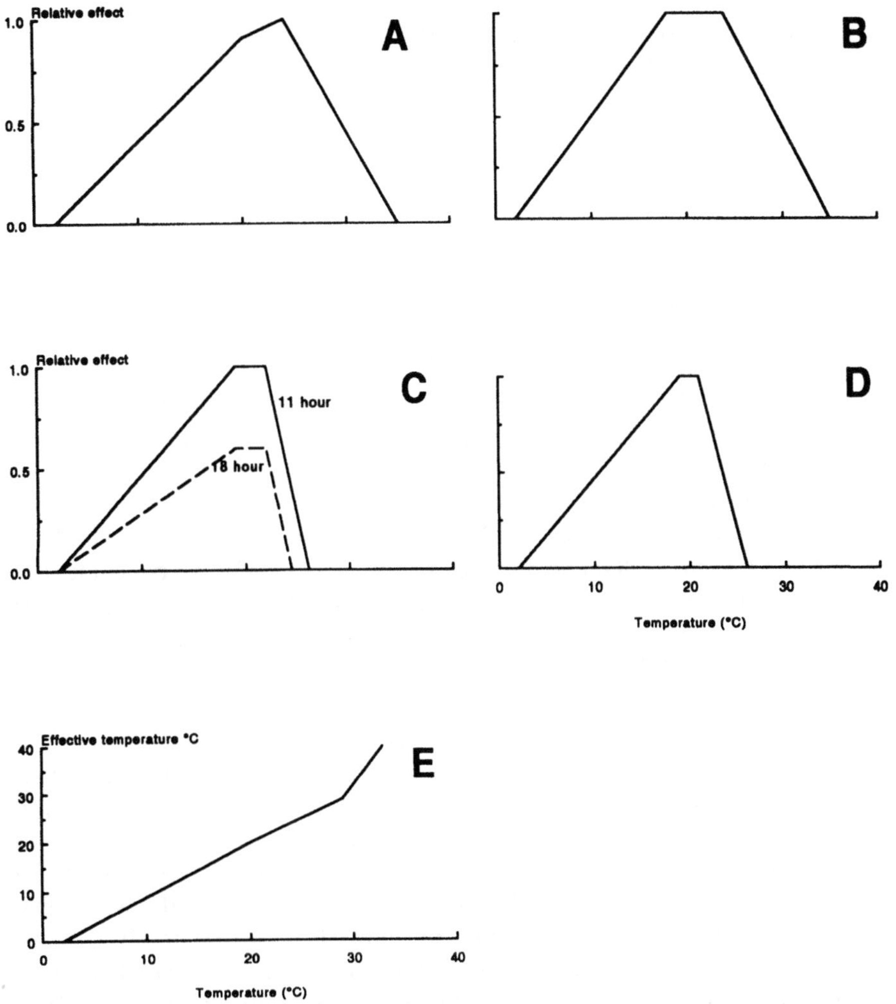

Figure 3. The relative effect of temperature in LINTUL-POTATO on the rates of sprout growth (A), light use efficiency (B), tuber initiation (C, —— at 11 h and --- at 18 h daylength) and tuber growth (D), and leaf senescence (E).

Ewing and Struik (1992) and Lorenzen and Ewing (1990) reported that short days reduce stem length, branching and weight, the number and weight of leaves and to increase the size of the leaflets and the leaf/stem weight ratio. Haverkort (1990) showed that shorter days reduce the vegetative growth and lead to earlier crop senescence. These observations agree with those of Demagante and Van der Zaag (1988) who observed that lengthening the photoperiod from 11.5 to 16 h increased branching, flowering and secondary growth. At the crop level, no clear influence on light interception and leaf growth was found by Kooman (1995) between the various climates and cultivars tested. Since the final maximum amount of foliage differed at the various locations, It can be concluded that the observations result from a shift in tuber initiation: shorter days increase the tuber initiation rate, thus reducing the length of Phase 1 in Figure 2 and bringing forward Phases 2 and 3 leading to an earlier crop. LINTUL-POTATO, therefore, has no relation linking foliar development to daylength.

Rate of foliar expansion: relation to temperature. Marinus and Bodlaender (1975) observed that foliar growth increased with temperature at low temperatures and Midmore (1984) and Van der Zaag et al. (1986) observed that foliar growth decreased with temperature at high temperatures. MacKerron and Waister (1985) linearly correlated leaf expansion from emergence to temperature above a base temperature of 2.4 °C, Spitters (1987) and Spitters and Schapendonk (1990) used a similar approach with a base temperature of 2 °C. Often minimum (0–7 °C), optimum (20–28 °C) and maximum temperatures for leaf growth have been recorded (e.g. Borah and Milthorpe 1962; Benoit et al. 1983; Ng and Loomis 1984; Manrique 1992). Maximum temperatures for leaf expansion have been reported as 30 °C by Ingram and McCloud (1984), 40 °C by Ng and Loomis (1984) and 35 °C by Benoit et al. (1983). We conclude from these literature data that at temperatures below 24 °C foliar expansion is linearly correlated to temperature and that it is zero again at 35 °C, as such it is simulated in LINTUL-POTATO with optimum rate shown in Table 2.

Development rate until tuber initiation: relation to daylength. Shorter days have been reported to increase the development rate until tuber initiation defined as the inverse of the number of days between emergence and the start of tuber growth. Some authors (e.g. Regel and Sands 1983) reported the existence of a 'critical photoperiod' 14.1 h for cv. Sebago and 15.5 h for cv. Kennebec above which tuber formation is retarded. Others (e.g. Lorenzen and Ewing 1990; Struik et al. 1988) reported that a shift from long (16–18 h) to short (10–12) days increased the development rate until tuber initiation and that short days compared to long days lead to earlier tuber initiation (Snyder and Ewing 1989; Wolf et al. 1990; Ewing and Struik 1992). This was confirmed under field conditions by Kooman (1995) who reported a negative relation between development rate until tuber initiation and the photoperiod prevailing at emergence of the crop. The relative effect of daylength (example of 11 h and 17 h) on development rate until tuber initiation as it is included in LINTUL-POTATO is shown in Figure 3d with optimum rate shown in Table 2.

Development rate until tuber initiation: relation to temperature. Under temperate conditions higher temperatures increase the tuberisation rate (Menzel 1983; Hay and Allen 1978) so that thermal time from planting with a base temperature of 0 °C describes the tuberisation rate well (Jefferies and MacKerron 1987). There is, however, an optimum as many reports on the relation between tuber initiation and temperature concern the observation that high temperatures retard tuber initiation (under tropical conditions) (Sands et al. 1979; Manrique and Hodges 1989; Midmore 1984; 1992; Struik et al. 1989a,b; Reynolds and Ewing 1989). Minimum (0–7 °C), optimum (15–20 °C) and maximum (25–30 °C) temperatures for tuber initiation rates have been reported by Borah and Milthorpe (1962), Sale (1979), Manrique and Hodges (1989) and by Ingram and McCloud (1984). The effect of photoperiod was found by Struik and Ewing (1992) to be more pronounced at higher temperatures. This was corroborated by Kooman (1995), who found a decrease of the length of the period of Phase 1 between emergence and tuber initiation (i.e. increased development rate until tuber initiation) at temperatures between 10 and 21 °C. Depending on cultivar and location, this period lasted between 8 and 40 days. These observations allow us to simulate the relative effect of temperature on development rate until tuber initiation in LINTUL-POTATO as shown in Figure 3d with an optimum rate as given in Table 2.

Phase 2

Tuber growth rate: daylength relation. Photoperiod has been reported to have an influence on the amount of dry matter allocated to the tuber after tuber initiation. Experimental results are often interpreted as 'short days favouring the partitioning of dry matter to the tubers' (e.g. Lorenzen and Ewing 1990; Ewing and Struik 1992; Menzel 1985b; Haynes et al. 1988). A shorter duration of Phase 2 under short days than under long days was found by Kooman (1995). He attributed this to an earlier tuber initiation (Phase 1) leading to higher tuber growth rates (not *relative* tuber growth rates) early in the season. This finding agrees with Wheeler and Tibbits (1987) as they reported similar dry matter allocation of tubers growing at 12 and 24 h daylength. No effect of daylength on tuber growth rate is included in LINTUL-POTATO.

Tuber growth rate: temperature relation. In LINTUL-POTATO, the relative tuber growth rate is only dependent on temperature. As with the development rate until tuber initiation, there are reports on the decrease of tuber growth rate at high temperatures leading to a shift in assimilate allocation from the tubers to the foliage (e.g. Krauss and Marschner 1984;, Menzel 1985b; Epstein 1966; Randeni and Caesar 1986; Midmore 1992). As with development rate until tuber initiation, so minimum (0–4.4 °C), optimum (15–22 °C) and maximum (25–35 °C) temperatures have been observed for tuber growth rates (Borah and Milthorpe 1962; Yandell et al. 1988; Manrique 1992; Hartz and Moore 1978; Spitters 1987; Ingram and McCloud 1984) The last, for instance assumed a maximum tuber growth rate between 16 and 24 °C. At 10 and 30 °C the growth rate was half of that at optimum and at 4 and 35 °C the tuber growth rates were 0. Kooman (1995) reported a shorter duration of Phase 2 with increasing temperature, up to the temperature of about 22 °C. Based on these observations we simulate the relative effect of temperature on the relative tuber growth rate in LINTUL-POTATO as shown in Figure 3e with maximum rate shown in Table 2.

Phase 3

Crop senescence rate: relation to temperature. Crop senescence takes place earlier at higher temperatures (Menzel 1985a) although the total length of the growing period may last longer because the total period of leaf formation may increase (Marinus and Bodlaender 1975). Leaf longevity has been recorded to last between 1440 and 1760 °Cd (Vos and Oyarzun 1987). A leaf formed was estimated to be functional for about 23 days at 20 °C and 12 days at 30 °C by Ingram and McCloud (1984) under conditions in Florida. Kooman (1995) estimated leaf longevity under conditions in the Netherlands to be between 1000 and 1700 °Cd, depending on cultivar. In LINTUL-POTATO we incorporated the effect of temperature as shown in Figure 3b: doubling the effective temperature for leaf senescence rate from 30 °C onward, adequately accommodates the reductions observed in foliar development at high temperatures.

Phases 1, 2 and 3

Light use efficiency: temperature relations. Daily crop growth per unit leaf area or unit soil covered by green foliage results from photosynthesis and respiration both of which are temperature dependent. Minimum (between 0 and 7 ° C), optimum (16–25 °C) and maximum (40 °C) temperatures for photosynthesis in potato crops have been reported by Ku et al. (1977), Dwelle et al. (1981), Dwelle (1985), Ng and Loomis (1984), Manrique (1992), Midmore (1992) and Hammes and Jager (1990). The highest total daily crop growth is reported between temperatures of 15 and 23 °C (Ben Kheder and Ewing 1985; Nishibe et al. 1988; Haverkort 1990). A strong increase in respiration with temperature, especially above about 25 °C (e.g. Sale 1974; Ng and Loomis 1984; Midmore 1992) could contribute to the difference between optimum for photosynthesis and for crop growth rate and to the often observed reduction of the light use efficiency at increasing temperatures. Haverkort and Harris (1987) and Manrique et al. (1991) presented linear correlations between the observed reduction of crop light use efficiencies and temperatures above 20 °C, with a slope of about -0.02 per °C. Combining the various evidence, in LINTUL-POTATO we simulate the effect of temperature on the light use efficiency as shown in Figure 3c with the optimum rate of 2.5 g MJ^{-1}. We assume that there is no influence of daylength on light use efficiency. Rather than using a negative linear relation between light use efficiency and total daily radiation to accommodate for light saturation of the canopy, LINTUL-POTATO only uses daily values up to a maximum of 12 MJ m^{-2} (Kooman 1995), disregarding radiation in excess of this value.

Table 1. Sequence of equations describing development and growth in LINTUL-POTATO ('V' = if)

Development	*Growth*
Emergence	*Leaf area*
$t_{em} = t_{pl} + \frac{P_d}{G_s} * T_{eff}$ (1)	$LAI = LAI_0 * e^{(R_L * \Sigma T_{LUE})}$ $LAI < 0.75 \vee \Sigma T_{LUE} < 450°Cd$ (2)
	$LAI = W_l SLA$ $LAI \geq 0.75$, $\Sigma T_{LUE} \geq 450°Cd$ (3)
	Light interception
	$I = I_0 * e^{(-k*LAI)}$ $I_0 \leq 12\ MJ\ m^{-2}$ (4)
	Total growth
	$dW_{Tot}/dt = I * LUE * E_{T,I}$ (5)
	Partitioning and daily growth of the foliage
	$dW_{sht}/dt = f_{sht} * dW_{Tot}/dt$ (6)
	$f_{sht} = 1 - f_{Tub}$ (7)
	$dW_l/dt = f_l * dW_{sht}/dt$ (8)
	$f_l = f_{l0} - (.5 - f_{Tub})$ (9)
	$dW_{st}/dt = f_{st} * dW_{sht}/dt$ (10)
	$f_{st} = 1 - f_l$ (11)
Tuber initiation	*Partitioning and daily growth of the tubers*
$R_{ti} = ST_T * T_{Tub} - ST_p * P + ST_c$ $t > t_{em}$ (12)	$dW_{Tub}/dt = f_{Tub} * dW_{Tot}/dt$ (13)
$f_{Tub} = \frac{rgr_{Tub} * W_{Tub} * E_{Tub}}{dW/dt + rgr_{Tub} * W_{Tub} * E_{Tub}}$ $W_{Tub} \geq W_{Tub_0}$	(14)
Leaf senescence	*Crop biomass*
$dW_{ld}/dt = \Sigma W_{l,i}$ $\forall_i$ $\Sigma T_{s,i} > S_t \vee S_{sh} > 1$ (15)	$W_{Tub}(t) = \int_{ti}^{t} dW_{Tub}/dt$ (16)
$S_t = A_{10} - dW_{Tub}/dt * S_{Tub}$ (17)	$W_l(t) = \int_{em}^{t} dW_l/dt$ (18)
$S_{sh,i} = R_{sh} * \frac{LAI - LAI_c}{LAI_c}$ $LAI > LAI_c$ (19)	$W_{st}(t) = \int_{em}^{t} dW_{st}/dt$ (20)
	$W_{Tot}(t) = W_l(t) + W_{st}(t) + W_{Tub}(t)$ (21)

Table 2. List of acronyms and equations used in LINTUL-POTATO

Acronym	Meaning and source	Value	Unit
A_{l0}	Initial longevity of a daily leaf class (Kooman 1995)	1000	°Cd
day	Julian day	Input	d
$E_{T,l}$	Relative effect of temperature on light use efficiency	Figure 3b	-
$E_{T,s}$	Relative effect of temperature on rate of leaf senescence	Figure 3e	-
$E_{T,ti}$	Relative effect of temperature on development rate until tuber initiation	Figure 3c	-
$E_{T,t}$	Relative effect of temperature on relative tuber growth rate	Figure 3d	
$E_{T,e}$	Relative effect of temperature on sprout growth rate	Figure 3a	
f_l	Fraction of daily shoot dry matter production allocated to the leaves	Output	-
f_{l0}	Initial fraction of daily shoot dry matter production allocated to the leaves (Kooman 1995)	0.8	-
f_{sht}	Fraction of daily total dry matter production allocated to the foliage	Output	-
f_{st}	Fraction of daily shoot dry matter production allocated to the stems	Output	-
f_{tub}	Fraction of daily total dry matter production allocated to the tubers	Output	-
G_s	Sprout growth rate (MacKerron and Waister, 1985)	1	mm °Cd^{-1}
i	Index daily leaf class	Index	d
I	Intercepted PAR	Output	MJ m^{-2}
I_0	Total incident PAR	Input	MJ m^{-2}
k	Extinction coefficient (Spitters and Schapendonk 1990)	1	-
LAI	Leaf Area Index	Output	m^2 m^{-2}
LAI_c	Critical leaf area Index (Kooman 1995)	6.0	m^2 m^{-2}
LAI_0	Initial leaf area Index (Spitters and Schapendonk 1990)	0.0155 × NPL	m^2 m^{-2}
LUE	Light use efficiency (Kooman 1995)	2.5	g MJ^{-1}
NPL	number of plants	4.4	# m^{-1}
P	Photoperiod	Input	h
P_d	Planting depth	Input	cm
rgr_{tub}	Relative growth rate of tubers (Ingram and McCloud 1984)	0.37	g g^{-1} d^{-1}
R_l	Relative increase rate leaf area (Spitters and Schapendonk 1990)	0.012	m^2 m^{-2} d^{-1}
R_{sh}	Senescence rate due to shade (Kooman 1995)	0.1	d^{-1}
R_{ti}	Development rate until tuber initiation	Calculated	
SLA	Specific leaf area (Van Oijen et al. 1995)	0.0023	m^2 g^{-1}
S_t	Leaf longevity	Output	
S_{sh}	Senescence due to shading	Output	
ST_C	Constant in calculating start tuber growth (Kooman 1995)	0.119	
ST_P	Daylength parameter calculating start tuber growth (Kooman 1995)	6.2 10^{-4}	h^{-1}
ST_T	Temperature parameter in calculating start tuber growth (Kooman 1995)	−5.8 10^{-3}	°C^{-1}

Table 2. Continued.

Acronym	Meaning and source	Value	Unit
t	time	Input	d
t_{em}	Day of emergence	Output	d
t_{pl}	Day of planting	Input	d
t_{st}	Day of start tuber growth	Output	d
T	Daily average temperature	Input	°C
T_0	Base temperature for potato growth (Spitters and Schapendonk 1990)	2	°C
T_{eff}	Effective temperature for potato growth	Output	°C
T_{lue}	Effective temperature for total growth	Output	°C
T_{tub}	Effective temperature for tuber growth	Output	°C
T_s	Effective temperature for senescence of foliage	Output	°C
W_l	Weight of green leaf dry matter	Output	kg m^{-2}
W_{ld}	Weight of dead leaf dry matter	Output	kg m^{-2}
W_{sht}	Weight of shoot dry matter	Output	kg m^{-2}
W_{st}	Weight of stem dry matter	Output	kg m^{-2}
W_{Tot}	Weight of total dry matter	Output	kg m^{-2}
W_{tub}	Weight of tuber dry matter	Output	kg m^{-2}

Model structure of LINTUL-POTATO

LINTUL-POTATO is partly based on dry matter accumulation with allocation governed by a dominant tuber sink (Spitters 1990; Kooman 1995; Spitters and Schapendonk 1990) using quantitative relationships from Moorby (1968), Sale (1974) and Gawronska et al. (1984). Part of the model was described by Kooman (1995) and the complete set of equations is given here in Table 1 with Table 2 describing the meaning of the acronyms and the (initial or optimum) values of the variables and parameters used.

The development of the crop starts at planting. The planting depth and the effective temperature (Figure 3a) determine the time between planting and emergence (Eq. 1). In the early stages of crop growth, temperatures rather than assimilates determine leaf area expansion and is therefore modeled according to Eq. 2. When the temperature sum from emergence exceeds 450 °Cd or the leaf area index exceeds 0.75, the increase of leaf area is determined by the availability of assimilates and is calculated as the product of leaf dry weight and specific leaf area (Eq. 3 from Spitters 1990; Spitters and Schapendonk 1990). Leaves are divided into classes based on the day in which the leaves are formed in. These daily classes senesce when the temperature sum integrated by the class exceeds the leaf longevity or when the leaf area above the leaf layer is such that shading is too strong (Eqs. 15 and 19). The leaf longevity is also influenced by tuber growth such that fast growing tubers reduce leaf longevity (Eq. 17). Crop growth ceases when the leaves in the latest class have senesced.

The daily amount of dry matter produced is calculated from the amount of photosynthetically active radiation intercepted by the canopy calculated from the leaf area index (LAI, Eqs. 3 and 4) and its conversion efficiency for dry matter production (Eq. 5, Figure 3c). The growth of the different plant parts is calculated from the total dry matter production and the partition coefficients to the parts (Eqs. 6–11, 13 and 14). To obtain the cumulative amount of dry matter produced, the daily growth is integrated over the growing period (Eqs. 16–21).

After emergence the development rate until tuber initiation is determined by temperature and daylength (Eq. 12, Figure 3d). Assimilate allocation to the tubers receives priority after tuber initiation: TI in Figure 2. Initially, tuber growth is sink-limited and results from the product of the relative growth rate and the tuber weight. Thereafter, tubers become an increasing sink as they become larger, however, in this phase the source is not large enough to deliver all assimilates needed to maintain simultaneously exponential tuber growth and leaf growth, resulting in an increasing part of the daily assimilates allocated to the tubers. Finally, tuber growth becomes source-limited and all assimilates produced go to the tubers. This course of tuber growth is described by Eq. 17 and the relative effect of temperature on the relative tuber growth rate is given in Figure 3e. Assimilates not allocated to the tubers are allocated to the shoot and divided between the leaves and stems (Eqs. 7–11). Root biomass is not taken into account as precise experimental measurements of roots in the field experiments are rare.

Simulation of temperature and daylength responses with LINTUL-POTATO

LINTUL-POTATO, as it is presented here, can be used for two main purposes.

Firstly, the values of variables and parameters in LINTUL-POTATO as they are presented here (Figure 3, Table 1) are mean values taken from literature and, where needed, adjusted by data given by Kooman (1995). Data presented by Kooman (1995) on parameter values of 8 *Solanum tuberosum* cultivars varying in earliness and grown in the Netherlands, Tunisia (winter, spring and autumn seasons) and Rwanda (2 altitudes) give a range of values for Dutch cultivars that are currently grown. Given a specific environment of temperature and daylength, desired cultivar characteristics for responses to these (especially on tuber initiation and tuber growth rates) can be evaluated by the model to calculate the potential tuber production in such an environment.

Secondly, the estimation of potential tuber yields in various regions (Figure 1) can be improved considerably by the use of a dynamic dry matter allocation as in the model, rather than assuming a fixed harvest index of about 0.75 at crop senescence. LINTUL-POTATO takes into account prohibitive low or high temperatures for sprout growth before emergence in Phase 0 (Figure 2). The parameters related to the tuber initiation rate (Phase 1), which is temperature and daylength dependent, can be given values such that the length of Phases 2 and 3 are sufficiently long so that the total length of the growing season fits in the period suitable for potato growth, between 2 and 30 °C.

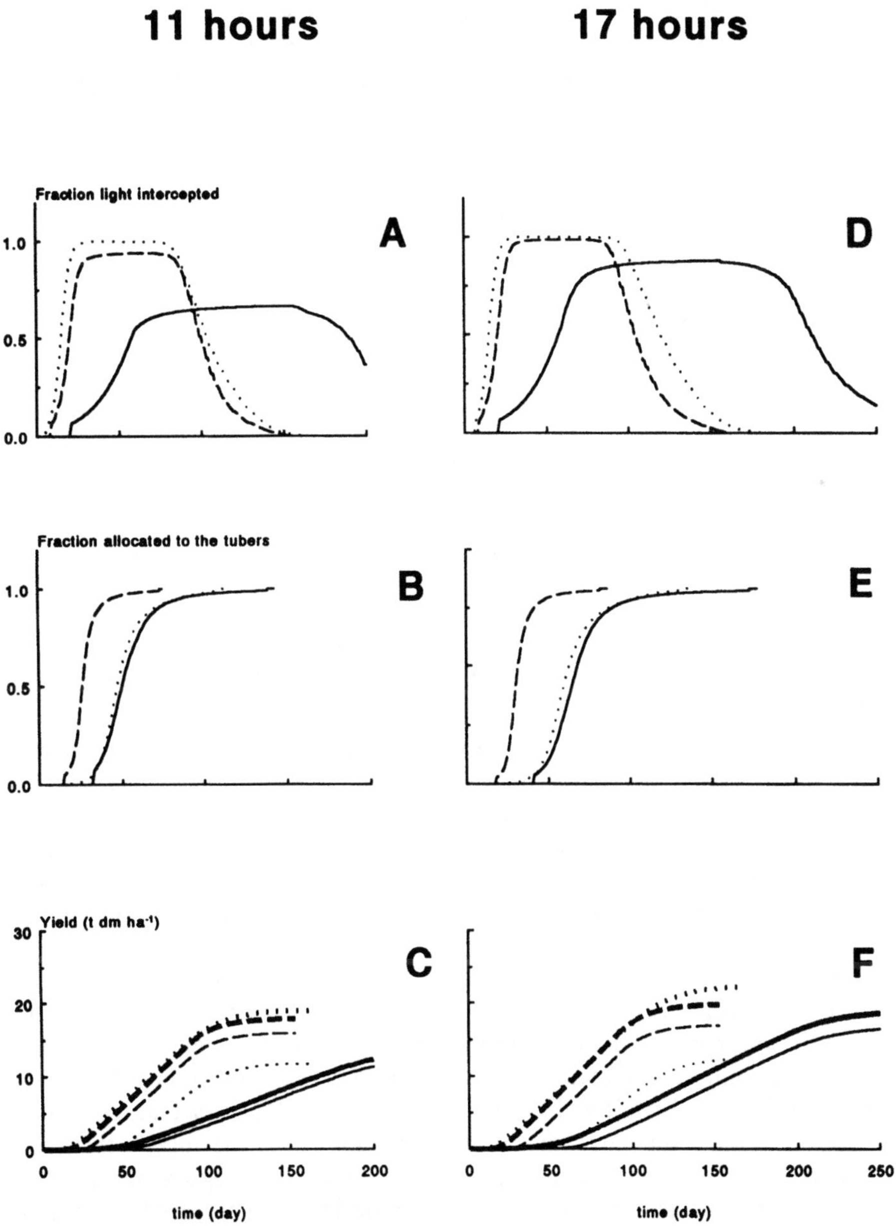

Figure 4. Simulated development of light interception (A,D), dry matter partitioning to the tubers (B,E), total (thick) and tuber dry matter production (thin) (C,F) at 10 (——), 17 (---) and 25 °C (...) and at daylengths of 11 (A,B,C) and 17 h (D,E,F).

The effect of different temperatures and daylengths on the length of the growing season, expressed as the period when the ground is (partially) covered by green foliage is shown in Figure 4 together with the development of the proportion of light interception, the fractional daily assimilates partitioning to the tubers, and tuber dry matter production, using the standard values of LINTUL-POTATO as presented in Table 1. and assuming a daily solar radiation level of 9 MJ m^{-2} throughout the season. The figure shows the effect of growing potatoes at mean daily temperatures of 10, 17 and 25 °C throughout the growing season at daylengths at emergence of 11 and 17 h.

Table 3. Simulated end of Phases 1, 2 and 3 (Days After Planting, DAP), total and tuber yield at the end of the growth cycle

Temperature	Phase 1 (DAP)		Phase 2 (DAP)		Phase 3 (DAP)		Total yield (t ha^{-1})		Tuber yield (t ha^{-1})	
	Photoperiod		Photoperiod		Photoperiod		Photoperiod		Photoperiod	
	11 h	17 h	11 h	17 h	11 h	17 h	11 h	17 h	11 h	17 h
10 °C	33	41	72	89	274	300	13.3	19.0	12.3	16.9
17 °C	15	18	32	42	153	162	17.8	19.7	15.9	16.9
25 °C	15	23	70	87	162	173	18.9	22.1	11.7	12.3

At 10 °C the rates of emergence and initial light interception are low. Emergence takes place on day 21 and the start of tuber growth takes place on day 33 which is too early to enable the crop to reach full ground cover. Ground cover reaches a maximum of about 60%. Because of the subsequent low total growth rate the fraction of dry matter allocated to the tubers increases rapidly and it takes about 40 days from tuber initiation until 90% of the dry matter is allocated to the tubers (Table 3). The rate of senescence of the crop is low, resulting in a relative long growth cycle of about 270 days. Because of the low ground cover and the low growth rate at 10 °C the tuber dry matter production is only 12.3 t ha^{-1}. At 17 °C, emergence and initial growth are faster than at 10 °C. Emergence is brought forward by 12 days and the start tuber growth is 18 days earlier. The higher crop growth rate at 17 °C allows the crop to reach 95% ground cover. Both tuber growth rate and leaf senescence rates are higher at 17 °C than at 10 °C leading to a reduction in the senescence period to 37 days. The total length of the growth cycle of the crop grown at 17 °C is 153 days. Due to the high growth rates and the early allocation of dry matter to the tubers, this crop produced 16 t ha^{-1} tuber dry matter by the end of the growing season. At 25 °C emergence takes place after 7 days but tuber initiation is not brought forward (see Figure 3d) but takes place two days later than at 17 °C. The initial tuber growth is very slow and, therefore, the end leaf growth (end of Phase 2, Figure 2) is only reached on day 70. The total growth cycle is as long as at 17 °C. The greater rate of senescence counteracts the effect of increased foliar development because of a reduced fraction of dry matter allocated to the tuber which also explains why tuber dry matter production does not exceed 12 t ha^{-1}.

Carrying out the simulation for 17 h (Figure 4D,E,F) rather than for a daylength of 11 h (Figure 4A,B,C) shows that the lengths of the growth cycles are increased and that a greater amount of light is intercepted. The increased growth cycle resulted from a delay in tuber initiation (Table 3), which delays dry matter allocation to the tubers. The increased growth cycle increased total dry matter production in t ha^{-1} from 13.3 (at 10 °C) and 18.9 (at 25 °C) at 11 h daylength to 19.0 and 22.1 at 17 h. The tuber dry matter production increased less than total dry matter production. The increase was largest at 10 °C (4.6 t) and smallest at 25 °C (0.6 t).

The model is not yet fully validated here. Comparison with independent data was carried out by Kooman (1995). LINTUL-POTATO simulates the trends in tuber dry matter production and crop behaviour as reported in literature: earlier tuber initiation under short days leads to earlier senescence (Haverkort 1990; Ewing and Struik 1992). At 10 °C total dry matter production was lower than at 17 °C which agrees with literature data from Ng and Loomis (1984), Ingram and McCloud (1984) and Spitters (1990) who reported optimal yields between 15 and 22 °C. Total dry matter differed only slightly between 17 and 25 °C as was also found by Marinus and Bodlaender (1975) comparing plant performance at 16, 22 and 28 °C. In long days, the growing season is longer at 25 than at 17 °C but the tuber dry matter production is lower due to reduced harvest indices at higher temperatures (Ben Kheder and Ewing 1985; Manrique 1992).

We also used LINTUL-POTATO to determine the optimum temperature for potato growth at varying daylengths. The range of temperatures tested throughout the growing season varied from 2 to 28 °C and the simulations were carried out for daylengths of 11, 14 and 17 h. The resulting total tuber dry matter production is presented in Figure 5. The highest tuber dry matter production at the 11 h daylength is achieved at about 17 °C. Longer days (e.g. 17 h) broaden the optimum range of temperatures and centre it at about 14 °C.

The range of optimum temperatures for potato production calculated by LINTUL-POTATO shown in Figure 5 includes the optimum temperatures reported in literature: 13 °C by Spitters (1987), 14–22 °C by Ingram and

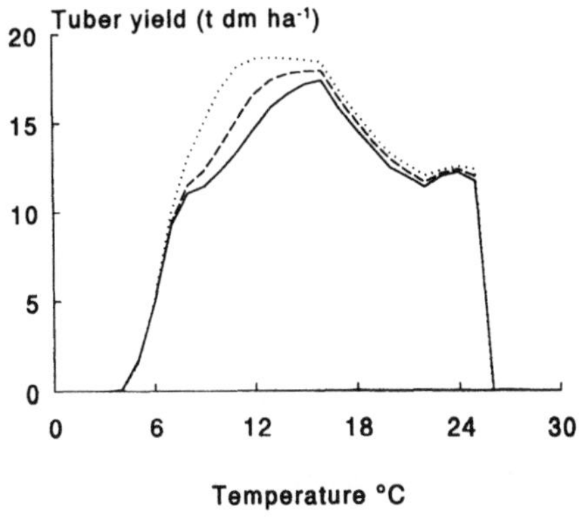

Figure 5. Simulated relation between tuber dry matter production by the time of crop senescence versus mean daily temperature throughout the growing season at daylengths of 11 (——), 14 (---) and 17 h (...).

McCloud (1984), 16–24 °C by Ng and Loomis (1984), 18 °C by Borah and Milthorpe (1962), 15–22 °C by Midmore (1992), 18–20 °C by Manrique (1992) and 21 °C by Sands et al. (1979).

The wider range of optimum temperatures in long days than in short days is explained as follows. The optimum crop growth rate is between 18 and 24 °C (Figure 3B) and is not affected by daylength. Short days bring forward tuber initiation and reduce the total length of the growth cycle and/or the maximum level of light interception, hence total and tuber dry matter production. Longer days at low temperatures however, retard tuber initiation to such an extent (Figure 3C) that it leads to longer growth cycles and/or higher maximum levels of light interception resulting in higher total and tuber production.

The phenomenon of a wider optimum temperature range for growth of the crop in longer days, may well explain the adaptability of potato to a wide range of climates. Besides its short day and low temperature origin, the crop also performs well in ecoregions with higher temperatures such as the Mediterranean area and the continental summer, because the prevailing daylength during the growing season here is longer than in the centre of origin.

Compared to experimental research, LINTUL-POTATO yields more commonly applicable information. In experiments the lateness of a crop is beside daylength, affected by other factors such as cultivar and seed age. The optimum temperature for tuber production, assessed in an experiment, therefore is always specific for that experiment. Which may explain the differences in optimum temperatures reported in literature.

The model LINTUL-POTATO simulated temperature and daylength responses similar to those reported in literature. The concept of dry matter allocation therefore is a valid tool to explain variation in potato production under different climatic conditions. Accurate quantification of the variables for cultivars differing in lateness and comparison of the model with independent data is reported by Kooman (1995).

Acknowledgements

Critical comments on the manuscript by Dr. M. van Oijen are gratefully acknowledged.

References

Allen E J, Scott R K (1980) An analysis of growth of the potato crop. Journal of Agricultural Science Cambridge 94:583–606.

Ben Kheder M, Ewing E E (1985) Growth analyses of eleven potato cultivars grown in the greenhouse under long photoperiods with and without heat stress. American Potato Journal 62:537–554.

Benoit G R, Stanley C D, Grant W J, Torrey D B (1983) Potato top growth as influenced by temperatures. American Potato Journal 60:489–501.

Borah M N, Milthorpe F L (1962) Growth of potato as influenced by temperature. Indian Journal of Plant Physiology 5:53–70.

Demagante A L, Van der Zaag P (1988) The response of potato (*Solanum* spp.) to photoperiod and light intensity under high temperatures. Potato Research 31 (1):73–83.

Dwelle R B, Kleinkof G E, Steinhorst R K, Pavek J J, Hurley P J (1981) The influence of physiological processes on tuber yield of potato genotypes (*Solanum tuberosum* L.): stomatal diffusive resistance, stomatal conductance, gross photosynthetic rate, leaf canopy, tissue nutrient levels and tuber enzyme activities. Potato Research 24:33–47.

Dwelle R E (1985) Photosynthesis and photoassimilate partitioning. Pages 35–58 in Li P H (Ed.) Potato Physiology. Academic Press, Orlando, USA.

Epstein E (1966) Effect of soil temperature at different growth stages on growth and development of potato plants. Agronomy Journal 58:169–171.

Ewing E E, Struik P C (1992) Tuber formation in potato: Induction, initiation, and growth. Horticultural Reviews 14:89–198.

Fahem M, Haverkort A J (1988) Comparison of the growth of potato crops grown in autumn and spring in North Africa. Potato Research 31:557–568.

Gawronska H, Dwelle R B, Pavek J J, Rowe P (1984) Partitioning of photoassimilates by four potato clones. Crop Science 24:1031–1036.

Hammes, P S, De Jager J A (1990) Net photosynthetic rate of potato at high temperatures. Potato Research 33:515–520.

Hartz T.K., Moore F.D. (1978) Prediction of potato yields using temperature and insolation data. American Potato Journal 55:431–436.

Haverkort A J (1990) Ecology of potato cropping systems in relation to latitude and altitude. Agricultural Systems 32:251–272.

Haverkort A J, Harris P M (1987) A Model for potato growth and yield under tropical highland conditions. Agricultural and Forest Meteorology 39:271–282.

Hay R K M, Allen E T (1978) Tuber initiation and bulking in the potato (*Solanum Tuberosum*) under tropical condition: The importance of soil and air temperatures. Tropical Agriculture 55:289–295.

Haynes K, Haynes F L, Swallow W H (1988) Temperature and photoperiod effects on tuber production and specific gravity in diploid potatoes. HortScience 23:562–565.

Ingram K T, McCloud D E (1984) Simulation of potato crop growth and development. Crop Science 24:21–27.

Jefferies R A, Mackerron D K L (1987) Thermal time as a non destructive method of estimating tuber initiation in potatoes. Journal of Agricultural Science Cambridge 108:249–252.

Kabat P, van den Broek B, Marshall B, Vos J and van Keulen H (1995) Modelling and parameterization of the soil – plant – atmosphere system, a comparison of potato growth models. Wageningen Pers, Wageningen, The Netherlands.

Klemke T, Moll A (1990) Model for simulation of potato growth from planting to emergence. Agricultural Systems 32:295–304.

Kooman P L (1995) Yielding ability of potato crops as influenced by temperature and daylength. Ph.D. Thesis, Wageningen Agricultural University, Wageningen, The Netherlands.

Krauss A, Marschner H (1984) Growth rate and carbohydrate metabolism of potato tubers exposed to high temperatures. Potato Research 27:297–303.

Ku S B, Edwards G E, Tanner C B (1977) Effects of light, carbon dioxide and temperature on photosynthesis, oxygen inhibition of photosynthesis and transpiration in *Solanum tuberosum*. Plant Physiology 59:868–872.

Lorenzen J H, Ewing E E (1990) Changes in tuberization and assimilate partitioning in Potato (*Solanum tuberosum*) during the first 18 days of photoperiod treatment. Annals of Botany 66:457–464.

MacKerron D K L, Waister P D (1985) A simple model of potato growth and yield. Part I. Model development and sensitivity. Agricultural and Forest Meteorology 34:241–252.

Manrique L A (1992) Potato production in the tropics: Crop requirements. Journal of Plant Nutrition 15:2679–2726.

Manrique L A, Kiniry J R, Hodges T, Axness D S (1991) Dry matter production and radiation interception of potato. Crop Science 31:1044–1049.

Manrique L A, Hodges T (1989) Estimation of tuber initiation in potatoes grown under tropical environments based on different methods of computing thermal time. American Potato Journal 66:425–436.

Marinus J, Bodlaender K B A (1975) Response of some potato varieties to temperature. Potato Research 18:189–204.

Menzel C M (1985a) Tuberization in potato at high temperatures: interaction between temperature and irradiance. Annals of Botany 55:35–39.

Menzel C M (1985b) The control of storage organ formation in potato and other species. Field Crop Abstracts 38:527–537.

Menzel C M (1983) Tuberization in potato at high temperatures: interaction between root and shoot temperatures. Annals of Botany 52:65–69.

Midmore D J (1992) Potato production in the tropics. Pages 728–793 in Harris P M (Ed.) The Potato Crop. Chapman & Hall, London, UK.

Midmore D J (1988) Potato in the hot tropics I. Soil temperature effects on emergence, plant development and yield. Field Crops Research 8:255–271.

Midmore D J (1984) Potato in the hot tropics I. Soil temperature effects on emergence plant development and yield. Field Crops Research 8:225–271.

Moorby J (1968) The influence of carbohydrate and mineral nutrient supply on the growth of potato tubers. Annals of Botany 32:57–68.

Ng E, Loomis R S (1984) Simulation of Growth and Yield of the Potato Crop. Simulation Monographs, Pudoc, Wageningen, The Netherlands.

Nishibe S, Iwama K, Nakaseko K (1988) Annual variation of growth and its relation to meteorological factors in potato plants. Japan. J. Crop Sci. 57:505–511.

Randeni G, Caesar K (1986) Effect of soil temperature on the carbohydrate status in the potato plant (*S. tuberosum* L.). Journal of Agronomy and Crop Science 156:217–224.

Regel P A, Sands P J (1983) A Model of the development and bulking of potatoes. IV daylength, plant density and cultivar effects. Field Crops Research 6:1–23.

Reynolds M, Ewing E E (1989) Effects of high air and soil temperature stress on growth and tuberization in *Solanum Tuberosum*. Annals of Botany 64:241–247.

Sale P J M (1979) Growth of potatoes (*Solanum Tuberosum*) to small tuber stage as related to soil temperature. Australian Journal of Agricultural Research 30:667–677.

Sale P J M (1974) Productivity of vegetable crops in a region of high solar input. III. Carbon balance of potato crops. Australian Journal of Plant Physiology 1:283–296.

Sands P J, Hacket C, Nix H A (1979) A model for the development and bulking of potatoes (*Solanum Tuberosum* L.) I. Derivation from well-managed field crops. Field Crops Research 2:309–331.

Snyder R G, Ewing E E (1989) Interactive effects of temperature, photoperiod, and cultivar on tuberization of potato cuttings. HortScience 24:336–338.

Spitters C J T (1990) Crop growth models: Their usefulness and limitations. Acta Horticulturae 267:349–368.

Spitters C J T (1987) An analysis of variation in yield among potato cultivars in terms of light absorbtion, light utilization and dry matter partitioning. Acta Horticulturae 214:71–84.

Spitters C J T, Schapendonk A H C M (1990) Evaluation of breeding strategies for drought tolerance in potato by means of crop growth simulation. Plant and Soil 123:193–203.

Struik P C, Ewing E E (1995) Crop physiology of potato (*Solanum tuberosum*): Responses to photoperiod and temperature relevant to crop modelling. Pages xxx in Haverkort A J, MacKerron D K L (Eds.) Ecology and Modelling of Potato Crops under Conditions Limiting Growth. Kluwer Academic Publishers, Dordrecht, The Netherlands.

Struik P C, Geertsema, J, Custers C H M G (1989a) Effects of shoot, root and stolon temperature on the development of the potato (*Solanum tuberosum* L.) plant. I. Development of the haulm. Potato Research 32:133–141.

Struik P C, Geertsema J, Custers C H M G (1989b) Effects of shoot, root and stolon temperature on the development of the potato (*Solanum tuberosum* L.) plant. II. Development of stolons. Potato Research 32:143–149.

Struik P C, Van Heusden E, Burger-Meijer K. (1988) Effects of short periods of long days on the development, yield and size distribution of potato tuber. Netherlands Journal of Agricultural Science 36:11–22.

Van Oijen M, De Ruijter F.J., Van Haren R.J.F., Responses of field grown Potato cultivar to cyst nematodes at three levels of soil compaction 1: leaf area dynamics, photosynthesis and crop growth (submitted).

Van der Zaag P, Demagante A L, Acasio R, Domingo A, Hagerman H (1986) Response of *Solanum* potatoes during the different seasons in an isohyperthermic environment in the Philippines. Tropical Agriculture 63:229–239.

Van Keulen H, Stol W (1995) Agro-ecological zonation for potato production. Pages xxx in Haverkort A J, MacKerron D K L (Eds.) Ecology and Modelling of Potato Crops under Conditions Limiting Growth. Kluwer Academic Publishers, Dordrecht, The Netherlands.

Vos J, Oyarzun P J (1987) Photosynthesis and stomatal conductance of potato leaves – effects of leaf age, irradiance and leaf water potential. Photosynthesis Research 11:253–264.

Wheeler R M, Tibbitts T W (1987) Utilization of potatoes for life support systems in space: III. Productivity at successive harvest dates under 12-h and 24-h photoperiods. American Potato Journal 64:311–320.

Wolf S, Marani A, Rudich J (1990) Effect of temperature and photoperiod on assimilate partitioning in potato plants. Annals of Botany 66:513–520.

Yandell B S, Najar A, Wheeler R, Tibbitts T W (1988) Modelling the effects of light, carbon dioxide, and temperature on the growth of potato. Crop Science 28:811–818.

4. Physiology of crop response to drought

R.A. JEFFERIES
Scottish Crop Research Institute, Invergowrie, Dundee, DD2 5DA, UK

Abstract. Effects of drought can be partitioned into effects on radiant energy interception, conversion of light into dry matter, partitioning of assimilate, and tuber dry matter concentration. The main effect of drought is to reduce radiant energy interception, by reducing both the rate and the duration of growth of leaves, and also the number of leaves produced. Rate of leaf expansion is inversely related to soil moisture deficit (SMD), and genotypes differ both in the threshold at which leaf expansion is constrained by soil moisture, and in their ability to maintain expansion with increasing SMD. Differences between genotypes appear to be related to the ability to maintain leaf water status, rather than in the ability for osmotic adjustment.

The efficiency of conversion of light into dry matter is less sensitive to drought stress than leaf growth. Drought has little effect on the functioning of Photosystem II, and effects on photosynthesis largely reflect stomatal limitation. Prolonged drought, though, may result in a reduction in light-harvesting complexes and down-regulation of carboxylating enzymes. Greater reduction in stomatal conductance to water vapour transfer than photosynthetic rate results in an increase in water use efficiency.

It is suggested that differences between experiments in the pattern of assimilate partitioning to tubers by droughted crops may reflect differences in timing and intensity of water-stress relative to tuber initiation. Tuber dry matter concentration can be related to thermal time accumulated from the time of plant emergence and to SMD. Increasing stability in dry matter concentration over a range of soil moisture conditions could contribute to improved drought tolerance. However, evidence for genotypic differences in stability in dry matter concentration is inconclusive.

The supply of water to shoots depends on the ability of roots to exploit available soil moisture. Compared with cereals, root systems of potatoes tend to be weak and shallowly distributed so that the ability to extract soil moisture effectively is restricted. Drought, while reducing dry matter production increases the root:shoot ratio indicating a shift in the balance of growth in favour of roots. Roots of plants grown in droughted conditions also tend to be thinner. Both responses enable droughted plants to exploit the available soil moisture more effectively. Genotypic differences in rooting depth and in amount of root growth have been found, but the significance of these differences for water extraction and growth of droughted crops is not clear.

Grafting experiments indicate that shoots largely determine changes in both stomatal conductance and partitioning of assimilates in response to drought. Interaction of hydraulic and non-hydraulic regulation of physiological processes provides a mechanism coupling long-term adaptation to soil moisture status with short-term response to evaporative demand. Future work must consider not only effects of water-stress on individual processes, but also the integration of response by the whole plant and the physiological processes underlying long-term responses to drought.

A.J. Haverkort and D.K.L. MacKerron (eds.), Potato Ecology and Modelling of Crops under Conditions Limiting Growth, 61–74.

Introduction

In comparison with other crops, potato is regarded as sensitive to water-stress (Salter and Goode 1967) with yields frequently constrained by drought in most of the environments in which the crop is grown. The effects of water-stress on a plant depends on the timing of the stress, its duration and severity. Transient water-stress, induced by high evaporative demand, may have acute, but only temporary, effects on plant growth. Such stress commonly occurs even in well-watered crops, particularly in hot, sunny conditions when evaporative demand exceeds the capacity of roots to supply water to shoots. In contrast, prolonged drought and gradual depletion of soil moisture, although affording greater opportunity for adaptation by the plant, may also induce chronic effects on plant growth and lead to premature senescence. Adaptation to water-stress may involve several different physiological and morphological characters, the relative importance of which may vary according to the type of water-stress experienced by the crop.

MacKerron and Waister (1985) proposed that total yield of the potato crop is the product of four processes: interception of light by the crop canopy, fixation of intercepted light into dry matter, partitioning of dry matter into tubers, and the dry matter concentration of the tubers. Each process is quantifiable, and can be analysed in more detail by looking at mechanisms underlying these processes. This chapter examines the effects of drought stress on crop growth and tuber yield by considering its influence on these processes and the underlying mechanisms.

Light interception and leaf growth

It has long been recognised that leaf growth is one of the first processes affected by water-stress (Hsiao 1973). In potato crops, the principal effect of drought is

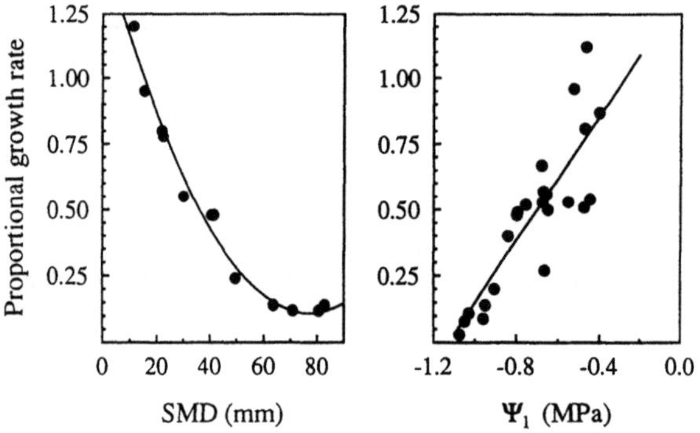

Figure 1. Relations between leaf extension rate in leaves of droughted plants, expressed as a proportion of the rate of the corresponding leaf of irrigated plants, and (a) soil moisture deficit and (b) leaf water potential (from Jefferies 1989).

on light interception. In crops droughted from the time of plant emergence, radiation interception is reduced as a result of slower and lesser canopy expansion, poorer light interception during the longer part of the growing season, and earlier senescence (Jefferies and MacKerron 1987; 1989). Examination of the growth of individual leaves (Jefferies 1989) revealed that drought resulted in a reduction in the both the rate and the duration of growth of leaves, and also in the number of leaves produced. Leaf expansion was closely related to soil moisture deficit (SMD) and declined rapidly when the SMD was greater than 16 mm to a minimum when the SMD was 77 mm (Figure 1). The sensitivity to increasing SMD was reflected in the relation between leaf expansion and leaf water potential (Ψ_l). Leaf expansion declined linearly when the mid-day Ψ_l was less than -0.3 MPa and ceased when Ψ_l was less than -1.1 MPa. However, these values of Ψ_l represent daily minimum values and actual threshold values are greater. In a glasshouse study, Gandar and Tanner (1976) found that leaf extension declined linearly with Ψ_l less than -0.2 MPa and ceased when Ψ_l was less than -0.5 MPa.

Table 1. Soil moisture deficit (mm) at which leaf expansion rate, in several potato cultivars, in the droughted treatment is equal to 1.0, 0.5, or 0.25 of that of fully irrigated plants (from Jefferies 1993b)

Proportional expansion rate	1.0	0.5	0.25
Baillie	< 20	38	51
Cara	< 35	49	62
Désirée	< 11	42	58
Maris Piper	< 22	40	55
Pentland Crown	< 35	61	76
Record	< 27	47	56
Russet Burbank	< 34	52	61
Spunta	< 23	51	69

Experiments comparing the effect of drought on 19 genotypes of potato (Jefferies 1993b), demonstrated genotypic differences both in the threshold SMD at which leaf expansion is constrained by soil moisture supply and in the ability to maintain expansion with increasing SMD (Table 1). Those cultivars considered to be drought susceptible (e.g. cv. Baillie, Maris Piper, and Record), generally had low thresholds and were sensitive to increasing SMD. Cultivars considered to be drought tolerant (e.g. cv. Cara, Pentland Crown, and Spunta) either had greater thresholds, or were less sensitive to increasing SMD, or both. Some of those cultivars that were able to sustain leaf expansion with increasing SMD are characterised by relatively large rooting systems (e.g. Cara), suggesting that the ability is linked to better exploitation of available soil moisture. However, there is no clear correlation between the ability to maintain leaf expansion and root dry matter production. Bansal and Nagarajan (1987) showed a negative correlation between reduction in stomatal conductance and reduction in leaf growth, suggesting that genotypic differences in the ability to maintain leaf growth when droughted may also be linked to stomatal control of leaf water status.

In some species, osmotic adjustment, the active accumulation of solutes within cells lowering osmotic potential so that turgor is maintained, is an important trait associated with drought tolerance (Ludlow and Muchow 1989). Where osmotic adjustment occurs, it allows turgor-dependent processes, such as leaf expansion, to continue at lower leaf water potentials. In the study by Jefferies (1993b), osmotic adjustment, measured as the difference in osmotic potential between rehydrated leaves from droughted and irrigated plants, proved to be limited (< 0.16 MPa) and was not correlated with the ability to maintain leaf growth at increasing SMD. Similarly, Vos and Groenwold (1988), in a study on three potato cultivars, found that although the osmotic potential of rehydrated leaves was influenced by both water-stress and ontogeny, osmotic adjustment was limited (≤ 0.25 MPa).

The expansion of cells depends not only on the water status of the cell but also on the ability of the cell walls to extend (Cleland 1986; Dale 1988), which is determined by biochemical processes. Turgor-dependent growth could be maintained with increasing water-stress by increasing the extensibility of cell walls. However, increased extensibility might also be a disadvantage as it would increase the amount of water required to rehydrate wilted leaves and could delay recovery from water-stress. Conversely, a reduction in cell wall extensibility would limit expansion and hence the evaporating surface so that soil moisture would be conserved initially and turgor maintained. Reduced extensibility would also reduce the amount of water required to rehydrate wilted leaves. The effects of water-stress on the extensible properties of cell walls has not been examined in potatoes. In rice, Turner et al. (1986) found no difference in the bulk modulus of elasticity of leaves, either among genotypes or between irrigated and droughted plants. In *Lolium perenne*, although drought increased the elasticity of cell walls (Thomas 1987), changes were small. Moreover, there was a large phenotypic variation and heritability was small, limiting the potential usefulness of the character for genetic selection.

Light conversion, photosynthesis, stomatal conductance and water use efficiency

The efficiency of conversion of light into dry matter is less sensitive to drought stress than is canopy expansion. Jefferies and MacKerron (1989), by fitting a quadratic curve to the relation between dry matter accumulation and intercepted radiation, estimated that the coefficient for the conversion of light into dry matter (L_c) was unaffected by SMD less than 47 mm (Figure 2). Although L_c is determined by a number of different processes, including canopy architecture, it does provide a measure of photosynthetic performance of the crop, and the results of Jefferies and MacKerron (1989) suggest that effects on photosynthetic performance are small compared with those on leaf expansion.

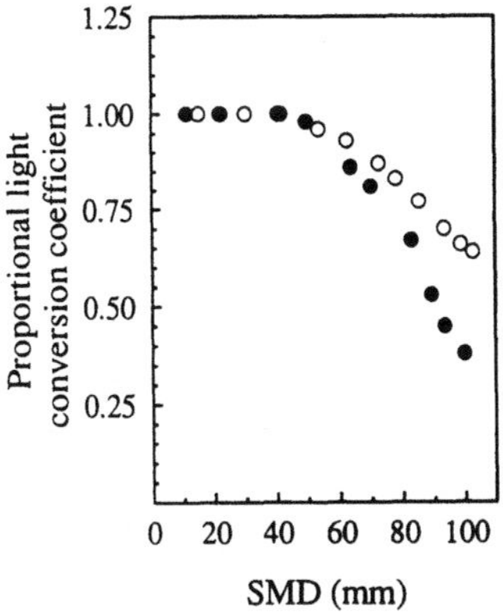

Figure 2. Relation between the coefficient for the conversion of light into dry matter in droughted crops, expressed as a proportion of that of irrigated crops, and soil moisture deficit (SMD) in 1986 (●) and 1987 (○) (from Jefferies and MacKerron 1989).

Analysis of chlorophyll fluorescence in field-grown potato (Jefferies 1994), revealed that drought had no significant effect on the functioning of Photosystem II (PSII). The ratio of variable to maximum fluorescence in either dark- or light-adapted leaves did not differ significantly between irrigated and droughted crops (Figure 3), indicating that the trapping efficiency of PSII was unaffected by drought. Analysis of fluorescence quenching, revealed no significant effect of drought on photochemical quenching (q_p). Consequently the quantum yield of PSII electron transport, which is the product of the ratio of variable to maximum fluorescence in the light-adapted state and q_p, was unaffected, indicating that electron transport was maintained in droughted plants. That, in turn, suggests that photosynthetic electrons are being used in photochemical processes in water-stressed leaves despite stomatal limitation of photosynthesis. Increased photorespiration, which is promoted by reduced CO_2 concentration in the sub-stomatal cavity, or increased oxidation by the pentose phosphate pathway may provide the process by which excess energy is dissipated in water-stressed plants. Both processes would generate CO_2 and provide a cycle which would consume excess energy and maintain high rates of electron transport, protecting the photosystem from damage. Such mechanisms may also enable rapid adaptation to and recovery from transient water-stress. However, prolonged drought results in a decline in minimum and maximum fluorescence (Jefferies 1992a; 1994), indicating premature senescence and a decline in the chlorophyll light harvesting complexes. In addition, prolonged drought may result in a reduction in photosynthetic capacity as a result of down-regulation of carboxylation enzymes (Chaves 1991).

Drought decreased the rate of quenching of maximum fluorescence following a dark-light transition (Jefferies 1992a), indicative of stomatal limitation of photosynthesis. However, there was only a poor correlation between the rate of quenching and Ψ_l. Moreover, the relation differed over time. In water-stressed plants, stomatal conductance to water vapour declined earlier, more rapidly,

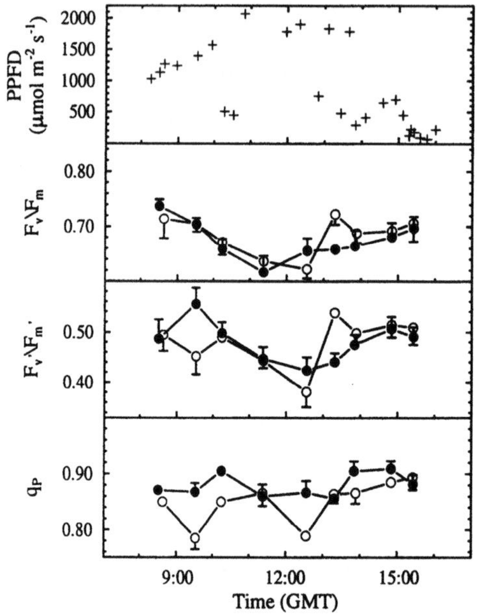

Figure 3. Daytime changes, 59 days after emergence, in photosynthetic photon flux density (PPFD), yield of variable fluorescence (F_v/F_m) in the dark-adapted state, yield of variable fluorescence in the light adapted state (F_v'/F_m') and photochemical quenching (q_P) in leaves of potato plants either irrigated (●) or droughted (○) from the time of plant emergence. Error bars represent standard errors of the mean (from Jefferies 1994).

and to a greater extent than photosynthetic rate measured by gas exchange (Vos and Oyarzún 1987; Vos and Groenwold 1989a). Although both stomatal conductance and photosynthetic capacity were related to Ψ_l, relations were not unique, differing with leaf position (Vos and Oyarzún 1987). In irrigated plants, Ψ_l fell to less than the pre-dawn values of droughted leaves without inducing changes in the ratios either between internal and external CO_2 concentration (C_i/C_a) or between stomatal conductance and photosynthetic rate (Vos and Groenwold 1989a), indicating that Ψ_l does not determine the response of stomata to drought.

Farquhar et al. (1982) showed theoretically that water use efficiency (the ratio between CO_2 assimilation and transpiration) is related to C_i/C_a, such that a decrease in C_i results in an increase in water use efficiency. Moreover, they also established a relation between C_i/C_a and the discrimination of naturally occurring stable isotope ^{13}C relative to ^{12}C, suggesting that increased water use efficiency would result in a decrease in carbon isotope discrimination. Vos and Groenwold (1989a), confirmed that, for potato, greater reduction in stomatal conductance to water vapour relative to photosynthetic rate, is associated with a decrease in C_i/C_a and an increase in water use efficiency. In well-watered plants, differences in water use efficiency between cultivars reflected differences in stomatal conductance rather than differences in photosynthetic capacity. Vos and Groenwold (1989b) also confirmed that, for potato, carbon isotope discrimination and water use efficiency are negatively correlated (one must be careful to note that these authors use the term fractionation, which has negative values, but Farquhar et al. (1982) use the term discrimination, which has positive values). Differences in water use efficiency were related to differences in stomatal conductance, but the correlation between stomatal conductance and isotope discrimination was poor. Recent work at the Scottish Crop Research Institute (R.A. Jefferies, unpublished) found, in well-watered diploid clones of potato, a significant correlation ($r^2 = 0.66$, $P < 0.001$) between isotope discrimination and stomatal conductance (Figure 4), but no correlation between isotope discrimination and photosynthetic capacity measured at 5% CO_2 concentration, confirming that differences in isotope discrimination largely reflect differences in stomatal- rather than mesophyll resistances.

Partitioning to tubers

The evidence on the effects of drought on partitioning of assimilate to tubers appears to be conflicting. In the cultivar Maris Piper, Jefferies and MacKerron (1989) found that drought initially increased assimilate partitioning to tubers so that harvest index was greater in droughted than in irrigated crops during mid-season. However, final harvest index at the end of the season was similar in both irrigated and droughted crops. Jefferies (1992b), in a comparison of seven potato genotypes, found drought significantly increased harvest index in one cultivar and decreased it in another. In a comparison of nineteen genotypes (Jefferies and MacKerron 1993) the harvest index of eight genotypes at the end of the season was significantly lower in droughted than in irrigated plants. Similarly, studies by MacKerron and Peng (1989) and Jefferies (1993a) found drought to reduce the partitioning of assimilate to tubers. These differences in response may reflect genotypic differences in development and also differences in the timing and severity of the water-stress imposed. Timing of water-stress

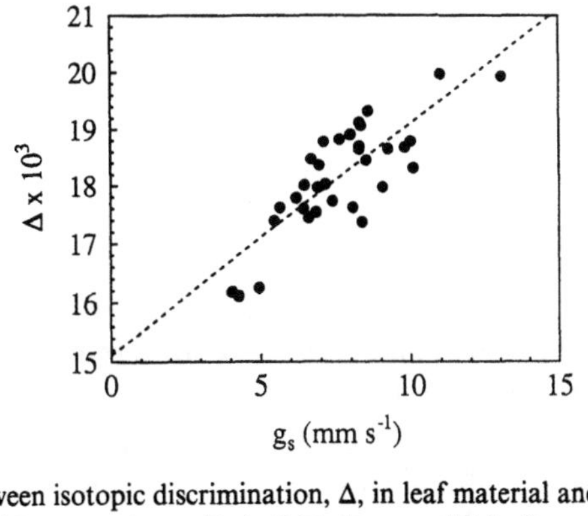

Figure 4. Relation between isotopic discrimination, Δ, in leaf material and stomatal conductance, g_s in well-watered diploid potato clones (R.A. Jefferies, unpublished).

relative to tuber initiation may be critical in determining patterns of assimilate partitioning. Water-stress early in the season reduces the numbers of stolons and tubers initiated (MacKerron and Jefferies 1986; Haverkort et al. 1990). Thus, severe and prolonged stress from early in the season may constrain tuber initiation causing assimilate to be partitioned to other organs. However, if water-stress occurs after tubers have been initiated, then partitioning to tubers is promoted and maturity is advanced. Further work is required to provide a quantified understanding of partitioning in relation to the timing and severity of water-stress. Labelling experiments using stable isotopes of carbon offer an elegant method of examining long term patterns of carbon partitioning, as has been demonstrated in *Brassica napus* (Deleens et al. 1989).

Tuber dry matter concentration

Tuber dry matter concentration, which is an important component of tuber yield, is dependent on both the dry matter accumulated into tubers and upon tuber water content. Tuber dry matter concentration tends to show a progressive increase from the time of tuber initiation until maturity, except when stress induces a remobilisation of starch. Tuber water content reflects plant water status and changes in it are associated with movement of water to and from the tuber in response to gradients in water potential. Tuber water potential is related to soil water potential (Epstein and Grant 1973) and exhibits diurnal fluctuations. In droughted crops, tuber dry matter concentration increases because tuber water content is affected to a greater extent than dry matter accumulation. As long as tubers are attached to an active root system, relief of water-stress results in a decrease in dry matter concentration as the uptake of water is increased (MacKerron and Jefferies 1985). There has been little attempt to quantify processes determining tuber dry matter concentration. Jefferies et al. (1989), using data for three potato genotypes, related tuber dry matter concentration to thermal time accumulated from plant emergence and SMD. Although validated against independent data, the relations are empirical and are unlikely to be appropriate for environments in which heat-stress reduces dry matter concentration.

For a given depression in the production of tuber dry matter caused by drought, cultivars that show the least change in tuber dry matter concentration will exhibit the greatest drought tolerance. Selection of genotypes for stability in dry matter concentration over a range of soil moisture conditions may contribute towards increased drought tolerance. Jefferies and MacKerron (1993) found that, although drought increased tuber dry matter concentration in all nineteen genotypes examined, the effect differed between genotypes, with drought increasing tuber dry matter concentration by 12.9 percentage points in clone 8906AC11, but by only 8.3 percentage points in the cultivar Up-to-Date. From the analysis of unreplicated trials, Killick and Simmonds (1974) concluded that genotype-by-environment effects on the specific gravity of tubers, and hence tuber dry matter concentration, were small, but they found a greater effect of years than of location, perhaps suggesting a greater effect of differences in water-stress between years than of other environmental effects differing between locations. Clearly, further work is required both to quantify the effects of environment on dry matter concentration to provide more mechanistic descriptions, and to determine the degree to which stability of tuber dry matter concentration differs between genotypes.

Root characteristics

The supply of water to shoots depends not only on soil moisture status, but also on the ability of roots to exploit available soil moisture. In a review of the literature, Gregory and Simmonds (1992) concluded that the quantity of roots produced by potatoes is generally similar to that produced by temperate grain legumes, only about one third of that produced by cereal crops, and that a greater proportion of the root system is in the surface layers. These rooting characteristics may limit the ability of potatoes to extract available soil moisture effectively and may contribute to the relative sensitivity of the crop to water stress. Unpublished studies reported by Gregory and Simmonds (1992), indicate that, in moist soils, between 65–75% of the total resistance to the flow of water between soil and leaf occurs below ground, and exceeds 80% in dry soils as the soil hydraulic resistance increases and hydraulic contact between roots and soil decreases. Moreover, in well-watered crops, hydraulic conductance between bulk soil and the base of the stem was proportional to root length in the soil, suggesting that the relative inability of potatoes to convey water to the shoots is a reflection of the relatively small root system, rather than of an inherently small conductance per unit length of root. Gregory and Simmonds (1992) also argued that, as the profile dries, the soil water potential (Ψ_S) experienced by the crop tends to decrease more rapidly in shallow-rooting crop species than in deeper rooting crop ones because the effective Ψ_S experienced by the crop is weighted according to root length density down the profile.

In a grafting experiment, Jefferies (1993) showed that drought, while reducing dry matter production, increased both root:shoot ratio and the ratio of root length to root weight. Both responses enable the plant to maximise the soil volume exploited and allow water uptake to be sustained. Thinner roots may also have smaller vessels and greater hydraulic resistance, which could lead to increased water-stress in the shoot, but might also conserve soil moisture. Although plants compensate to sustain water uptake, if evaporation exceeds the rate at which the roots can grow and exploit the potential rooting volume, roots may fail to extract all the available water in the profile. Thus, MacKerron and Peng (1989) found that, in plants grown in 1.5 m pipes, drought resulted in an upward shift in the proportional distribution of roots within the soil profile, and that soil was moist below the lower limit of root growth. This effect may be further exacerbated if drying increases soil strength and inhibits root extension.

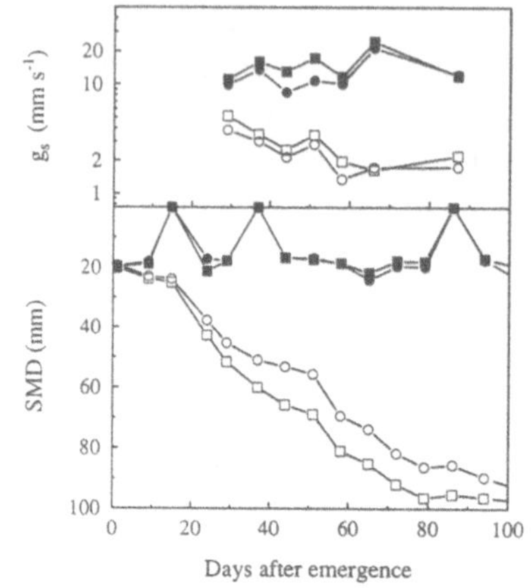

Figure 5. Soil moisture deficit (SMD) and stomatal conductance (g_s) in irrigated (closed symbols) and droughted (open symbols) crops of potato cultivars Cara (●, ○) and Désirée (■, □) (R.A. Jefferies and D.K.L. MacKerron, unpublished).

Genotypic differences have been found for both rooting depth (Steckel and Gray 1979) and in amount of root growth (MacKerron and Peng 1989; Jefferies 1993). However, the importance of these characters for crop water use and yield under drought is not clear. In a comparison of four cultivars, Steckel and Gray (1979) were unable to relate differences in yield to small in differences in rooting depth or water extraction. Similarly, experiments conducted at the Scottish Crop Research Institute (R.A. Jefferies and D.K.L. MacKerron, unpublished) comparing growth and water use of two potato cultivars, Désirée and the more prolific-rooting Cara, showed that although SMD eventually were similar in droughted crops of both cultivars, the increase in SMD with time was faster in Désirée than in Cara (Figure 5) as stomatal conductance tended to be lower in Cara. Differences in the pattern of water use in the two cultivars did not lead to significant differences in dry matter production in droughted conditions. Clearly, work is required to quantify the dynamics of root growth and water use in potato crops. These results also emphasise the need to consider the integrated response of the plant to drought stress.

Integration of response to drought

Over the past 30 years, it has generally been regarded that plant response to water-stress is controlled by plant water status. In particular, water potential has been promoted as the determinant of physiological response to water-stress because of its physical basis, its ease of measurement, and its suitability for modelling using electrical analogues. However, although relations between various plant processes and water potential have been established, the relations are not unique and are empirical rather than causal. Consequently, although these relations enable models of plant response to water-stress to be constructed, such models should not necessarily be regarded as mechanistic simply because they are based on physical concepts. Sinclair and Ludlow (1985) argued that relative water content is more important than leaf water potential as a determinant of physiological activity. This is supported by the fact that, for a variety of physiological processes, differences between species in response to water deficits, disappear when the response is related to relative water content rather than to water potential. However, although relative water content may be a more important determinant of physiological activity and can easily be measured experimentally, it is not as amenable as water potential is to modelling, except through the use of empirical relations between it and water potential.

Although plant water status is an important determinant of physiological response, it has become increasingly apparent that plant response to drought is also regulated by non-hydraulic signals produced by roots in response to drying. Split-root experiments (Blackman and Davies 1985; Blum et al. 1991) have demonstrated that drying part of the root system can reduce stomatal conductance and leaf growth without any measurable effect on plant water status. Passioura (1988) found that, in wheat seedlings, soil drying reduced leaf expansion even though roots were pressurised to maintain leaf turgor. Although some studies (Munns 1990; Munns et al. 1993) have suggested that other substances may be involved, the increase in concentration of abscisic acid (ABA) in xylem sap with soil drying, and negative correlations between the concentration of ABA in xylem sap and either the rate of leaf expansion or stomatal conductance (Zhang and Davies 1990a,b) have implicated ABA as the major substance involved. Moreover, feeding ABA into plant stems artificially to increase the concentration of ABA in the xylem resulted in a decrease in stomatal conductance (Tardieu et al. 1993), suggesting that the relation between stomatal conductance and the concentration of ABA in xylem sap is causal. Recent results have shown that sensitivity of stomata to ABA increases with decreasing Ψ_l (Tardieu and Davies 1992; Tardieu et al. 1993) indicating an interaction between non-hydraulic and hydraulic regulation of stomatal conductance. Tardieu and Davies (1993) argued that the interaction of hydraulic and non-hydraulic effects provides a mechanism coupling long-term adaptation to soil moisture status with short term response to evaporative demand, and they proposed a water-flux-dependent model integrating hydraulic and non-hydraulic effects. Work on non-hydraulic root-to-shoot communication has been confined to species other than potato, and its significance for water relations and crop growth ought to be examined in potato. Moreover, genetic differences in root-to-shoot communication and their possible significance for selecting for drought tolerance have yet to be explored in detail.

Work on non-hydraulic root-to-shoot communication has tended to emphasise the importance of the root system in sensing soil moisture status and integrating the response of the plant to drought. However, it is evident that long term integration of the response of the plant to drought is more complex.

Table 2. Effects of scion and root stock on dry matter partitioning between shoot, tubers and roots in irrigated and droughted potato (from Jefferies 1993a)

Scion/Stock	Shoot/Total	Root/Total	Tuber/Total	Root/Shoot
Cara/Cara				
Irrigated	0.45	0.04	0.51	0.10
Droughted	0.86	0.14	0.00	0.20
Cara/Désirée				
Irrigated	0.46	0.03	0.51	0.07
Droughted	0.81	0.12	0.07	0.19
Désirée/Cara				
Irrigated	0.25	0.02	0.73	0.11
Droughted	0.62	0.10	0.28	0.18
Désirée/Désirée				
Irrigated	0.27	0.02	0.71	0.08
Droughted	0.49	0.08	0.43	0.21
SED (d.f. = 26)	0.034	0.007	0.038	0.015

Jefferies (1993a) used reciprocal grafting experiments to examine the balance between root and shoot in determining the response of potato to drought. In these experiments, stomatal conductance was largely determined by the scion in both irrigated and droughted plants. Scions also had a dominant effect in determining partitioning of dry matter between shoot, root and tubers (Table 2). Thus scions of the cultivar Cara increased root growth in root stocks of the cultivar Désirée. Differences between grafts in response to drought were determined largely by the scion with those of Cara showing a greater shift in partitioning in favour of shoots and roots than did scions of Désirée. These effects of shoots on dry matter partitioning to roots in response to drought may have implications for the exploitation and uptake of soil water, and root-to-shoot communication. Future work must consider not only effects of water-stress on individual processes, but also both the integration of response by the whole plant and the physiological processes underlying long term responses to drought.

Acknowledgements

The financial support of the Scottish Office Agriculture and Fisheries Department is acknowledged.

References

Bansal K C, Nagarajan S (1987) Reduction of leaf growth by water stress and its recovery in relation to transpiration and stomatal conductance in some potato (*Solanum tuberosum* L.) genotypes. Potato Research 30:497–506.

Blackman P G, Davies W J (1985) Root to shoot communication in maize plants of the effects of soil drying. Journal of Experimental Botany 36:39–48.

Blum A, Johnson J W, Ramseur E L, Tollner E W (1991) The effect of drying topsoil and a possible non-hydraulic root signal on wheat growth and yield. Journal of Experimental Botany 42:1225–1231.

Chaves M M (1991) Effects of water deficits on carbon assimilation. Journal of Experimental Botany 42:1–16.

Cleland R E (1986) The role of hormones in wall loosening and growth. Australian Journal of Plant Physiology 13:93–103.

Dale J E (1988) The control of leaf expansion. Annual Review of Plant Physiology and Plant Molecular Biology 39:267–295.

Deleens E, Marcotte L, Schwebel-Dugue N, Vartanian N (1989) Stable isotope carbon study: long-term partitioning during progressive drought stress in *Brassica napus* var. *oleifera*. Plant, Cell and Environment 12:615–620.

Epstein E, Grant W J (1973) Water stress relations of the potato plant under field conditions. Agronomy Journal 65:400–404.

Farquhar G D, O'Leary M H, Berry J A (1982) On the relationship between carbon isotope discrimination and the intercellular carbon concentration in leaves. Australian Journal of Plant Physiology 9:121–137.

Gandar P W, Tanner C D (1976) Leaf growth, tuber growth and water potential in potatoes. Crop Science 16:534–538.

Gregory P J, Simmonds L P (1992) Water relations and growth of potatoes. Pages 214–246 in P. Harris (Ed.) The Potato Crop: The Scientific Basis for Improvement, 2nd Edition. Chapman & Hall, London, UK.

Haverkort A J, Van de Waart M, Bodlaender K B A (1990) The effect of early drought stress on numbers of tubers and stolons of potato in controlled and field conditions. Potato Research 33:89–96.

Hsiao T C (1973) Plant responses to water stress. Annual Review of Plant Physiology 24:519–517.

Jefferies R A (1989) Water-stress and leaf growth in field-grown crops of potato (*Solanum tuberosum* L.). Journal of Experimental Botany 40:1375–1381.

Jefferies R A (1992a) Effects of drought on chlorophyll fluorescence in potato (*Solanum tuberosum* L.). I. Plant water status and the kinetics of chlorophyll fluorescence. Potato Research 35:25–34.

Jefferies R A (1992b) Effects of drought on chlorophyll fluorescence in potato (*Solanum tuberosum* L.). II. Relations between plant growth and measurements of fluorescence. Potato Research 35:35–40.

Jefferies R A (1993a) Cultivar responses to water stress in potato: effects of shoot and roots. New Phytologist 123:491–498.

Jefferies R A (1993b) Responses of potato genotypes to drought. I. Expansion of individual leaves and osmotic adjustment. Annals of Applied Biology 122:93–104.

Jefferies R A (1994) Drought and chlorophyll fluorescence in field-grown potato (*Solanum tuberosum*). Physiologia Plantarum 90:93–97.

Jefferies R A, Heilbronn T D, MacKerron D K L (1989) Estimating tuber dry matter concentration from accumulated thermal time and soil moisture. Potato Research 32:411–417.

Jefferies R A, MacKerron D K L (1987) Aspects of the physiological basis of cultivar differences in yield of potato under droughted and irrigated conditions. Potato Research 30:201–217.

Jefferies R A, MacKerron D K L (1989) Radiation interception and growth of irrigated and droughted potato (*Solanum tuberosum*). Field Crops Research 22:101–112.

Jefferies R A, MacKerron D K L (1993) Responses of potato genotypes to drought. II. Leaf area index, growth and yield. Annals of Applied Biology 122:105–112.

Killick R J, Simmonds N W (1974) Specific gravity of potato tubers as a character showing small genotype-environment interactions. Heredity 32:109–112.

Ludlow M M, Muchow R C (1989) A critical evaluation of traits for improving crop yields in water-limited environments. Advances in Agronomy 43:107–153.

MacKerron D K L, Jefferies R A (1986) The influence of early soil moisture stress on tuber numbers in potato. Potato Research 29:299–312.

MacKerron D K L, Jefferies R A (1985) Observations on the effects of the relief of late water stress in potato. Potato Research 28:349–359.

MacKerron D K L, Peng Z Y (1989) Genotypic comparisons of potato root growth and yield response to drought. Pages 199–206 in Aspects of Applied Biology 22, Roots and the Soil Environment. Association of Applied Biologists, UK.

MacKerron D K L, Waister P D (1985) A simple model of potato growth and yield. I. Model development and sensitivity analysis. Agricultural and Forest Meteorology 34:241–252.

Munns R (1990) Chemical signals moving from roots to shoots: the case against ABA. Pages 175–183 in Davies W J, Jeffcoat B (Eds.) British Society for Plant Growth Regulation, Monograph 21. Importance of Root to Shoot Communication in Responses to Environmental Stress. BSPGR, UK.

Munns R, Passioura J B, Milborrow B V, James R A, Close T J (1993) Stored xylem sap from wheat and barley in drying soil contains a transpiration inhibitor with a large molecular size. Plant, Cell and Environment 16:867–872.

Passioura J B (1988) Root signals control leaf expansion in wheat seedlings growing in drying soil. Australian Journal of Plant Physiology 15:687–693.

Salter P J, Goode J E (1967) Crop responses to water at different stages of growth. Research Reviews of the Commonwealth Bureau of Horticulture, East Malling 2:93–97.

Sinclair T R, Ludlow M M (1985) Who taught plants thermodynamics? the unfulfilled potential of plant water potential. Australian Journal of Plant Physiology 12:213–217.

Steckel J R, Gray D (1979) Drought tolerance of potatoes. Journal of Agricultural Science, Cambridge 47:770–775.

Tardieu F, Davies W J (1992) Stomatal response to abscisic acid is a function of current plant water status. Plant Physiology 98:540–545.

Tardieu F, Zhang J, Gowing D J G (1993) Stomatal control of both [ABA] in the xylem sap and leaf water status: a test of a model for droughted or ABA-fed field-grown maize. Plant, Cell and Environment 16:413–420.

Thomas H (1987) Physiological responses to drought of *Lolium perenne* L.: measurement of, and genetic variation in, water potential, solute potential, elasticity and cell hydration. Journal of Experimental Botany 38:115–125.

Turner N C, O'Toole J C, Cruz R T, Yambao E B, Sayeed A, Namuco O S, Dingkuhn M (1986) Responses of seven diverse rice cultivars to water deficits. II. Osmotic adjustment, leaf elasticity, leaf extension, leaf death, stomatal conductance and photosynthesis. Field Crops Research 13:273–286.

Vos J, Groenwold J (1988) Water relations of potato leaves. I. Diurnal changes, gradients in the canopy, the effects of leaf insertion number, cultivar and drought. Annals of Botany 62:363–371.

Vos J, Groenwold J (1989a) Characteristics of photosynthesis and conductance of potato canopies and the effects of cultivar and transient drought. Field Crops Research 20:237–250.

Vos J, Groenwold J (1989b) Genetic differences in water use efficiency, stomatal conductance and carbon isotope fractionation in potato. Potato Research 32:113–121.

Vos J, Oyarzún P J (1987) Photosynthesis and stomatal conductance of potato leaves – effects of leaf age, irradiance, and leaf water potential. Photosynthesis Research 11:253–264.

Zhang J, Davies W J (1990a) Changes in the concentration of ABA in xylem sap as a function of changing soil water status can account for changes in leaf conductance and growth. Plant, Cell and Environment 13:277–285.

Zhang J, Davies W J (1990b) Does ABA in the xylem control the rate of leaf growth in soil-dried maize and sunflower plants? Journal of Experimental Botany 41:1125–1132.

5. Modelling water and nitrogen uptake of a potato crop growing on a ridge

P. DE WILLIGEN[1], M. HEINEN[1] and B.J. VAN DEN BROEK[2]

[1] *DLO Research Institute for Agrobiology and Soil Fertility (ABDLO), P.O. Box 129, 9750 AC Haren, The Netherlands*

[2] *DLO Winand Staring Centre for Integrated Land, Soil and Water Research (SC.DLO), P.O. Box 125, 6700 AC Wageningen, The Netherlands*

Abstract. The potato crop is usually grown on ridges of height 20–25 cm and about 75 cm apart. It is to be expected that not only will vertical movement of water and nutrients occur in the soil but, also, due to the form of the ridge and the spatial variation of root distribution, horizontal movement will occur as well. Therefore, a two-dimensional model for soil water movement was constructed which could be used to simulate water transport within a ridge for a crop with a non-regular distribution of roots. The model was extended with modules describing nutrient transport, water and nutrient uptake. Data taken from literature were used to demonstrate the possibilities of this model. For identical environmental conditions and nitrogen supply it was shown that in a sandy clay loam nitrogen uptake was larger than in a loamy sand. This could be explained by considering the differences in the hydraulic properties of the two soils, the sandy clay loam having a higher water content at the same pressure head.

Introduction

Plants require large amounts of water, removing it from the soil and transpiring nearly all of it into the atmosphere. For the production of 1 kg of dry matter, depending on climate, plant type and fertility, 300 to 2000 kg of water may be required (Wesseling 1976).

Pressure head gradients between the soil and the plant roots tend to draw water into the roots, through the plant and into the atmosphere. If a critical pressure head is reached in the soil, plants can no longer remove water at the required rate and crop growth will eventually cease.

Next in importance to the availability of water in the soil for satisfying the evaporative demand, water in soil also plays a prominent role in many other agronomically important processes (e.g. mineralization, precipitation, solubilization) and determines to a large degree the possible rate of uptake of nutrients. Firstly, because uptake of water by the plant generates a flow of solutes towards the root, i.e. transport of nutrients by mass flow. Secondly, because even in the absence of any water flow, the water content of the soil determines the possible rate of diffusion of solutes, as the diffusion coefficient is a function of the water content. Thus one can expect a considerable interaction between nutrient uptake and water status of the soil, as indeed has

A.J. Haverkort and D.K.L. MacKerron (eds.), Potato Ecology and Modelling of Crops under Conditions Limiting Growth, 75–88.

been found in the case of nitrate for potato (Asfary et al. 1983, Clutterbuck and Simpson 1978).

The potato crop is considered to be sensitive to water-stress (Harris 1978; Loon 1981 in Jefferies 1992). Vos and Groenwold (1986) mention that root development of potato is often found to be less than that of other arable crops e.g. cereals. This may partly explain the sensitivity of potatoes for drought.

A potato crop is usually grown in rows, where the row distance is about 75 cm, so that one can expect considerable horizontal gradients perpendicular to the row. Moreover, potatoes are generally grown on ridges about 20–25 cm above soil surface. The systematic variation associated with planting pattern and ridge building may lead to different demands for the spatial supply of water and nutrients than those for crops which are sown broadcast and grown on a even surface. To take the effects of rows and ridges into account a two-dimensional model was designed to simulate water and nitrate transport in the soil and their uptake by the plant.

Description of the model

The modules for water transport, water uptake, nutrient transport and nutrient uptake will be briefly discussed in turn.

Water transport

The core of the model is a module describing transport of water in a rectangular domain. This module is called FUSSIM2, which is an acronym for *F*low of water in *U*nsaturated (vadose) *S*oil *S*imulation *M*odel in two (*2*) dimensions (Heinen and De Willigen 1992). We will present and discuss several aspects of this module. These aspects are, succesively: the governing flow equation, boundary conditions, the numerical (finite difference) discretization steps for the flow equation and the boundary conditions, possible ways to obtain average hydraulic conductivities, the representation of a soil ridge and the solution procedure.

Governing flow equation

Generally, the governing flow equation, which is valid for saturated-unsaturated, heterogeneous, isotropic, non-hysteretic and rigid soils, reads (Koorevaar et al. 1983):

$$\frac{\partial \Theta}{\partial t} = \Delta \cdot \{K(h)\Delta h\} - \frac{\partial K(h)}{\partial z} + S \qquad (1)$$

where h represents the pressure head (cm), and z represents the gravitational head (cm). The minus sign in Eq. 1 appears because the vertical coordinate (also denoted by z) is chosen positive downward. The pressure head and volumetric water content are related by the water retentivity function $\Theta(h)$. The hydraulic conductivity K (cm • d^{-1}) is a function of h and consequently of Θ, so it is denoted by $K(h)$ or $K(\Theta)$, respectively. S (cm^3 • cm^{-3} • d^{-1}) represents a sink and/or source term.

As mentioned before we consider two-dimensional flow in a rectangular flow domain of depth L and width X. Then Eq. 1 is written as:

$$\frac{\partial \Theta}{\partial t} = \frac{\partial}{\partial x} K(h) \frac{\partial h}{\partial x} + \frac{\partial}{\partial z} K(h) \frac{\partial h}{\partial z} - \frac{\partial K(h)}{\partial z} + S \qquad (2)$$

where x (cm) represents the horizontal coordinate. In the remainder of this paper we will use K to represent $K(h)$. As Eq. 1, Eq. 2 is a non-linear partial differential equation, for which analytical solutions can be found for special $K(h)$ and S relationships. For any other situation, the governing flow equation can only be solved numerically. Here an implicit finite difference scheme is used.

Boundary conditions

The governing flow equation can only be solved when the boundary conditions are known. For our purposes the following set of conditions is appropriate. The boundary condition at the upper boundary $z = 0$, the soil atmosphere interface, is that of prescribed flux (rain and/or evaporation):

$$z = 0,\ F_z = R - E_p \qquad (3)$$

where F_z is the flux in the z-direction, R is rain (cm • d^{-1}), and E is the actual evaporation. Potential evaporation is calculated from the potential evapotranspiration taking into account the leaf area index of the crop (Belmans et al. 1982):

$$E_p = ET_p \,.\, e^{-0.6 * LAI} \qquad (4)$$

where E_p is potential evaporation, ET_p is potential evapotranspiration, which is input to the model. Evaporation is then calculated from the potential evaporation and the pressure head in the relevant control volume. The reduction of evaporation is a function of the pressure head at the soil surface, the functional relation being taken from Van Keulen (1975). The pressure head at the upper surface of the control volumes forming the top boundary of the ridge was found by extrapolation of the pressure head in the center of the control volume, by assuming a steady-state profile.

At the lower boundary $z = L$ the condition is that of prescribed pressure head:

$$z = L,\ h = h_L \qquad (5)$$

The prescribed fluxes and prescribed pressure heads can be a function of x and t. Moreover mixed boundary conditions can occur, e.g. prescribed flux at some points and prescribed pressure head at others.

The boundary conditions at the boundaries $x = 0$, and $x = X$, are those of no flow (vanishing flux):

$$x = 0,\ x = X,\ \frac{\partial h}{\partial x} = 0 \qquad (6)$$

Discretization of flow equation
The formulation of discretization equations is as recommended by Patankar (1980). The rectangular domain is divided by a grid in so-called control volumes, in the centre of which the grid points are situated. The pressure heads at these points are to be calculated as well as the fluxes over the faces of the control volumes. Discretisation is achieved by integrating the flow equation over the control volume and the time step. Details of the procedure can be found in Heinen and De Willigen (1992).

Averaging the hydraulic conductivity
In the discretisized equations we need the hydraulic conductivity at the interfaces between the control volumes. However, only the values of K at the nodes are known. Therefore, K at the interface between two control volumes must be obtained as some kind of an average using the K values of the two nodes on either side of the interface. Different possibilities of averaging exist: the arithmetic average, the geometric average, and the reciprocal of the average of the reciprocals, known as the harmonic average. Whenever the hydraulic conductivity at a certain node is zero, the geometric and the harmonic average cause K to be zero at all the faces of the control volume. This can be used to model irregular boundaries as explained below.

Representation of the ridge
Though a zero value for the hydraulic conductivity is physically impossible, allocating a zero value to the conductivity for part of the soil domain can be used to 'block off' some of the control volumes of the regular grid, so that the remaining active control volumes form the irregular domain, by which the ridge is represented. The hydraulic conductivity of the hatched control volumes in Figure 1 is set to zero, the boundary of the ridge is approximated by the

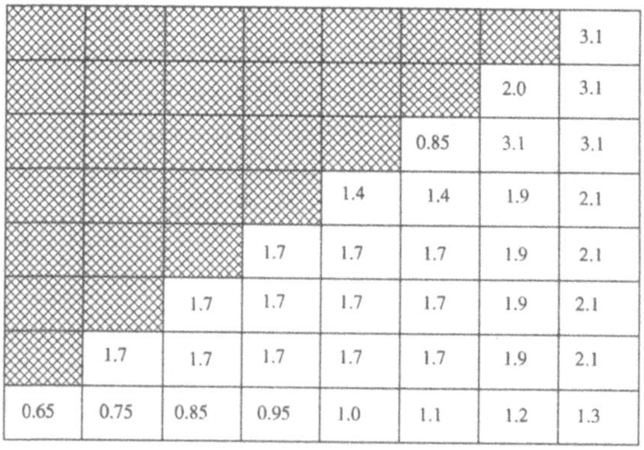

Figure 1. Representation of the centre and one side of a potato-ridge in the model FUSSIM2. The ridge is divided into 64 control volumes. The hatched control volumes form the blocked-off region, the saturated conductivity K_s (cf. Eq. 8) is set at zero. The figures in the control volumes indicate the local root length density in cm • cm^{-3}.

'staircase' of the boundaries of the uppermost active control volumes. In doing so one can handle an irregularly shaped boundary, while maintaining an orthogonal grid (Patankar 1980).

Solution

After discretization of the flow equation for all grid points one obtains $M \times N$ equations, which can be solved directly by for example the Gauss-Seidel elimination method. But according to Patankar (1980) convergence of this method is rather slow especially for high numbers of grid points. We used the Alternating Direction Implicit (ADI) method (Peaceman and Rachford 1955), which is often recommended. It has been used in connection with the two-dimensional Richards' equation for saturated and unsaturated soil conditions (e.g. Rubin 1968; Wang and Anderson 1982). This method consists of two steps: first the equations are written implicitly in one direction and explicitly in the other direction, and then vice versa. Solving the equations obtained in step one yields intermediate values for h at the grid points. These intermediate values of h are used in the second step. Solution of the equations obtained in the second step then yields the new h values at $t+\Delta t$. Because of the non-linearity of Eq. 2 an iterative ADI method was used.

Water uptake

Water uptake is calculated as described by De Willigen and Van Noordwijk (1995). It is based on a microscopic numerical uptake model, describing the flow of water to a root to consist of two components: one describing the flux of water from the bulk soil to the root surface with Darcy's equation and the second describing the flux over the root surface into the root as the product of the root conductivity and the difference of pressure head of the soil at the root surface and the pressure head in the root. Earlier De Willigen and Van Noordwijk (1987) showed that the pressure head distribution in the soil around the root could be very well approximated by calculating it from a steady-rate distribution of the matric flux potential, the latter being defined as:

$$\Phi = \int_h^0 K \mathrm{d}h \qquad (7)$$

By equating the flux to the root and the flux into the root, the pressure head at the surface of the root and that in the plant can be calculated. This procedure can be used to calculate uptake by a root system distributed over different control volumes. By iteration, then, the pressure head at the root surface in each control volume and the pressure head in the plant, which is assumed to have the same value everywhere in the plant, can be computed. The relevant equations can be found in De Willigen (1990) and De Willigen and Van Noordwijk (1995). The assumption that the pressure head in the plant is everywhere the same, leads under some conditions to the phenomenon of hydraulic lift, i.e. in some control volumes water will exude from the root into the soil. The actual transpiration is taken to be a function of the pressure head in the plant and potential transpiration, the latter being found as the difference between potential evapotranspiration and potential evaporation. The reduction of the transpiration due to low values of the pressure head in the plant was as proposed by Campbell (1985):

$$f(hp) = \frac{1}{\left(1+\frac{hP}{hP_{1/2}}\right)^a}$$

where h_P is the pressure head in the plant, $h_{P,1/2}$ is the pressure head in the plant where the transpiration is half of the potential transpiration, and the exponent a is a species-dependent constant.

Nutrient transport

Transport of solutes is by mass flow and diffusion/dispersion. The former is calculated by multiplying the flux of water over a face of two adjacent control volumes by the concentration in that control volume for which the direction of the flux is outward. For a nutrient like nitrate the concentration is simply the amount in the control volume divided by the amount of water. For other nutrients e.g. phosphate the concentration has to be calculated from the amount of labile phosphate and the adsorption isotherm (Van Noordwijk et al. 1990). The dispersion coefficient is calculated as the product of the absolute value of the flux of water and a characteristic length: the dispersion length. The dispersion coefficient is added to the diffusion coefficient, which is calculated as the product of the diffusion coefficient in water, the water content and a tortuosity factor, which depends on the water content (Barraclough and Tinker 1981). The tortuosity factor is a roughly linear function of the water content.

Nutrient uptake

This module is based on the assumption that uptake is determined by plant demand, as long as the transport of the nutrient towards the root allows the concentration at the root surface to be greater than zero, otherwise the roots take up with the maximum rate the nutrient can de transported to the root. This maximum rate can be shown to be proportional to the average concentration and the diffusion coefficient, whereas the proportionality constant is a function of root length density. Derivation of the equations used can be found in De Willigen and Van Noordwijk (1994a,b). With these equations as a basis, a procedure has been constructed by which uptake by a non-regular distributed root system can be calculated, in which also eventual compensatory uptake by roots in favourable situations is taken into account. The procedure is described in detail in De Willigen and Van Noordwijk (1987; 1989).

Parameters and functions

Most of the values, necessary parameters, and initial conditions were taken from Van den Broek (1992). These pertain to an experiment with potatoes conducted in 1986 and 1987 on two locations: Rusthoeve where the soil profile consists of a sandy clay loam overlying a loamy sand and Vredepeel with a profile of two types of loamy sand. The simulation was done for a period of seven weeks, starting at June 9 (calendar day number 160). Except for the soil profile the parameters and functions used were identical for both locations.

Dimensions of the ridge and root distribution

The ridge has a width of 75 cm and a height of 20 cm (Vos and Groenwold 1986). Because of the assumed symmetry, the width of the domain considered in the model was taken to be 40 cm, horizontally divided in 8 columns of width 5 cm. The total depth of the domain was 100 cm, divided (top down) in eight layers of 3 cm thickness, one of 6 cm thickness, and seven with 10 cm thickness.

Since root distribution was not measured in the experiments at Rusthoeve and Vredepeel, the data of Vos and Groenwold (1986) were used. They did detailed measurements on root length density distribution in horizontal and vertical directions in the soil profile. In Figure 1 the numbers in the control volumes indicate the root length density as determined by Vos and Groenwold (1986). When necessary, root length densities were found by interpolation as the grid used in the experiment was more coarse than that employed in the model. Figure 2 gives a two-dimensional view of the distribution of root length density in the whole profile. The root system was taken to be stationary, i.e. no change occurred in the root length density distribution. Both Asfary et al. (1983) and Vos and Groenwold (1986) found only small changes in the root length density distribution once the crop had been established.

Initial and boundary conditions

Meteorological data of 1986 for the Rusthoeve location were used. This was a dry year as can be seen from Figure 3 showing precipitation and evapotranspiration in the growing season of 1986. It was assumed that the pressure head at the lower boundary of the soil domain was in equilibrium with the ground water table as measured at Rusthoeve. The depth of the ground water table was in the range 125–175 cm below the top of the ridge.

Soil hydraulic properties

The soils were assumed to have the properties of soils mentioned in the so-called Staring series (Wösten et al. 1994). In Rusthoeve the upper 30 cm was a sandy clay loam (ZavelB8) and the lower part of the profile (30–100 cm) a loamy sand (SandO2). In Vredepeel the upper 30 cm was a loamy sand (Sandb1), and the

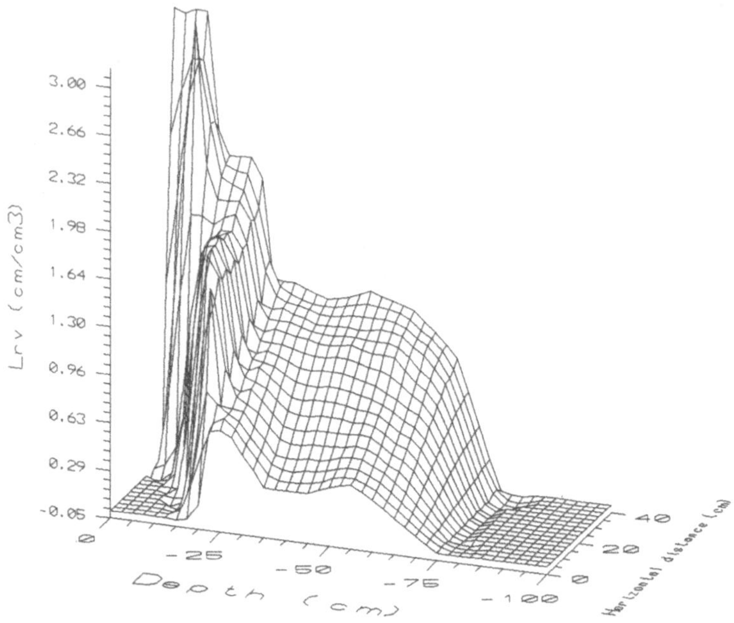

Figure 2. Two-dimensional distribution of root length density in cm • cm^{-3} over the soil profile.

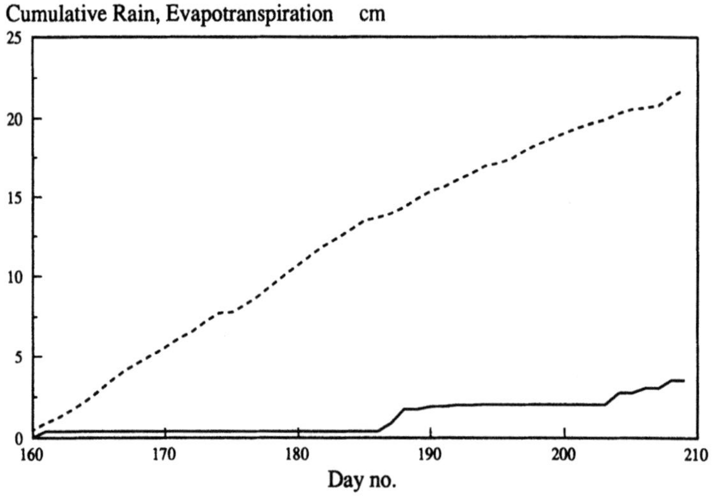

Figure 3. Rain (solid line) and potential evapotranspiration (dashed line) at Rusthoeve in 1986.

subsoil another loamy sand (SandO1) (Van den Broek 1992). The experimental data of these soils have been fitted to Mualem-Van Genuchten functions (Van Genuchten 1980; Mualem 1976) by Wösten et al. (1994). The relation between water content and pressure head is given by:

$$\Theta = \Theta_r + \frac{\Theta_s - \Theta_r}{(1+|\alpha h|^n)^{m(l+2)}}$$

where Θ is the water content, the subscript s indicates the water content at saturation, and the subscript r the residual water content, h is pressure head in cm, α and n are curve shape parameters and m is defined by $m = 1 - \frac{1}{n}$ The relation between hydraulic conductivity and pressure head is given by:

$$K = K_s + \frac{((1+|\alpha h|^n)^m - |\alpha h|^{n-1})^2}{(1 + |\alpha h|^n|^{m(l+2)})} \tag{10}$$

Here again the subscript s indicates saturated conductivity and l is a curve-shape parameter. The values of the parameters are given in Table 1. For the blocked control volumes (the hatched ones in Figure 1) the same functions were used but with $K_s = 0$.

Table 1. Parameters of the Van Genuchten-Mualem equations for the top- and subsoil of the locations at Rusthoeve and Vredepeel

Rusthoeve	Θ_r $cm^3 \bullet cm^{-3}$	Θ_s $cm^3 \bullet cm^{-3}$	K_s $cm \bullet d^{-1}$	α cm^{-1}	l	n
Topsoil						
ZavelB8	0.0	0.43	2.25	0.0096	−2.733	1.284
Subsoil						
SandO2	0.02	0.38	15.56	0.0214	0.039	2.075
Vredepeel						
Topsoil	0.01	0.43	17.46	0.0249	−0.14	1.507
Subsoil						
SandO1	0.01	0.36	13.21	0.0224	0.0	2.167

Nitrogen supply and demand

The nitrogen demand is set at 4.1 $kg \bullet ha^{-1} \bullet d^{-1}$ (Asfary et al. 1983). The nitrogen supply is 210 $kg \bullet ha^{-1}$, of which 200 $kg \bullet ha^{-1}$ is located in the ridge, and another 10 $kg \bullet ha^{-1}$ in the profile downwards. Calculations were made for two distributions within the ridge: one where the nitrate is concentrated in the uppermost control volumes and one where it is evenly distributed over all control volumes in the ridge.

Water uptake and transpiration

The root conductivity was set at $5 \bullet 10^{-6}$ $cm \bullet d^{-1}$, an average value for different species, taken from De Willigen and Van Noordwijk (1987). The values of $h_{P,1/2}$ and a (Eq. 8) were chosen such that reduction of transpiration begins when the plant water potential falls below −0.35 MPa, as found by Campbell et al. (1976).

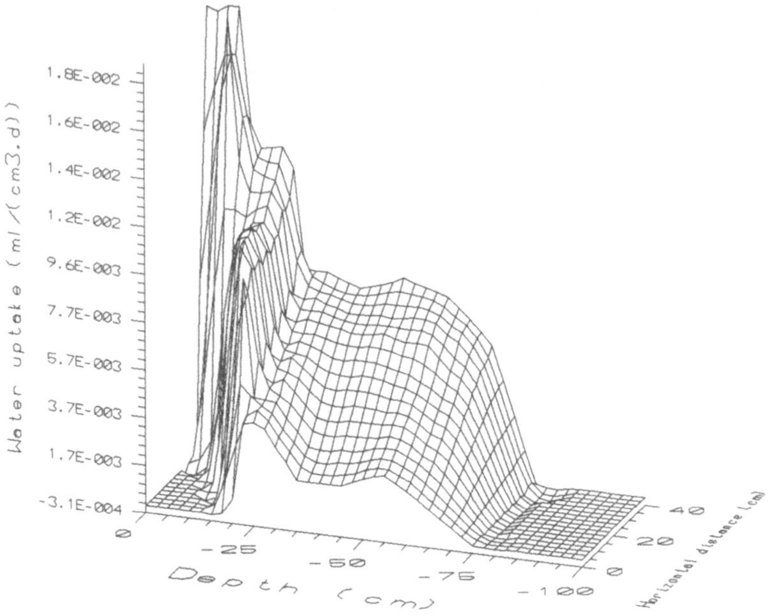

Figure 4. Distribution of water uptake in ml • cm^{-3} • d^{-1} at first day of simulation period in the sandy clay loam.

Results and discussion

Figure 4 gives the distribution of water uptake at the first day of the simulation for the Vredepeel location. In a fairly moist soil the dominant resistance in the pathway soil-rhizosphere-root is the root resistance. Accordingly the distribution of water uptake in such a situation is to a large degree determined by the distribution of root length density. Comparing Figure 4 with Figure 2 one can see the similarity between distribution of water uptake and root length density, uptake being higher where root length density is higher.

Campbell et al. (1976) give the resistance of the root system of a potato crop in the field in days. To compare their value with ours the total resistance is calculated as follows: at the first day the soil is at field capacity which means an average pressure head in the soil of -0.05 MPa (= -50 cm), the transpiration was 0.354 cm • d^{-1}, and according to the model the plant water potential -0.126 MPa (= -1260 cm). The transpiration can be given as the difference between soil and plant potential divided by the total resistance of the root system. This total resistance therefore is (1260 − 50)/0.354 = ca. 3400 day, which agrees with the value calculated by Campbell et al. (1976).

When the upper layers dry out due to transpiration and evaporation, the soil resistance increases, and gradually more of the required water is taken up from lower layers with lower root density but with higher water content.

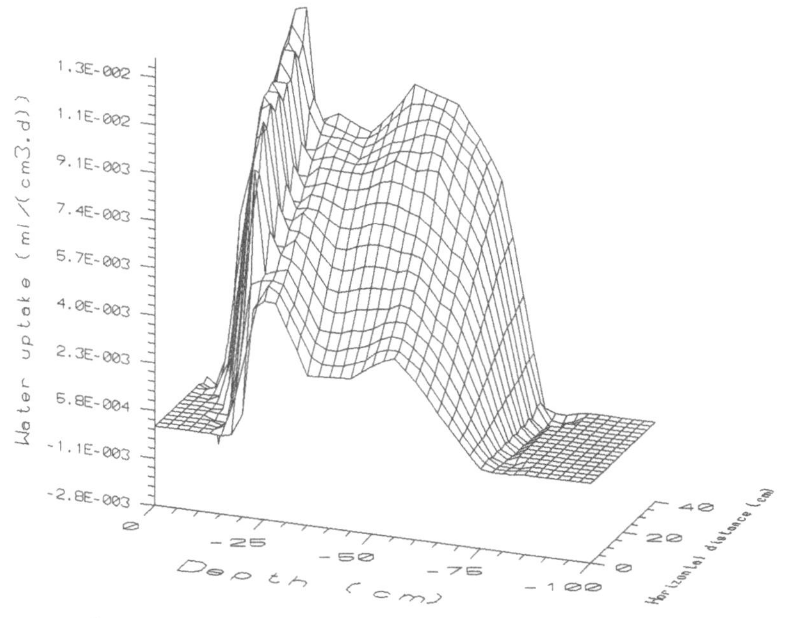

Figure 5. As Figure 4, but after 14 days.

This is shown in Figure 5 where the distribution of water uptake is shown after a dry spell of 14 days. In fact in the top control volume in the middle of the ridge there is negative water uptake, i.e. here some water flows from the root to the soil.

The total nitrogen uptake in the seven weeks simulation period was not influenced by the initial distribution of nitrogen in the case of the sandy clay loam. For the parameter values and environmental conditions here, both distributions allowed an optimum uptake, amounting in total to 198 kg • ha^{-1}, during the complete period. This was different for the loamy sand where the uniform distribution gave an uptake of 173 kg • ha^{-1}, whereas the distribution concentrated in the edge of the ridge gave 185 kg • ha^{-1}. The differences in uptake between the two locations are due to the differences in hydraulic properties. To illustrate this the situation at day 202, after a dry spell of about 12 days, will be considered. The average pressure head in the ridge of the sandy clay loam was about -0.45 MPa, and of the loamy sand -0.7 MPa. The volumetric water content differed much more, being 0.13 (sandy clay loam) and 0.04 (loamy sand). This last low water content causes a low value of the diffusion coefficient. At the beginning of day 202 the nitrate uptake amounted to 4.1 kg • ha^{-1} • d^{-1} and 0.36 kg • ha^{-1} • d^{-1}, respectively and the water uptake to 0.22 and 0.17 cm • d^{-1}, respectively, at a potential transpiration rate of 0.26 cm • d^{-1}. In the case of the loamy sand the nitrate uptake is much more reduced than the water uptake, due to dryness. The effect can be illustrated in more detail by considering uptake from a specific control volume, that in the

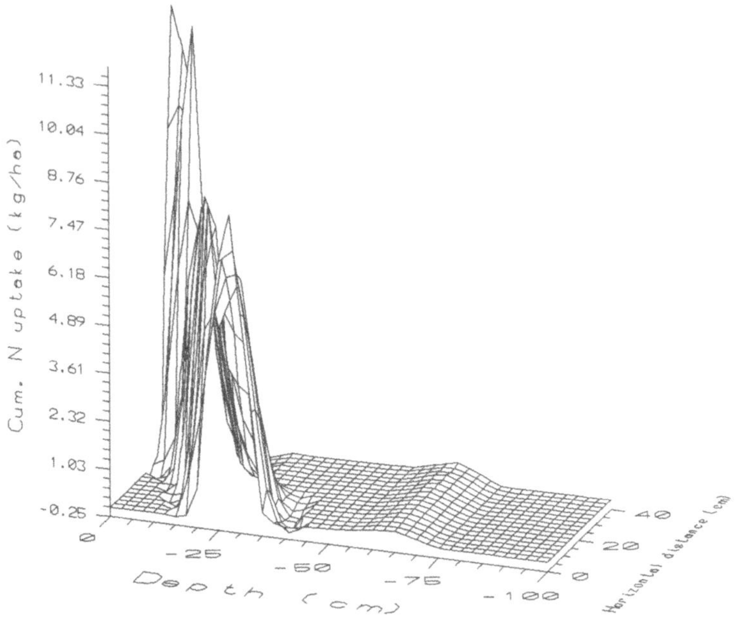

Figure 6. Distribution of cumulative uptake after seven weeks of uptake. Initially 200 kg • ha^{-1} of nitrogen was uniformly distributed over all active control volumes of the ridge.

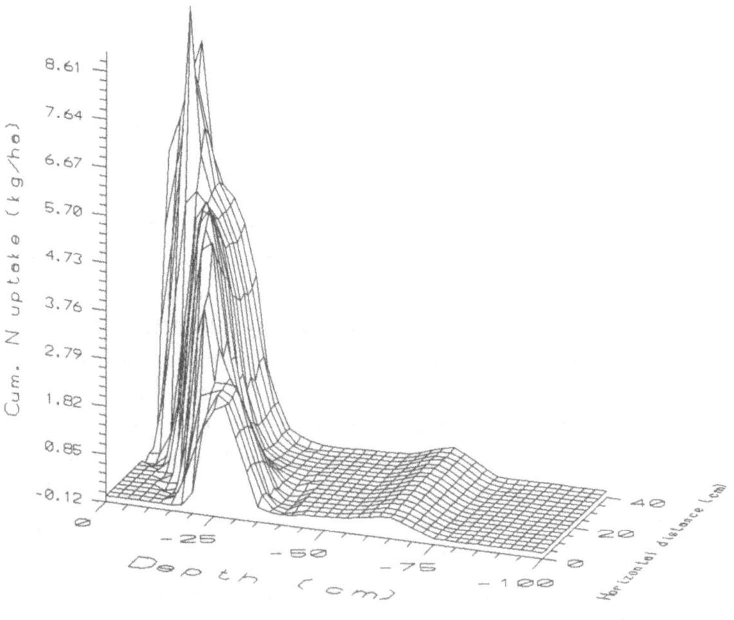

Figure 7. As Figure 6, but initially 200 kg • ha^{-1} distributed over the uppermost control volumes of the ridge.

middle of the ridge at depth 10 cm. Here, the amount of nitrogen present at day 202 in the case of the sandy clay loam was 1.6 kg • ha^{-1}, and in the case of the loamy sand it was 3.5 kg • ha^{-1}, the concentrations being 0.02 and 0.168 mmol • ml^{-1}, respectively. However, the uptake from this control volume was from the sandy clay loam 0.22 kg • ha^{-1} • d^{-1} and 0.021 kg • ha^{-1} • d^{-1} from the loamy sand. The uptake of water was negative in both cases so that the mass flow of nitrate towards the roots contributed negatively to the total transport to the root. The uptake then is dominated by transport by diffusion, and due to the low water content in the loamy soil the possible diffusion rate towards the plant root will be much smaller than in the sandy clay loam.

Generally the concentration is proportional to the reciprocal of the water content, the diffusion coefficient is linearly related to the square of the water content, so the maximum diffusion rate, which is proportional to the product of the diffusion coefficient and the concentration decreases in a linear fashion with the water content.

The distribution of the cumulative uptake of nitrate over the profile is given in Figure 6 for initial uniform distribution of nitrate in the ridge and in Figure 7 for initial concentration of nitrate in the top layer of the ridge.

The model discussed here deals exclusively with processes in soil. Plant demand for water or nutrient is not computed, it requires input of potential values of transpiration and nutrient uptake. The model then calculates whether conditions in the soil profile are such that these demands can be met. It offers the opportunity to estimate the effects of horizontal gradients and of crops growing on ridges. Its predictions, however, should be handled with caution as only some parts of the model have been validated.

References

Asfary A F, Wild A, Harris P M (1983) Growth, mineral nutrition and water use by potato crops. Journal of Agricultural Science Cambridge 100:87–101.

Barraclough P B, Tinker P B (1981) The determination of ionic diffusion coefficients in field soils. I. Diffusion coefficients in sieved soils in relation to water content and bulk density. Journal of Soil Science 32:225–236.

Belmans C, Dekker L W, Bouma J (1982) Obtaining soil physical data for simulating soil moisture regimes and associated potato growth. Agricultural Water Management 5:319–333.

Campbell M D, Campbell G S, Kunkel R, Papendick R I (1976) A model describing soil-plant-water relations for potato. American Potato Journal 53:431–441

Campbell G S (1985) Soil Physics with BASIC. Elsevier, Amsterdam, The Netherlands.

Clutterbuck B J, Simpson K (1978) The interactions of water and fertilizer nitrogen in effects on growth pattern and yield of potatoes. Journal of Agricultural Science Cambridge 91:161–172.

De Willigen, P (1990) Calculation of Uptake of Nutrients and Water by a Root System. Nota 210, Institute for Soil Fertility, Haren, The Netherlands.

De Willigen P, Van Noordwijk M (1987) Roots, Plant Production and Nutrient Use Efficiency. Ph.D. Thesis, Agricultural University Wageningen, The Netherlands.

De Willigen P, Van Noordwijk M (1989) Rooting depth, synchronization, synlocalization and N-use efficiency under humid tropical conditions. Pages 145–156 in Van der Heide J (Ed.) Nutrient

Management for Food Production in Tropical Farming Systems. Institute for Soil Fertility, Haren, The Netherlands, and Universitas Brawijaya, Malang, Indonesia.

De Willigen P, Van Noordwijk M (1995) A model for interactions between water and nutrient uptake. (In press) in Kabat, P., B. Marshall, B.J. van den Broek, J. Vos, H. van Keulen (Eds.). Modelling and Parameterization of the Soil-Plant-Atmosphere System. A Comparison of Potato Growth Models. Wageningen Press, Wageningen.

De Willigen P, Van Noordwijk M (1994a) Mass flow and diffusion of nutrients to a root with constant or zero-sink uptake I. Constant uptake. Soil Science 157:162–170.

De Willigen P, Van Noordwijk M (1994b) Mass flow and diffusion of nutrients to a root with constant or zero-sink uptake II. Zero-sink uptake. Soil Science 157:171–175.

Harris P M (1978) Water. Pages 244–277 in Harris P M (Ed.) The Potato Crop: The Scientific Basis for Improvement. Chapman & Hall, London, UK.

Heinen M, De Willigen P (1992) FUSSIM2. A Simulation Model for Two-Dimensional Flow of Water in Unsaturated Soil. Nota 260, Institute for Soil Fertility, Haren, The Netherlands.

Jefferies R A (1992) Use of a simulation model to asses possible strategies of drought tolerance in Potato (*Solanum tuberosum* L.). Agricultural Systems 41:93–104.

Koorevaar P, Menelik G, Dirksen C (1983) Elements of Soil Physics. Development in Soil Science 13. Elsevier Science Publishers, Amsterdam, The Netherlands.

Mualem Y (1976) A new model for predicting the hydraulic conductivity of unsaturated porous media. Water Resources Research 12:513–522.

Van Noordwijk M, De Willigen P, Ehlert P A I, Chardon W J (1990) A simple model of P uptake as a possible basis for P fertilizer recommendations. Netherlands Journal of Agricultural Science 38:317–322.

Patankar S V (1980) Numerical Heat Transfer and Fluid Flow. Hemisphere Publishing Corporation, New York, USA.

Peaceman D W, Rachford H H (1955) The numerical solution of parabolic and elliptic equations. Journal of the Society for Industrial and Applied Mathematics 3:28–41.

Rubin J (1968) Theoretical analysis of two-dimensional, transient flow of water in unsaturated and partly unsaturated soils. Soil Science Society of America Proceedings 32:607–615.

Van den Broek B J (1992) Predicting Times of Irrigation with a Simulation Model. M.Sc. Thesis, University of Alberta, Canada.

Van Genuchten M T H (1980) A closed form equation for predicting the hydraulic conductivity of unsaturated soils. Soil Science Society of America Proceedings 32:607–615.

Van Keulen H (1975) Simulation of Water Use and Herbage Growth in Arid Regions. PUDOC, Wageningen, The Netherlands.

Vos J, Groenwold J (1986) Root growth of potato crops on a marine clay soil. Plant and Soil 94:17–33.

Wang H F, Anderson MP (1982) Introduction to Groundwater Modelling. Finite Difference and Finite Element Methods. W.H. Freeman and Company, New York, USA.

Wesseling J (1976) De betekenis van water voor de landbouw. (The role of water in agriculture.) Polytechnisch Tijdschift (editie Bouwkunde Weg- en Waterbouw nr 2) 2–4 (in Dutch).

Wösten J H M, Veerman G J, Stolte J (1994) Waterretentie- en Doorlatendheidskarakteristieken van Boven- en Ondergronden in Nederland: De Staringreeks. Technical Document 18, DLO-Staring Centrum, Wageningen, The Netherlands.

6. Physiological effects of changes in atmospheric carbon dioxide concentration and temperature on growth and water relations of crop plants

SIEBE C. VAN DE GEIJN and PAUL DIJKSTRA
AB-DLO, P.O. Box 14, Wageningen, The Netherlands

Abstract. The effects of elevated concentrations of carbon dioxide on physiological processes are discussed. The higher photosynthesis in general leads to a higher plant biomass and yield (10 to 40% at double present concentrations). An important aspect is the change in plant water relations due to the partial closure of the stomata. This results in an enhanced water use efficiency and improved water status, especially during periods of transient water shortage. Although instantaneous transpiration efficiency (actual net photosynthesis per unit water transpired) is increased by about 40%, total water use per unit ground area per season usually remains about the same. This is due to a higher leaf area index and the many feed-back mechanisms in the soil-plant canopy-atmosphere system.

For potato, only few experiments have been published to date. Results point to a complex interaction between plant nutrition, light level, daylength and growth enhancement by elevated [CO_2]. Further experimental research and model evaluations are needed.

Introduction

Public concern regarding climate change is growing, calling for a better understanding of the impact of changed conditions on (agro)-ecological systems. In this context it is necessary to consider possibilities for adaptation and mitigation. In most cases the stress resulting from environmental conditions, imposed on the biotic system, does not come as a single well understood and predictable event. Frequently the timing of events in the growing season determines the effects of environmental factors on plant and crop fitness or functional status. Information on the physiological mechanisms behind the interactions is in general inadequate for the estimation of quantitative effects for specific crops and conditions.

The aim of this paper is to discuss the physiological mechanisms that determine the effects of changes in atmospheric carbon dioxide concentrations on agricultural crops in general and, more specifically, on potato crops. Attention will be given to the effects on biomass production, water relations and yield formation of crops.

A.J. Haverkort and D.K.L. MacKerron (eds.), Potato Ecology and Modelling of Crops under Conditions Limiting Growth, 89–100.

Climate change parameters: State of the art

Climate change may on a short to medium term affect agricultural production mainly through four factors: temperature, radiation, water relations and atmospheric CO_2 concentration. The change in atmospheric carbon dioxide concentration is the most predictable and best documented factor. At present concentrations have reached 356 μmol/mol from pre-industrial levels of about 280 μmol/mol, rising at a rate of about 1.8 μmol/mol (i.e. 0.5%) per year. There is considerable uncertainty about the mean values for temperature and precipitation, and their seasonal patterns, as projected by Global Circulation Models (GCM), particularly at scales relevant to agricultural practice. This is limiting the possibility for the evaluation of future trends in crop production and food security in various regions. Also the technological advancement and management flexibility of agronomic practice as it may adjust and adapt to future climate situations (best technical means) (McKenney et al. 1992) is adding to the uncertainty. At present most papers are restricted to extrapolations to future conditions using present crops and crop management with the changed temperature regime, water availability and $[CO_2]$.

Few studies are both geographically explicit and scientifically complete and allow precise estimates even without additional uncertainty introduced by climate and weather parameters. Most studies do depend on the reliability of crop growth models applied in conditions where datasets for testing and validating the model performance are scarce, or lacking altogether. This area of research is getting growing attention from the scientific community in International programmes, like the IGBP-GCTE-Crops Network (Steffen et al. 1992; Anonymous 1994).

Carbon dioxide effects on photosynthesis and water relations

Photosynthesis

As a consequence of higher $[CO_2]$ in the atmosphere the entry of the primary substrate for photosynthesis into the substomatal cavity is enhanced (Kirschbaum 1994). Crops exposed to elevated $[CO_2]$ generally respond with positive but variable increases in biomass productivity and yield (Kimball 1983; Baker and Allen 1993). Production of C3-plants is increased by between 10% and 40% at double-present $[CO_2]$ (700 μmol/mol) (Kimball 1983; Rogers and Dahlman 1993), depending on the crop species and climatic and nutritional conditions. Variations in growth enhancement both between crops and over years are high (Dijkstra et al. 1993; Grashoff and Nonhebel 1993; Rogers and Dahlman 1993), and interactions between $[CO_2]$ and nutrient and water availability are complex. Direct effects of elevated $[CO_2]$ on productivity are negligible in C4-plants (Wong 1979; Goudriaan and Unsworth 1990), as a result of the different biochemical pathway for photosynthesis. In conditions of (temporary) water shortage growth enhancement can be considerable even in C4 plants however, because of improved water status as will be explained.

The extent and occurrence of physiological adaptations of the C3-photosynthetic apparatus to long-term exposure to high [CO_2] is still unresolved (Cure and Acock 1986; Wolfe 1994). Some recent papers point to the interference of the experimental set-up, imposing an effect of physical limitation of the root volume (pot sizes) leading to limitation of sink strength (Arp 1991; Thomas and Strain 1991), although this has not been confirmed by others (McConnaughay et al. 1993; Visser and Rozema, pers. comm.). Such a feed-back, if it arises from a limitation of the sink-strength for carbohydrates, should be dependent on the growth stage, and disappear in periods of vigorous growth of fruits, seeds or tubers. At a canopy level, photosynthetic acclimation is not always detected (Dijkstra et al. 1993). The sensitivity of C3 plants to higher [CO_2] increases with temperature, and the optimal temperature for canopy photosynthesis may therefore shift slightly upwards (Long 1991; Kirschbaum 1994).

Misinterpretation of data from growth analyses may lead to the conclusion that a drop in relative growth rate (RGR) at elevated [CO_2] is observed. In fact this may be the result of the initially higher exponential growth of plants, grown at elevated [CO_2], of which the growth changes earlier to linear growth when the canopy is closing because of self-shading of leaves. As such the phenomenon is due to a shift of the growth curve in time (Van de Geijn et al. 1993). Similar ambiguities may appear in the interpretation of the enhancement of growth by [CO_2] when growth rates are compared between closed canopies of annual crops and free standing individual citrus trees (Baker and Allen 1993).

For herbacious species it has been shown that a strong negative correlation exists between net assimilation rate (NAR) and the growth response to elevated [CO_2] (Roumet et al. subm.). From the same analysis it appeared that cultivated perennial species react strongly to [CO_2], in the same way as annuals, and in contrast to wild perennials.

Water use efficiency and seasonal water use

The effect of elevated [CO_2] in modifying the gas-exchange properties of leaves and their water use, is of special interest in field grown crops. Many studies have reported an increase in the water use efficiency, i.e. the amount of dry matter produced per unit of water transpired. Stomatal conductivity may drop by 40 to 60% with a doubling of the ambient [CO_2] (Morison 1987). Although the stomatal reaction of a single leaf to elevated [CO_2] has been unequivocally shown, its effect cannot be extrapolated directly to a larger scale, and over the whole season. Changes in the micro-environment at the leaf surface in the canopy may modify transpiration and alter vapour pressure deficit.

Effects at the level of stomata

The opening status of the stomata is a compromise between water loss and uptake of CO_2 from ambient air (Collatz et al. 1991). In line with this proposition, stomatal response to elevated [CO_2], (Ca), in general is reflected in partial stomatal closure (Hendrey et al. 1993; Morison 1987). The biochemical reactions leading to this stomatal closure are not yet clear. The resulting effect has been described assuming that plants regulate the internal [CO_2] (Ci) in the substomatal cavity leading to a constant ratio (Ci/Ca) with the atmospheric concentration (Kirschbaum 1994). Such a regulation matches the partial closure at elevated CO_2 observed in many studies using porometers (Morison and Gifford 1984a; Morison 1987). The required supply of carbon dioxide by diffusion through the stomata to stabilise Ci is coupled to the rate of photosynthesis and therefore stomatal conductance is well correlated with light level. This relation can be modified by the presence of environmental factors such as stress by air pollution or drought. The ratio Ci/Ca at light saturation is about two-thirds for C3 plants but about one-third for C4 plants. The lower value for C4 plants reflects the higher affinity for CO_2 of the C4 photosynthetic pathway, and leads to a more efficient use of water by those plants (Kimball et al. 1993). Both plant types profit from this response to [CO_2] because of reduced sensitivity in the event of a transient water shortage (Rogers et al. 1984; Chaudhury et al. 1990; Morison 1993).

Effects at the level of the leaf

Water loss by transpiration is not only affected by the conductivity of the stomata, but also by the driving forces for exchange of the water vapour from the leaf surface to the surrounding atmosphere. In most instances the dominant factors are the radiative energy flux and advection. Also changes in area of individual leaves have a direct effect on water loss at a per leaf basis. Leith et al. (1986) found a 5% increase in leaf area for soybeans with a doubling [CO_2]. In general however, leaf thickness or specific leaf weight increases more than leaf area.

Apart from the effect on leaf properties, the gradient in partial pressure of water vapour at the leaf surface is important. When transpiration is reduced by elevated [CO_2], the existing vapour pressure deficit (VPD) between the stomatal cavity and the air in the immediate vicinity of the leaf surface, in the boundary layer, will increase, and stimulate transpiration, all other factors being equal. On the other hand, this same reduced transpiration would cool the leaf less, and consequently a rise in leaf temperature may occur. Due to this latter phenomenon the temperature at the leaf surface may rise with about 1 °C, apart from the rise in leaf temperature due to the greenhouse effect (Idso et al. 1987; Morison 1987; Van de Geijn et al. 1993). The resulting homeostatic effects tend to stabilise water loss.

A rise of growing-season temperatures due to the global greenhouse effect will tend to stimulate transpiration. Depending on changes in humidity it may affect VPD at the leaf surface and thereby the energy balance determining the fluxes of sensible and latent heat. Increased water availability may, however, offset the warming at the leaf surface, especially if, through water saving, partial stomatal closure alleviates water stress during periods of high water requirements in the diurnal sequence (Rogers et al. 1984). It requires detailed microclimatological data and complex calculations to integrate these various mechanisms in operation at the leaf level (McNaughton and Jarvis 1991).

Effects at the canopy level

The aspects considered at the leaf level also have an effect at the canopy level. Light distribution in the canopy, leaf age, humidity gradients and the degree of coupling with the atmosphere at specific positions in the canopy all have their effect.

The leaf area index (leaf surface per unit soil surface) of a crop in general will also increase, as a result of earlier and more rapid leaf production (Jones et al. 1985; Ackerly et al. 1992; Morison and Gifford 1984a,b; Chaudhury et al. 1990). This applies especially for indeterminate growing species, but also perennials will have an early start. Early development of the canopy, e.g. due to enhanced tillering, will lead to an earlier full ground cover, and may thus limit water loss by direct soil evaporation. Depending on the local precipitation and soil water reserves such early, enhanced plant development may also be profitable for the full use of water resources (Rogers et al. 1984). The higher transpiration early in the season may however also lead to an earlier depletion of soil water reserves (Morison and Gifford 1984a,b; Chaudhury et al. 1990).

In their FACE (Free Air Carbon dioxide Enrichment) experiments Hendrey et al. (1993) found that leaf area of cotton increased by 25% in dry plots, but only by 11% in irrigated plots as compared with their respective controls. It has been shown that plant water status is improved at elevated [CO_2], and some of the effect can be ascribed to partial alleviation of the water stress (Rogers et al. 1984; Chaudhury et al. 1990). Also for plants with a normal water supply pre-dawn water potentials are less negative.

Most available data on water relations have been collected from single plants grown in pots or hydroponically, and using porometer measurements supplemented with weight determinations. Values for water use efficiency and total water use in (semi-)field conditions over the full season, or over an extended period are scarce (Dijkstra et al. 1993).

It should be realised that water use efficiency (WUE) is often confused with instantaneous transpiration efficiency (ITE) (Eamus 1991; Morison 1993). The latter expression is best described as the ratio between the instantaneous CO_2 fixation (actual net photosynthesis) and water loss by transpiration (both measured with a leaf chamber). Water use efficiency is the ratio of the net gain in dry matter (mostly as measured above ground) in a given period, divided by the water loss in the same period. In general the ITE is higher than the WUE because of the variations in ITE over the day and season, and as carbohydrate losses by respiration in the dark are not included. Changes in water use efficiency and total (seasonal) water use per unit ground surface are more important for effects on agricultural production. Table 1 illustrates seasonal water use efficiencies for some crops and water supply conditions. In general the (seasonal) crop water use efficiency is further increased in the case of water deficiency. Although the water use efficiency at elevated [CO_2] is increased by

Table 1. Seasonal crop water use efficiency (WUE in g DM kg^{-1} water) (Data from Morison 1993 and ref. therein)

	Ambient [CO_2]	Double [CO_2]	Ratio	
sorghum	3.08	4.13	1.34	
wheat well-watered	5.1	6.3	1.23	
wheat water short.	6.2	8.9	1.43	
wheat	2.62	3.45	1.31	
wheat well watered	1.58	2.14	1.35	grain only
wheat water short.	1.27	1.86	1.46	grain only
faba beans	4.91	7.82	1.54	
water hyacinth	1.4	2.6	1.85	

25–50%, total seasonal water use by the crop (per unit soil surface) quite often is hardly affected (Dijkstra et al. 1993; Chaudhury et al. 1990). This is caused by the simultaneous increase in biomass. The magnitude of the positive effect on yield in water stress conditions depends on the timing of water availability relative to the stage of crop development.

Table 2. Daytime canopy water use efficiency (photosynthesis per unit water transpired: mmol CO_2 per mol H_2O) (Data from Baker and Allen 1993)

	[CO_2]	Temp. (°C)	WUE	High/low ratio
Soybean	330	28	2.95	
		31	2.63	
		33	2.37	
		35	2.33	
	800	28	6.00	2.03
		31	5.11	1.94
		33	4.95	2.09
		35	4.71	2.02
citrus	330	25	2.30	
		34	1.25	
	840	25	5.40	2.34
		34	3.52	2.81

Transpiration increases and WUE drops with increasing temperature (Table 2). This is caused by changes in the energy balance and VPD at the leaf surface. Data by Baker and Allen (1993) suggest that the relative improvement of the water use efficiency does not or only little depend on temperature.

A complication for the correct estimation of the real effect of elevated [CO_2] and temperature changes is that almost all experiments to study the impact of elevated [CO_2] and/or temperature changes have been performed in environmentally controlled and generally well mixed and ventilated experimental set-ups (enclosures or Open Top Chambers (OTCs)), where the indirect effects may not show up so prominently as in a real future climate (Unsworth et al. 1984; Leuning and Foster 1990; McNaughton and Jarvis 1991). Similar problems of interpretation have been seen in research on air pollution. The few larger scale FACE experiments are therefore extremely important for validation and extrapolation.

Temperature

The projected change in mean annual temperature will also affect crop yield. For most crops, plant development can be described as a function of an effective temperature sum (Carter et al. 1990; Ellis et al. 1990), and a rise in temperature will reduce the duration of various crop development stages e.g. the tuber initiation period and crop maturation. This type of effect is treated in crop growth modelling studies that describe changes in crop productivity due to climate change (Wolf and Van Diepen 1992; Grashoff and Nonhebel 1993; Yandell et al. 1988; Schapendonk et al., this volume). Temperature will also influence the partitioning of dry matter and the growth rate. If shortening the growing season is not fully compensated by a change in ontogenetic development and higher growth vigour at the higher temperature (Ellis et al. 1990; Struik and Ewing, this volume), a net yield loss may occur as the duration of the vegetative growth stage and the integral of light interception during maturation is of overriding importance. Especially in regions where present growing season temperatures are near maximum (above optimal) levels for the crop, serious yield reductions may occur even with a temperature rise of only 1–2 °C.

Some studies have considered a shift of agro-ecological production zones to higher latitudes, following a rise in mean seasonal temperature (Parry et al. 1988; Wolf and Van Diepen 1992). An earlier start of the growing season because of higher spring temperatures means in cool temperate climates that the growing season locally starts at shorter daylength. This requires a shift of daylength sensitive cultivars from lower to higher latitudes (Morison and Butterfield 1990). Daylength sensitive cultivars from lower latitudes, adapted to present conditions, however, are selected for a different duration of the growing season, and may therefore not take full advantage of the possibilities of the future climate.

Changes in precipitation pattern may have consequences for the field operations during the season. Late spring rains or early autumn wet periods may seriously hamper access to the field, and therefore limit possibilities to take advantage of the early season higher temperatures, and expose spring crops to higher temperature in present daylength conditions.

Carbon dioxide effects on potato growth

Only a limited number of studies have directly addressed the physiological response and changes in productivity of potato to changes in [CO_2] (Yandell et al. 1988; Wheeler and Tibbitts 1989; Wheeler et al. 1991). Therefore, only very preliminary and partly speculative statements can be made. On the basis of the observation that plants exposed to elevated [CO_2] in general tend to accumulate more starch in their foliage it can be postulated that specific starch accumulators like potato would profit by storing more starch in their tubers. Experiments to confirm this have not been conclusive. Preliminary results show that in some instances the foliage of potato plants shows unexpected changes in shape and colour during the season (e.g. yellowing, necrotic leaf margins), resembling the symptoms of mineral deficiencies, but not unequivocally attributable to one or more elements (Dijkstra et al., unpublished). This observation could mean that the proper maintenance of a balanced nutrition throughout the season requires special attention.

Table 3. Total plant weight of potato in response to [CO_2], photoperiod, and light level (PPF) for three cultivars (Data from Wheeler et al. 1991)

		Photoperiod			
		12h		24h	
	[CO_2]	PPF, umol m^{-2} s^{-1}		PPF, umol $m^{-2}s^{-1}$	
Cultivar	μmol/mol	400	800	400	800
Norland	350	329 %	397 %	463 %	462 %
	1000	412 *(+25.2)*	445 *(+12.1)*	493 *(+ 6.5)*	430 *(−6.9)*
Russet	350	322	374	587	504
Burbank	1000	443 *(+37.6)*	431 *(+15.2)*	662 *(+12.8)*	418 *(−17)*
Denali	350	352	454	556	679
	1000	488 *(+38.6)*	583 *(+20.4)*	592 *(+6.4)*	667 *(−1.8)*

A complex picture emerges also from the experiments of Wheeler et al. (1991) and Wheeler and Tibbitts (1989). In an experiment on potato, including effects of irradiance, [CO_2] and photoperiod they found that under 24 h light, a change from 350 to 1000 μmol/mol CO_2 gave enhanced growth (+8.6%) at an irradiance of 400 μmol m^{-2} s^{-1} but reduced growth at 800 μmol m^{-2} s^{-1}, averaged over three cultivars (Table 3). Under a 12-h photoperiod, the [CO_2] always enhanced growth but the increase was less at the higher irradiance (+15.9%) than at the lower irradiance (+33.8%). The interactions present in the data are not clear and the data is too meagre a basis to draw firm and general conclusions. They certainly indicate the confounding effect of assimilate availability on the effects of [CO_2] on crop growth. Recent experiments in Open Top Chambers showed that the yield at double-present [CO_2] may increase by 0–24% (Dijkstra and Jansen, unpublished).

Concluding remarks

Although the general principles of plant reactions to changes in temperature and [CO_2] are quite well understood, the quantitative estimation of the effect on yield for a complex system such as a potato crop is not yet justified. Several factors seem to interfere. The down-regulation of photosynthetic rate in the conditions of high light and elevated [CO_2] may play a role, as well as damage to the tissues by enhanced accumulation of starch. The few results of experiments reported so far with potato also suggest that the nutritional status is very important.

The observation that both light level and photoperiod are of importance in determining the response to elevated [CO_2] suggests that also interactions may exist with the light level at different positions in the canopy. This would add to the complexity of the system, especially in the period of senescence of the foliage and rapid tuber bulking. Too little experimental evidence is available to come to firm conclusions. Computer simulation models may help us however to design the experiments most likely to give us answers to the many questions (Schapendonk et al., this volume). The establishment of an international network (e.g. an IGBP-GCTE-Potato Network) to exchange experimental datasets and compare results from such modelling exercises may give an important push to the developments in this area.

References

Ackerly D D, Coleman J S, Morse S R, Bazzaz F A (1992) CO_2 and temperature effects on leaf area production in two annual plant species. Ecology 73:1260–1269.

Anonymous (1994) GCTE Focus 3 Wheat Network: 1993 Model and Experimental Meta Data. IGBP-GCTE Report No. 2, Canberra, Australia.

Arp W J (1991) Effects of source-sink relations on photosynthetic acclimation to elevated CO_2. Plant, Cell and Environment 14:869–876.

Baker J T, Allen Jr L H (1993) Contrasting crop species responses to CO_2 and temperature: rice, soybean and citrus. Vegetatio 104/105:239–260.

Carter T R, Parry M L, Porter J H (1990) Climate change and future crop suitability in Europe. Pages 3–6 in Goudriaan J, Van Keulen H, Van Laar H H (Eds.) The Greenhouse Effect and Primary Productivity in European Agro-Ecosystems. PUDOC, Wageningen, The Netherlands.

Chaudhury U N, Kirkham M B, Kanemasu E T (1990) Carbon dioxide level effects on yield and water use of winter wheat. Agronomy Journal 82:637–641.

Collatz G J, Ball J T, Grivet C, Berry J A (1991) Physiological and environmental regulation of stomatal conductance, photosynthesis and transpiration: a model that includes a laminar boundary layer. Agricultural and Forest Meteorology 54:107–136.

Cure J D, Acock B (1986) Crop responses to carbon dioxide doubling: a literature survey. Agricultural and Forest Meteorology 38:127–145.

Dijkstra P, Schapendonk A H C M, Groenwold J (1993) Effects of CO_2 enrichment on canopy photosynthesis, carbon economy and productivity of wheat and faba bean under field conditions. Pages 23–41 in Van de Geijn S C, Goudriaan J, Berendse F (Eds.) Climate Change: Crops and Terrestrial Ecosystems. Agrobiologische Thema's 9, AB-DLO, Wageningen, The Netherlands.

Eamus D (1991) The interaction of rising CO_2 and temperatures with water use efficiency. Plant, Cell and Environment 14:843–852.

Ellis R H, Headley P, Roberts E H, Summerfield R J (1990) Quantitative relations between temperature and crop development and growth. Pages 85–115 in Jackson M, Ford-Lloyd B V, Parry M J (Eds.) Climate Change and Plant Genetic Resources. Belhaven, London, UK.

Goudriaan J, Unsworth M H (1990) Implications of increased carbon dioxide and climate change for agricultural productivity and water resources. Pages 111–130 in Kimball B A, Rosenberg N J, Allen Jr L H (Eds.) Impact of Carbon Dioxide, Trace Gases, and Climate Change on Global Agriculture. ASA Spec. Publ. no. 53.

Grashoff C, Nonhebel S (1993) Effect of CO_2 increase on the productivity of cereals and legumes; model exploration and experimental evaluation. Pages 43–57 in Van de Geijn S C, Goudriaan J, Berendse F (Eds.) Climate Change: Crops and Terrestrial Ecosystems. Agrobiologische Thema's 9, AB-DLO, Wageningen, The Netherlands.

Hendrey G R, Lewin K F, Nagy J (1993) Free air carbon dioxide enrichment: development, progress, results. Vegetatio 104/105:17–31.

Idso S B, Kimball B A, Mauney J R (1987) Atmospheric carbon dioxide enrichment effects on cotton midday foliage temperature: Implications for plant water use efficiency. Agronomy Journal 79:667–672.

Jones P, Allen Jr L H, Jones J W (1985) Responses of soybean canopy photosynthesis and transpiration to whole-day temperature changes in different CO_2 environments. Agronomy Journal 77:242–249.

Kimball B A (1983) Carbon dioxide and agricultural yield: an assemblage and analysis of 770 prior observations. Agronomy Journal 75:779–788.

Kimball B A, Mauney J R, Nakayama F S, Idso S B (1993) Effects of increasing atmospheric CO_2 on vegetation. Vegetatio 104/105:65–75.

Kirschbaum M U F (1994) The sensitivity of C3 photosynthesis to increasing CO_2 concentration: a theoretical analysis of its dependence on temperature and background CO_2 concentration. Plant, Cell and Environment 17:747–754.

Leith J H, Reynolds J F, Rogers H (1986) Estimation of leaf area of soybeans grown under elevated carbon dioxide levels. Field Crops Research 13:193–203.

Leuning R, Foster L J (1990) Estimation of transpiration by single trees: comparison of a ventilated chamber, leaf energy budgets and a combination equation. Agricultural and Forest Meteorology 51:63–68.

Long S P (1991) Modification of the response of photosynthetic productivity to rising temperature by atmospheric CO_2concentrations: Has its importance been underestimated? Plant, Cell and Environment 14:729–739.

McConnaughay K D M, Berntson G M, Bazzaz F A (1993) Limitations to CO_2-induced growth enhancement in pot studies. Oecologia 94:550–557.

McKenney M S, Easterling W E, Rosenberg N J (1992) Simulation of crop productivity and responses to climate change in the year 2030: the role of future technologies, adjustments and adaptations. Agricultural and Forest Meteorology 59:103–127.

McNaughton K G, Jarvis P G (1991) Effects of spatial scale on stomatal control of transpiration. Agricultural and Forest Meteorology 54:279–301.

Morison J I L, Butterfield R E (1990) Climatic variability and crop development. Pages 7–10 in Goudriaan J, Van Keulen H, Van Laar H H (Eds.) The Greenhouse Effect and Primary Productivity in European Agro-Ecosystems. PUDOC, Wageningen, The Netherlands.

Morison J I L, Gifford R M (1984a) Plant growth and water use with limited water supply in high CO_2 concentrations. I. Leaf area, water use and transpiration. Australian Journal of Plant Physiology 11:361–374.

Morison J I L, Gifford R M (1984b) Plant growth and water use with limited water supply in high CO_2 concentrations. II. Plant dry weight, partitioning and water use efficiency. Australian Journal of Plant Physiology 11:375–384.

Morison J I L (1987) Intercellular CO_2 concentration and stomatal response to CO_2. Pages 229–

251 in Zeiger E, Farquhar G D, Cowan I R (Eds.) Stomatal Function. Stanford University Press, CA, USA.

Morison J I L (1993) Response of plants to CO_2 under water limited conditions. Vegetatio 104/105:193–209.

Parry M L, Carter T R, Konijn N T (1988) The Impact of Climatic Variations on Agriculture. Vols. I and II. Kluwer Academic Publishers, Dordrecht, The Netherlands.

Rogers H H, Dahlman R C (1993) Crop responses to CO_2 enrichment. Vegetatio 104/105:117–131.

Rogers H H, Sionit N, Cure J D, Smith J M, Bingham G E (1984) Influence of elevated carbon dioxide on water relations of soybeans. Plant Physiology 74:233–238.

Roumet C, Bel M P, Roy J (1995) Response of 11 grass species to CO_2 (submitted).

Schapendonk A H C M, Pot C S, Goudriaan J (1995) Simulated effects of elevated carbon dioxide concentration and temperature on the productivity of potato. Interaction with cultivar differences for earliness. Pages 101–114 in Haverkort A J, MacKerron D K L (Eds.) Potato Ecology and Modelling of Crops under Conditions Limiting Growth. Kluwer Academic Publishers, Dordrecht, The Netherlands.

Steffen W L, Walker B H, Ingram J S I, Koch G W (1992) Global Change and Terrestrial Ecosystems: The Operartional Plan. IGBP Report 21.

Struik P, Ewing E E (1995) Crop physiology of potato (*Solanum tubersoum*): Responses to photoperiod and temperature relevant to crop modelling. Pages 19–40 in Haverkort A J, MacKerron D K L (Eds.) Potato Ecology and Modelling of Crops under Conditions Limiting Growth. Kluwer Academic Publishers, Dordrecht, The Netherlands.

Thomas R B, Strain B R (1991) Root restriction as a factor in photosynthetic acclimation of cotton seedlings grown in elevated carbon dioxide. Plant Physiology 96:627–634.

Unsworth MH, Heagle AS, Heck WW (1984) Gas exchange in open top field chambers. I. Measurement and analysis of atmospheric resistances. Atmosph. Environ. 18:373–380.

Van de Geijn S C, Goudriaan J, Van der Eerden L J, Rozema J (1993) Problems and approaches to integrating the concurrent impacts of elevated carbon dioxide, temperature, ultraviolet-B radiation and ozone on crop production. Pages 333–338 in Buxton et al. (Eds.) International Crop Science I, CSSA, Madison, WI, USA.

Wheeler R M, Tibbitts T W (1989) Utilization of potatoes for life support systems in space. IV. Effect of CO_2 enrichment. American Potato Journal 66:25–34.

Wheeler R M, Tibbitts T W, Fitzpatrick A H (1991) Carbon dioxide effects on potato growth under different photoperiods and irradiance. Crop Science 31:1209–1213.

Wolf J, Van Diepen C A (1992) Effects of Climate Change on Crop Production in the Rhine Basin. Report no. 52. Winand Staring Centre for integrated Land, Soil and Water Research and Institute for Inland Water Management and Waste Water Treatment, SC-DLO, Wageningen, The Netherlands.

Wolfe D W (1994) Physiological and growth responses to atmospheric carbon dioxide concentration. Pages 223–242 in Pessarakli M (Ed.) Handbook of Plant and Crop Physiology. Chapter 10. Marcel Dekker, New York, USA.

Wong S C (1979) Elevated atmospheric partial pressure of CO_2 and plant growth. I. Interactions of nitrogen nutrition and photosynthetic capacity in C3 and C4 plants. Oecologia 44:68–74.

Yandell B S, Najar A, Wheeler R, Tibbitts T W (1988) Modelling the effects of light, carbon dioxide, and temperature on the growth of potato. Crop Science 28:811–818.

7. Simulated effects of elevated carbon dioxide concentration and temperature on the productivity of potato

Interaction with Cultivar Differences for Earliness

A.H.C.M. SCHAPENDONK[1], C.S. POT[1] and J. GOUDRIAAN[2]

[1]*Research Institute for Agrobiology and Soil Fertility (AB-DLO), P.O. Box 14, 6700 AA Wageningen, The Netherlands*

[2]*Agricultural University, Department of Theoretical Production Ecology, P.O. Box 430, 6700 AK Wageningen, The Netherlands*

Abstract. The effects of doubled carbon dioxide concentration and elevated temperature were evaluated by means of a crop growth model. Total dry matter production and tuber yield of potato cultivars, growing on a sandy loam were simulated for Dutch weather conditions, considering cultivar differences for earliness and soil moisture content as variables. Averaged over the years 1988–1991, a simulated increase of the carbon dioxide concentration from 350 vpm to 700 vpm increased tuber dry matter production by 22% for late cultivars and 29% for early cultivars. The effects were smaller for late cultivars irrespective of the occurrence of a drought period.

Elevated temperature reduced the positive effect of carbon dioxide because the stimulation of leaf area expansion in the juvenile stage, was upset later in the season due to an early start of foliage senescence. Raised temperature had only a net positive effect for conditions of optimum water supply and high irradiance during the juvenile stage, combined with a severe late drought. Then, the larger foliage prevented a rapid decrease in light interception during senescence.

Introduction

Agricultural productivity will respond to changes in the atmospheric CO_2 concentration and to possibly related changes in temperature. Various studies have shown a positive effect of CO_2 and a counteracting effect of a concomitant elevated temperature. The effects cannot be traced back to one or a few critical physiological and morphological components because the link between plant productivity and physiological characteristics is concealed by feed-back control mechanisms and by the large variance introduced by interactions with other environmental and developmental factors. Simulation models can be helpful to quantify the impact of a hypothetical environment on plant production and to evaluate the best strategy for further experimental research. Several crop growth models have been published for potato, differing in approach and detail (Fishman et al. 1984; MacKerron and Waister 1985; Ng and Loomis 1984). At present, the physiological models have mainly been used as a research tool, especially as a framework in the analyses of experimental results, and in studying the effect of individual processes on crop growth in relation to the environment.

In the present paper, the approach of crop growth modelling will be

A.J. Haverkort and D.K.L. MacKerron (eds.), Potato Ecology and Modelling of Crops under Conditions Limiting Growth, 101–114.

illustrated, first by describing the model in general terms and secondly by showing its usefulness and limitations for management and cultivar choice and its explanatory value for research purposes, especially in the stage prior to expensive experimental activities.

Material and methods

Experimental design

Experimental results were obtained form a field experiment. Sixteen cultivars were grown on a sandy soil in Renkum, near Wageningen, in 1988. The experiment was designed as a split plot in 6 replicates. The experiments were performed to check the general assumption that tuber yield and total biomass are closely correlated with the cumulative intercepted light (Spitters et al. 1989a). Pre-sprouted tubers were planted on 19 April in ridges, 0.75 m apart with a spacing of 0.35 m between plants in a row. Each plot consisted of 40 plants. There were 3 harvest dates: one at the beginning of June to determine the differences in time of tuber initiation, one at the end of July to have an estimate of the maximum haulm weight of each variety and a final harvest in September. At each harvest, dry weight and fresh weight of haulms and tubers were determined for each cultivar. During the whole growth period, the percentage soil cover by green foliage was estimated visually every week as an indicator of light interception.

Description of model for potential production

The light profile within the canopy is calculated from LAI and the extinction coefficients for both the flux of direct solar radiation and that of diffuse sky light. This procedure is followed to calculate the rates of photosynthesis at various heights within the canopy, discriminating between shaded leaf area receiving diffuse radiation only, and sunlit leaf area receiving both diffuse and direct radiation (Spitters et al. 1986).

Leaf area index
Leaf area increment (ΔLAI) is calculated from simulated leaf growth rate (ΔW_{lv}) and specific leaf area (SLA):

$$\Delta LAI = SLA \times \Delta W_{lv} \quad (1)$$

To account for the phenomenon that the early growth of leaf area ($LAI < 0.75$) is limited by temperature rather than by the supply of assimilates (and thus by radiation), early leaf area growth is described by an exponential function of the temperature sum after plant emergence or after planting (Spitters et al. 1989b):

$$LAI_t = N \times L_o \times \exp(RGR_L \times \Sigma\ °Cd) \quad (2)$$

where the LAI after some temperature sum (Σ °Cd) is calculated from the plant density (N in plants m^{-2}), the initial leaf area per plant at emergence (L_o in m^2 $plant^{-1}$) and the relative growth rate of leaf area (RGR_L in $°C^{-1}d^{-1}$). Foliage senescence is calculated from the simulated weight of green leaves (W_{lv}) and a relative death rate (RDR):

$$\Delta LAI = -W_{lv} \times RDR \tag{3}$$

The relative death rate (RDR in d^{-1}) is a function of crop development stage, earliness and temperature (Spitters and Schapendonk 1990).

Dry matter production

Daily dry matter growth rate is calculated from rates of photosynthesis and respiration:

$$\Delta W = C_f \times (P_g - R_m) \tag{4}$$

where ΔW is the growth rate (g DM $m^{-2}d^{-1}$), R_m the maintenance respiration rate (g CH_2O g^{-1} DM d^{-1}), C_f the conversion efficiency (g DM g^{-1} CH_2O) derived from average synthesis costs according to the known biochemical synthesis pathways of the chemical components of the dry matter (Penning de Vries and Van Laar 1982). Daily gross canopy assimilation rate (P_g) is obtained from momentaneous photosynthetic rates of individual leaves according to the model of Farquhar et al. (1980) and the light profile in the canopy,integrated by Gaussian integration over LAI and day length (Goudriaan 1986).

Dry matter partitioning

Partitioning factors:

$$\Delta W_i = p_i \times \Delta W \tag{5}$$

Daily dry matter growth of individual tissues (i), expressed on a soil area basis (W_i, g $m^{-2}d^{-1}$) is calculated from total growth (W, g $m^{-2}d^{-1}$) using a partitioning factor, which is a function of crop development stage (in °C d after plant emergence or planting).

Total biomass (W) is obtained by integrating the daily growth rates of the tissues (ΔW_i) over time. The harvest index (HI), being the fraction of tuber dry matter in total dry matter, is calculated by dividing tuber dry matter yield by total biomass.

Leaf photosynthesis

Leaf gross CO_2 assimilation was simulated according to the biochemical model described by Farquhar et al. (1980) and Von Caemmerer and Farquhar (1981). The model, based on Michaelis Menten kinetics, simulates the kinetics of electron flow and the carboxylation rates. From these relations, effects of temperature, photosynthetically active photon flux density (PPFD) and CO_2 concentration on gross leaf photosynthesis can be derived.

Gross leaf photosynthesis is determined by two rate-limiting processes, i.e.

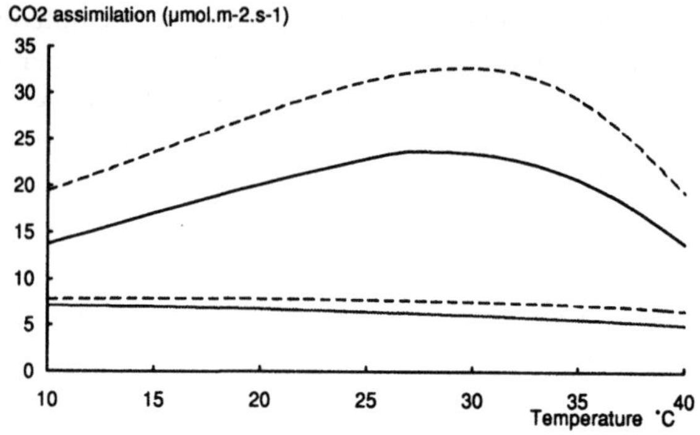

Figure 1. Simulated temperature dependence of gross CO_2 assimilation at 112 and 900 umol PPFD, lower and upper curves respectively and ambient CO_2 concentrations of 350 (——) and 700 vpm (---).

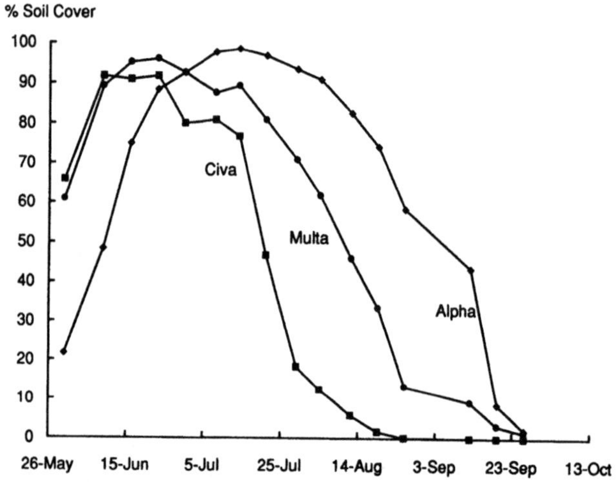

Figure 2. Foliage development of three potato cultivars, plotted as soil cover, estimated from visual observations in the experiment of 1988.

the production of reducing equivalents in the electron-transport chain and the rate of carbon fixation in the Calvin cycle. At low PPFD, the electron-transport rate, equivalent to energy delivery is limiting. A high CO_2 concentration has a positive effect due to more efficient energy utilization by suppression of photorespiration. At high PPFD the positive effect is enhanced because the carboxylation rate, linked with the availability of CO_2, is rate-limiting under high light conditions.

The effects of temperature on photosynthesis are complex. Temperature operates on the rates of electron transport and the rates of carboxylation and oxygenation. In addition the affinities of Rubisco for CO_2 and O_2 are affected. Due to a lower activation energy, higher temperatures will promote O_2 binding at the expense of CO_2 binding. At low PPFD, the effect of temperature on gross assimilation will be negative (Figure 1, lower part) due to the increase in photorespiration. However, at high PPFD, photosynthesis becomes less dependent on energy efficiency and more on the maximum turnover rates of electron transport and carboxylation, which are increased by a higher temperature. CO_2 amplifies this trend (Figure 1).

Earliness

Effects of earliness (m) are expressed by the initiation of tuberization and the onset of leaf senescence. Simulations were made for all cultivars of the 1988 experiment. Figure 2 shows the visual estimates of percentage of ground covered by green foliage for three cultivars. A more accurate but time-consuming method is based on counting the number of grids that were at least filled by more than half with projections of leaves (Khurana and McLaren 1982).

Table 1. Correlation matrix between yield components, measured in 1988

	Eff.	W	HI	Ydry	%DM	Yfresh	m	%CVe	%CVg	Mean	Units
Attribute											
PAR	0.07	0.89	−0.58	0.77	0.71	0.17	−0.96	2.4	13.5	4.9	GJ/ha
Efficiency	1	0.51	0.4	0.65	0.36	0.51	−0.12	4.2	6.1	2.93	g/MJ
Biomass		1	−0.32	0.96	0.78	0.38	−0.88	3.4	16.2	14.3	t/ha
Harvest index			1	−0.04	−0.32	0.38	0.62	1.2	4.6	0.807	-
Tuber yield (dry)				1	0.72	0.51	−0.75	4.1	15.2	11.6	t/ha
Dry matter content					1	−0.22	−0.74	2.4	13.0	0.214	-
Tuber yield (fresh)						1	−0.11	2.7	10.2	54.2	t/ha
Earliness Variety list							1	.	.	.	-

Eff: efficiency of intercepted radiation; W: total dry matter; HI: harvest index; Ydry: dry weight tubers; Yfresh: fresh weight tubers; %DM: dry matter concentration in tubers; m: earliness (Dutch cultivar list); %CVe: percentage environmental variation; %CVg: percentage genetic variation.

Cumulative light interception has been proven to be a reliable determinant of plant production. A correlation matrix for the observed yield components, and the calculated light interception for the same field experiment, is given in Table 1. Most of the variation in cumulative light interception can be attributed to differences in time of foliage senescence as indicated by earliness.

The correlation between cumulative light interception and earliness, rated according to the Dutch cultivar list (Cv. list) was −0.96. Not only total plant productivity, but also tuber dry matter production were well correlated with cumulative light interception and thus with earliness (Figure 3). This, however, was not true for tuber fresh weight because there was a compensating negative correlation between tuber dry matter concentration and earliness (Table 2).

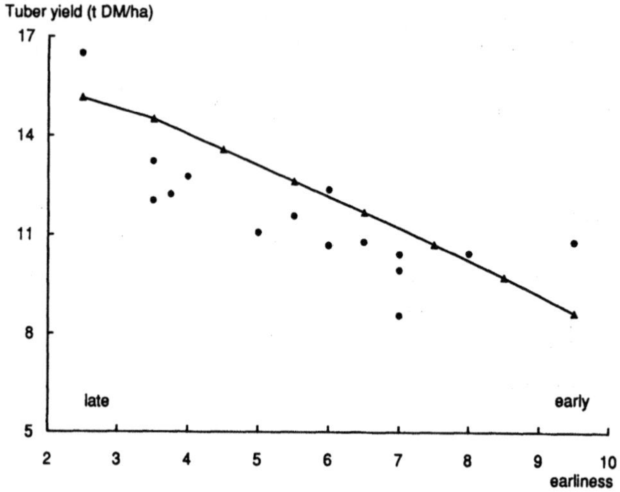

Figure 3. Effect of earliness on tuber yield of 16 potato cultivars in the experiment of 1988. Measured (dots) and simulated data (triangles) at the final harvest.

Table 2. Yield and dry matter concentrations. Harvest 1988

Cultivar	Yfresh	Ydry	%DM	m
Alcmaria	60.23	10.44	17%	8
Alpha	58.09	12.74	22%	4
Bintje	53.65	10.78	20%	6.5
Civa	59.28	10.79	18%	9.5
Karnico	63.83	16.48	26%	2.5
Katahdin	46.54	8.55	18%	7
Kennebec	55.5	10.68	19%	6
Krostar	45.29	10.41	23%	7
Maritta	50.8	12.21	24%	3.75
Multa	61.19	13.2	22%	3.5
Pimpernel	50.3	12.02	24%	3.5
Saturna	50.12	11.56	23%	5.5
Spunta	58.31	9.93	17%	7
Veenster	50.17	12.35	25%	6
Woudster	49.21	11.06	22%	5

Death rate of leaves due to developmental senescence is given by Spitters and Schapendonk (1990):

$$R_{dv} = (T - T_{base}) \exp(-11.7 + 0.48\,m) \exp((\text{sum}(T - T_{base}) - 725)(0.0068 - 0.0005\,m)) \qquad (6)$$

T (average day temperature);
T_{base} (effective temperature for leaf development) = 2 °C;
m (earliness).

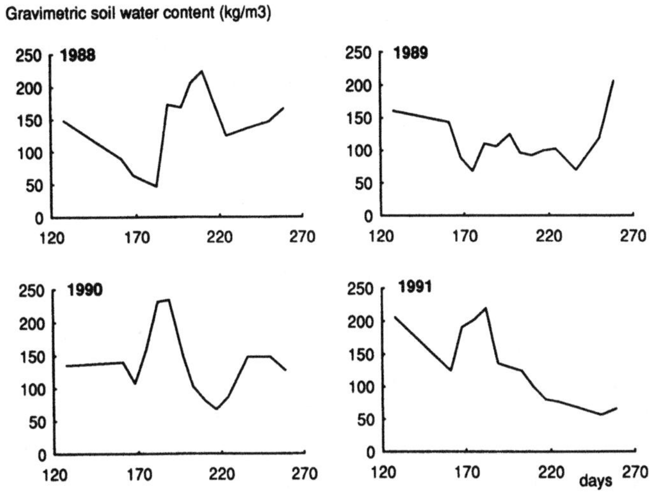

Figure 4. Simulated time courses of the gravimetric soil water content in four successive years.

Soil water relations

The soil moisture balance is tracked by a separate subroutine, largely the same as described by Van Keulen (1986). Rainfall is set against percolation of water to the soil layers below the rooting zone and the evapotranspiration rate. When the soil moisture content exceeds field capacity, the surplus water flows to the subrooting zone. Figure 4 shows the simulated soil moisture contents over the years 1988–1991. Evapotranspiration is derived by using a calculated reference value, that quantifies the evaporation from a short grass cover. This reference value marks the potential transpiration and evaporation from daily radiation, temperature and the fractional light interception by the canopy. Next, the actual values for soil evaporation and crop transpiration are calculated. Based on soil moisture balance studies of a sandy soil, the actual soil suction can be derived. Next, an estimated critical volumetric soil water content with a concomitant soil suction is used as a threshold to derive a reduction factor for transpiration (Spitters and Schapendonk 1990).

$$f_{wt} = (p_{wp} - p)/(p_{wp} - p_{cr})$$
$$0 < f_{wt} < 1 \tag{7}$$

p_{wp} (wilting point);
p_{cr} (critical suction; suction of soil moisture above which stomates close and transpiration is reduced);
p (actual suction).

Actual growth of the crop is then given by the product of potential growth and the reduction factor f_{wt} (Haverkort and Goudriaan 1993):

$$\Delta W_a = \Delta W_p \, f_{wt} \, (q - 1)/\{(q - 1) \, f_{wt} + 1\} \tag{8}$$

ΔW_p: potential dry matter production rate;
q: parameter that determines the improvement of the water use efficiency under drought conditions. If q = 1 there is no effect, if q = 2 the maximum increase of the water use efficiency, due to the drought condition, is about 100%.

Carbon partitioning

The partitioning of the assimilated carbon is dependent on the developmental stage of the crop and earliness. Late varieties reach a higher LAI during the season and they maintain a green foliage for a longer time. However, tuber initiation starts later. The effect of stress on dry matter partitioning is often small. However, in seed crops, stress around flowering usually causes an additional yield reduction due to a reduction of sink capacity (empirical relations by Doorenbos and Kassam 1979), which can substantially affect the harvest index. Shortage of soil moisture and/or nutrients favours allocation of assimilates to the below-ground parts at the expense of the above-ground parts, which may significantly affect yield of tuber crops (Seligman and Van Keulen 1981).

For the present application, effects of drought on the specific leaf area and leaf death rate are not implemented, but c.f. Spitters and Schapendonk (1990). In addition the effects of drought and day length on tuber initiation and bulking are neglected despite the evident influence on carbon partitioning (Kooman and Haverkort, implemented in this volume). However, because we only intend to show the basic effects, these interactions are omitted but can easily be inserted into the model.

Interaction with water stress

Water stress induces closure of stomata. The greater resistance of the stomata is reflected in a inhibited rate of photosynthesis. However, photosynthesis is less inhibited than transpiration because the concentration gradient for CO_2 is increased by drought, while that for water remains the same. Therefore, the water-use efficiency is enhanced. However, after a few days, water stress reduces the photosynthetic capacity directly, either by down-regulation of electron transport or by effects on the carboxylation enzymes that lower the mesophyll conductance (Schapendonk et al. 1989) again lowering the water use efficiency. Thus, closure of stomata and the lower mesophyll resistance show a correlated response to water stress, that leads to a constant ci/ca ratio. Therefore, photosynthesis and transpiration are equally reduced and water use efficiency remains the same.

Effects of drought can now be translated into production losses by multiplication of the potential production by the ratio between actual and potential transpiration (Ta/Tp), given a steady-state condition after a few days of a drought period. In general it appears that the immediate effect of drought on physiological parameters is much greater than the integrated seasonal effect, because of the close relation between whole crop transpiration and leaf area (i.e.

growth). In other words, an impaired development of leaf area will save water for later consumption and, therefore, smooth the temporary effects.

Seasonal effects
Photosynthetic light-use efficiency measured early in the season, was low. This was partly due to low photosynthetic rates of individual leaves and partly because a low LAI leads to more light-saturated leaves. Later in the season the light-use efficiency of the canopy increases but gross photosynthesis of single leaves (P_g) gradually decreases mainly due to senescence. The efficiency of net photosynthesis (P_n) decreased even more due to the increasing respiration needed to maintain the standing biomass. This is reflected in an increased respiration/photosynthesis ratio. The declining trend in efficiency of above-ground biomass production (W), is counterbalanced by a gradually decreasing allocation to roots. As a result of all these interactions, the efficiency for above-ground dry matter is rather constant.

Effects of elevated CO_2

Light use efficiency
We measured a higher light use efficiency, for a canopy with a higher leaf area index (LAI). In accordance with these results, Figure 5a shows the calculated light use efficiencies of cultivars varying in earliness. Thus late cultivars are more efficient, mainly due to the overall higher leaf area that leads to less light-saturation in the canopy. The relative effect of elevated CO_2 is just opposite. It is slightly higher for early than for late cultivars. This is due to the CO_2 effect on the light response curve which is higher in the light saturated region and thus under conditions with low LAI, i.e. early cultivars, where individual leaves, on average, receive more light. Caution should be taken to generalize this statement because analyses showed that the light use efficiency varies considerably from day to day. However for longer periods of time (e.g. a 14-day period), the application of an average light use efficiency for above-ground dry matter production is satisfactory.

The relative effect of CO_2 on tuber yield is greater then on the light use efficiency because tuber bulking of early cultivars has an additional benefit from CO_2 due to the enhanced rate of leaf expansion, so that the tuber filling later in the season becomes less vulnerable to drought stress (Figure 5b).

Basic trends
Sink-source interaction. The reported effects of CO_2-enrichment on the yield of potato vary between a slightly negative effect (Goudriaan and De Ruiter 1983) and a maximum positive effect of 39% for a doubling of the CO_2 concentration from 350 to 700 vpm (Wheeler et al. 1991). The yield increase, predicted by the present model is on average 26%, which is in agreement with Wheeler et al. (1991). When the yield of potato is limited by CO_2-assimilation, i.e. when the source of carbohydrate is limited, effects of CO_2 on

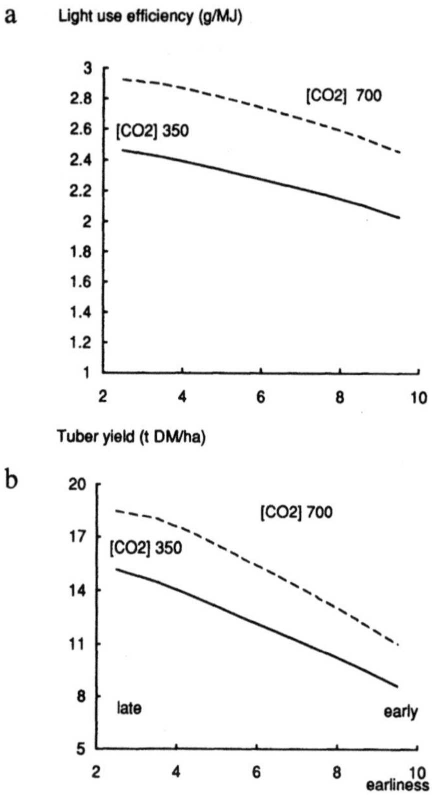

Figure 5. The efficiency of absorbed irradiance (a) and the tuber yield (b) plotted as a function of earliness at two different CO_2-concentrations.

photosynthesis will be important. However, in the case of sink limitation, increasing the CO_2 concentration may have no effect. Wheeler et al. (1991) demonstrated that doubling the CO2 concentration resulted in an increase of 32% in tuber dry matter under 12-h photoperiods but no increase under 24-h photoperiods. This difference was attributed to earlier tuberization under the shorter photoperiod, providing a greater sink for assimilates. Sink limitation may decrease photosynthetic rate or even result in an irreversible damage of the chloroplasts due to starch accumulation (Goudriaan and De Ruiter 1983) and thus it can inhibit photosynthesis physically.

Photosynthesis in general, responds to a sink limitation by a decreased quantum efficiency. In that way assimilate supply and demand are tuned. However, there are exceptions. For instance, the mesophyll resistance of the late cultivar Kennebec, in our experiments, hardly responded to drought-induced growth limitation. This implies a poor operational control of photosynthesis. Indeed it has been noticed that carbohydrate levels in the leaves of this cultivar increased to injurious levels under elevated CO_2 and high irradiance (Wheeler and Tibbitts 1985).

The broad variation of responses to elevated CO_2 emphasizes the existence of genetic variation in both source and sink limited processes. On the other hand, the close correlation between the cumulative light absorption and dry matter production (Table 1), favours the idea of a prevailing source limitation.

Interaction with temperature

Gross CO_2 assimilation responds to temperature by changes in electron-transport, turn-over of Rubisco and the affinity of O_2 and CO_2 to Rubisco. Temperature enhances the CO_2-effect because it induces higher rates of enzyme turn-over, that are amplified by CO_2-induced suppression of photorespiration (Figure 1). This is valid for high and low PPFD. At high PPFD, limitation of carboxylation will be less at elevated CO_2 concentrations, which provides an additional advantage.

At temperatures above 35 °C, irreversible damage of the electron carriers in the chloroplasts and denaturation of Calvin cycle enzymes cause a sharp decline of the maximum electron transport- and carboxylation rates. It has been shown that the combined effects of high temperature and drought severely impair photosynthetic electron transport (Schapendonk et al. 1989).

The effects of temperature on respiration depend on the developmental stage of the crop. Initially respiration is proportional to growth and therefore it remains a fixed fraction of photosynthesis. However, with increasing biomass, maintenance respiration of the crop will take an increasing share of the total respiratory costs. At leaf area ratios above 4, the maintenance may consume substantial amounts of energy, and higher temperatures will have a negative effect on net photosynthesis and increase the light-compensation point of the crop. The simulated results show losses of about 11 and 15% for total biomass production and tuber yield, respectively when the temperature rises 3 °C (Table 3). The year 1991 was an exception. High temperature and abundant water in spring caused rapid leaf expansion, that compensated partly for the late drought in the summer and autumn. Further, the temperature was low in the summer of 1991 and so the simulated yield increased by 6–7%, when the temperature was raised by 3 °C. In general there was no interaction between CO_2 concentration and temperature on yield or evaporation (Table 3). The momentaneous effects are apparently counteracted by long-term effects, most likely through indirect effects on the leaf area duration (LAD).

Conclusions

- A doubling of the CO_2 concentration is expected to stimulate the yield of potato, ranging from 20% for the late varieties to 30% for the early varieties. The effects can be attributed to increased light use efficiency and a positive effect on the amount of intercepted light.
- The positive effect of CO_2 is greatest for early varieties and early-drought conditions. The effect of drought on late cultivars is smaller and independent of the period in the year. However one should recall the fact that effects of

Table 3. Simulated data on transpiration, total dry weight and tuber yield. The figures between brackets show the relative effect of a temperature increase by 3 °C

year	350 vpm	T = (+3 °C) (%)	700 vpm	T = (+3 °C) (%)
Cumulated transpiration (mm)				
1988	139	(−10.4)	144	(−9.9)
1989	134	(−3.7)	145	(−4.5)
1990	177	(−1.4)	186	(−1.5)
1991	138	(+4.3)	142	(+3.6)
Cumulated total dry weight (t/ha)				
1988	13.14	(−14.1)	16.31	(−12.9)
1989	14.17	(−10.0)	17.92	(−11.0)
1990	16.83	(−11.4)	20.81	(−11.5)
1991	12.04	(+4.3)	14.88	(+3.6)
Cumulated tuber dry weight (t/ha)				
1988	11.83	(−17.7)	14.78	(−17.0)
1989	11.10	(−17.1)	14.14	(−17.2)
1990	13.79	(−14.3)	17.14	(−14.3)
1991	8.59	(+7.0)	10.72	(+6.0)

- CO_2 and temperature on time of tuber initiation are not accounted for in the presented model.
- The amount of water transpired over the whole growth period is not changed by an elevated CO_2 concentration because an increased water use efficiency is compensated by an increased LAI.
- A temperature increase of 3 °C has a negative effect on the yield. This is mainly due to an effect on leaf senescence. The simulated data for both high and low CO_2 are practically identical, which means that in the long term there is no interaction between CO_2 concentration and temperature.

Acknowledgements

We are grateful to D.K.L. MacKerron and A.J. Haverkort for critical comments on the manuscript.

References

Doorenbos J, Kassam A H (1979) Yield Response to Water. FAO Irrigation and Drainage Paper 33, FAO, Rome, Italy.

Farquhar G D, Von Caemmerer S, Berry J A (1980) A biochemical model of photosynthetic CO_2 assimilation in leaves of C3 species. Planta 149:78–90.

Fishman S, Talpaz H, Dinar M, Levy M, Arazi Y, Rozman Y, Varshavsky S (1984) A phenomenological model of dry matter partitioning among plant organs for simulation of potato growth. Agricultural Systems 14:159–169.

Goudriaan J, De Ruiter H E (1983) Plant growth in response to CO_2 enrichment, at two levels of nitrogen and phosphorus supply. 1. Dry matter, leaf area and development. Netherland Journal of Agricultural Science 31:157–169.

Goudriaan J (1986) A simple and fast numerical method for the computation of daily totals of crop photosynthesis. Agricultural and Forest Meteorology 38:249–254.

Haverkort A J, Goudriaan J (1993) Fysiologische aanpassing aan droogte:perspectieven voor verbetering van droogtetolerantie. Pages 49–65 in Van Keulen H, Penning de Vries F W T (Eds.) Agrobiologische Thema's 8. Watervoorziening en Gewasproduktie.

Khurana S C, McLaren J S (1982) The influence of leaf area, light interception and season on potato growth and yield. Potato Research 25:329–342.

Kooman P L, Haverkort A J (1995) Modelling development and growth of the potato crop influenced by temperature and daylength: Lintul-potato. Pages xxx in Haverkort A J, MacKerron D K L (Eds.) Ecology and Modelling of Potato Crops under Conditions Limiting Growth. Kluwer Academic Publishers, Dordrecht, The Netherlands.

MacKerron D K L, Waister P D (1985) A simple model of potato growth and yield. Part I: Model evaluation and sensitivity analysis. Agricultural and Forest Meteorology 34:241–252.

Ng N, Loomis R S (1984) Simulation of Growth and Yield of the Potato Crop. Simulation Monographs. Pudoc, Wageningen, The Netherlands.

Penning de Vries F W T, Van Laar H H (Eds.) (1982) Simulation of Plant Growth and Crop Production. Simulation Monographs. Pudoc, Wageningen, The Netherlands.

Schapendonk A H C M, Spitters C J T, Groot P J (1989) Effects of water stress on photosynthesis and chlorophyll fluorescence of five potato cultivars. Potato Research 32:17–32.

Seligman N G, Van Keulen H (1981) PAPRAN: A simulation model of annual pasture production limited by rainfall and nitrogen. Pages 192–221 in Frissel M J, Van Veen J A (Eds.) Simulation of Nitrogen Behaviour of Soil-Plant Systems. Pudoc, Wageningen, The Netherlands.

Spitters C J T, Toussaint H A J M, Goudriaan J (1986) Separating the diffuse and direct component of global radiation and its implications for modelling canopy photosynthesis. I. Components of incoming radiation. Agricultural and Forest Meteorology 38:231–242.

Spitters C J T, Neele A E F, Schapendonk A H C M (1989a) Crop physiological components of tuber yield and their use in potato breeding. Pages 51–60 in Parental Line Breeding and Selection in Potato Breeding. Proceedings EAPR Breeding Section and EUCARPIA Potato Section Conference. Pudoc, Wageningen, The Netherlands.

Spitters C J T, Van Keulen H, Van Kraalingen D W G (1989b) A simple and universal crop growth simulator: SUCROS87. Pages 147–181 in Rabbinge R, Ward S A, Van Laar H H (Eds.) Simulation and Systems Management in Crop Protection. Simulation Monographs 32. Pudoc, Wageningen, The Netherlands.

Spitters C J T, Schapendonk A H C M (1990) Evaluation of breeding stategies for drought tolerance in potato by means of crop growth simulation. Plant and Soil 123:193–203.

Van Keulen H (1986) A simple model of water-limited production. Pages 130–152 in Van Keulen H, Wolf J (Eds.) Modelling of Agricultural Production: Wheather, Soils and Crops. Simulation Monographs, Pudoc, Wageningen, The Netherlands.

Von Caemmerer S, Farquhar G D (1981) Some relations between the biochemistry of photosynthesis and the gas exchange of leaves. Planta 153:376–387.

Wheeler R M, Tibbitts T W, Fitzpatrick A H (1991) Carbon dioxide effects on potato growth under different photoperiods and irradiance. Crop Science 31:1209–1213.

Wheeler R M, Tibbitts T W (1985) Growth of potato (*Solanum Tuberosum* L.) under continuous light. Plant Physiology 80:801–804.

8. Nitrogen and the growth of potato crops

J. VOS

Wageningen Agricultural University, Department of Agronomy, Haarweg 333, 6709 RZ Wageningen, The Netherlands

Abstract. The lengthening of the period of full soil cover in response to increasing nitrogen supply can be explained from the effects of nitrogen on the rate of leaf expansion and on stem branching. Nitrogen supply modifies most other developmental processes or plant attributes only marginally or transiently. Specific effects of nitrogen on leaf photosynthesis and light use efficiency (gram dry matter per Mega joule intercepted radiation) are only small. In crops provided with 'near optimal' rates of nitrogen, the 'Greenwood equations' (Greenwood et al. 1985a; 1990) provide a reasonable description of the change in the average nitrogen concentration as a function of the total dry matter per unit area. Experiments with several rates of nitrogen, supplied as one single dose in spring, showed a fairly uniform relation between nitrogen uptake and dry matter yield (at maturity) for cultivars of the same maturity class. This seems to imply a fairly close association between the accumulation of dry matter and nitrogen. Observations from a pot experiment, however, showed that both processes are not necessarily closely linked. Suggestions are made for modelling nitrogen uptake, and the effects of nitrogen on plant development and crop photosynthesis.

Introduction

Quantitative understanding of responses to nitrogen of potato crops is important for two reasons. One is the diminishing response in terms of yield to increased rates of supply; this poses the question of finding the economic optimum rates of supply. The second, and only recently acknowledged, reason is losses of nitrogen to the environment, posing the question of the ecologically optimum rate of nitrogen supply. Neither the economic optimum rate, nor the ecological optimum rate is fixed. The economic optimum is particularly affected by prices of input and product, and the ecologically optimum depends on the definition of environmental constraints and the criteria that one wants to respect. Both optima will depend on weather conditions (e.g. rainfall distribution), properties of the soil (e.g. water storage capacity, mineralization potential, rootability) and on the regime of application of nutrients (e.g. application of a single dressing in spring, split applications, organic manures supplemented with chemical fertilizer). Hence, there is no universally applicable answer to the question 'what is the optimal rate of supply', be it in the economic sense or in the ecological sense. Modelling can be helpful in those cases in order to calculate probabilities of attaining the optimal yield and probabilities of risks

A.J. Haverkort and D.K.L. MacKerron (eds.), Potato Ecology and Modelling of Crops under Conditions Limiting Growth, 115–128.

of violating environmental standards. The objective of this contribution is to discuss the nitrogen responses of potato with a modeler's requirement of data and relations in mind and with the aim to apply a model to find answers to the question of the best application strategy. The focus will be on plant and crop responses exclusively; transformations and transport of nitrogen in the soil, which need to be modeled to examine nutrient management strategies under various conditions, are left beyond the scope of the paper. It is noted that much of the current knowledge of nitrogen responses of plants and crops is based on experiments, either in the field, or in containers or pots where plant and crop attributes were studied in response to single doses. However, for risk analyses it is important that models can describe the responses to delayed and split nitrogen applications (cf. Vos and Marshall 1994).

Soil cover and light interception

A common response to a variable dose of nitrogen at planting is that there is no difference between treatments in the initial rate of change of soil cover (green tissue, measured with a grid, cf. Burstall and Harris 1983). However, as nitrogen becomes limiting (which is earlier in time the lower the dose), the rate of canopy expansion declines towards zero. At severe nitrogen limitation this occurs even before full soil cover was reached, followed by a stage of declining green cover. The duration of the period with full soil cover is longer and the stage of decline starts later in time the higher the nitrogen dose (Figure 1). Assuming that the proportion of light intercepted is equal to the proportion of soil cover, one can calculate seasonal light interception, using actual meteorological data (Figure 2). Figures 1 and 2 show that the length of the period of full soil cover is affected by the nitrogen dose and that the advantage in yield from larger supplies of nitrogen result from a later end to light interception. An explanation of this general response to nitrogen can be obtained from an understanding of the effects of nitrogen on the dynamics of branching and of growth and senescence of leaves.

Uptake of nitrogen and dry matter yield

The relation between uptake of nitrogen and total dry matter yield (haulm plus stolons and tubers, fibrous roots excluded) appears to be fairly fixed. Figure 3 provides an example for potatoes in the maturity class that matures end of September in the Netherlands (maturity class 4.5 in the Dutch List of Recommended Varieties). The response of yield to uptake is curvilinear: each additional unit of N taken up is converted less 'efficiently' into dry matter. As a consequence, the average concentration of nitrogen increases as more nitrogen is absorbed. The maximum yield not limited by insufficient nitrogen supply, may depend on the maturity class, and can be expected to be lower for early maturing cultivars.

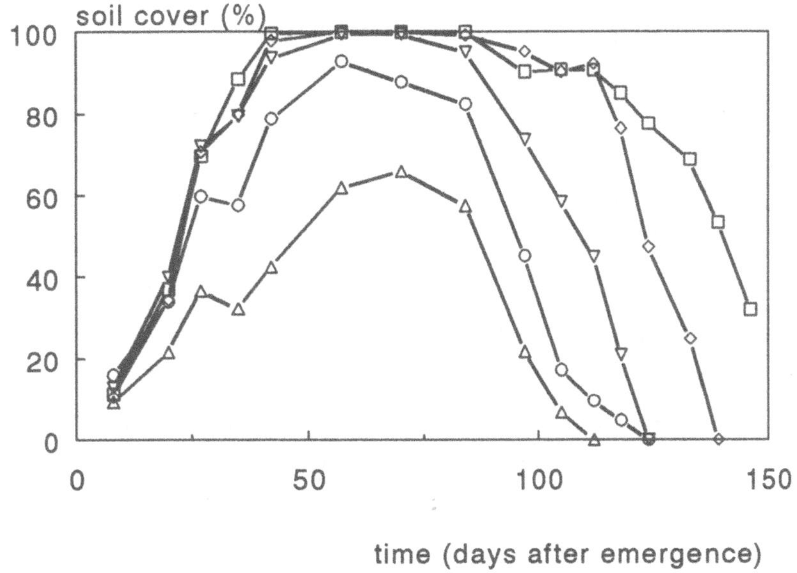

Figure 1. Progress of percentage soil cover, measured with a grid (Burstall and Harris 1983) as influenced by nitrogen supply; symbols for the rates of nitrogen supply (single doses in spring in kg ha^{-1}) represent: triangles pointing upwards: 0 N, circles 50 N, triangles down 100 N, diamonds 200 N, squares 300 N (Vos, unpublished).

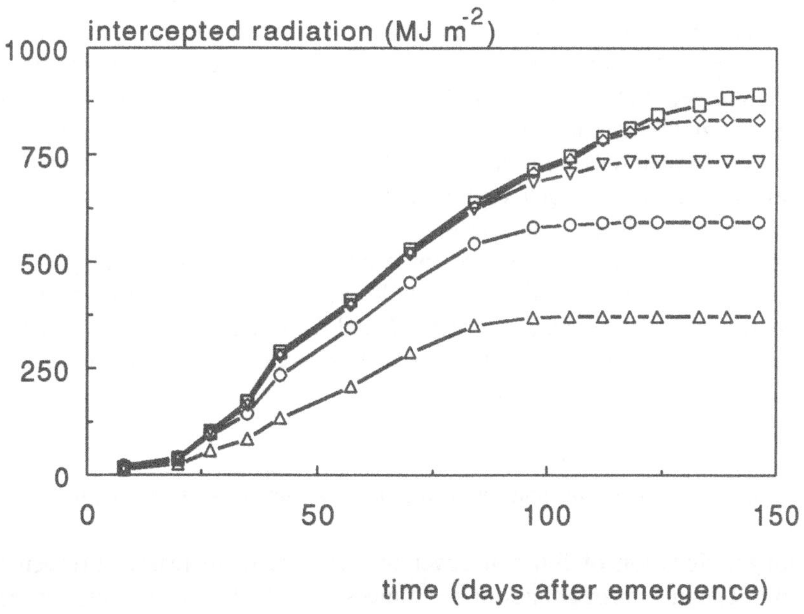

Figure 2. Progress of the cumulative radiation interception of crops provided with different rates of nitrogen, for symbols see caption to Figure 1.

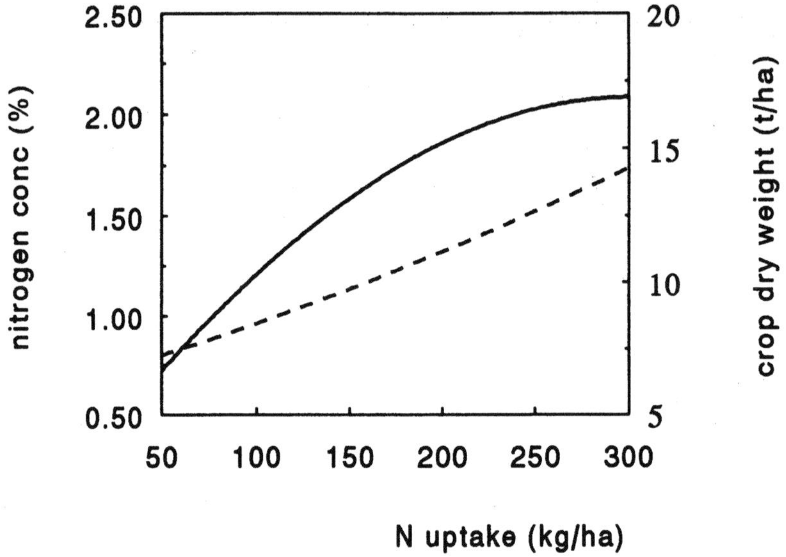

Figure 3. The average concentration of nitrogen in the dry matter of potato crops at maturity (broken line) and the dry weight of the mature crops (foliage plus tubers; full drawn line) as functions of the nitrogen taken up at maturity; based on data from four seasons with different rates of nitrogen supply varying between 0 and 400 kg ha^{-1} (Vos, unpublished).

Another important general relation is that between nitrogen supply and apparent nitrogen recovery, where the latter is defined by the nitrogen that is taken up by a fertilized crop as a proportion of nitrogen fertilizer applied, correcting for the nitrogen uptake by an unfertilized control. Neeteson et al. (1987) showed that the recovery clearly declined with increase in nitrogen dose in spite of much variation in recovery for a given level of nitrogen supply; Figure 4 is based on their data and shows only the maximum recovery that was attained at several rates of nitrogen supply. The average recovery (based on total dry matter) of 96 fertilizer trials was 46–50% (Neeteson 1989).

So far, a fairly general picture has been drawn of responses of potatoes to nitrogen. From these relations one could arrive at first approximations of the amount of dry matter that can be produced from a given amount of available nitrogen, but a more mechanistic approach is needed for modelling to be useful to explore solutions to real problems.

Developmental processes and plant attributes affected by nitrogen supply

A longer duration of full soil cover in response to increased nitrogen dose (Figure 1) can be brought about by changes in several developmental processes and plant attributes including the rate of appearance of leaves, the life span of leaves and the degree of branching. In Figure 5 illustrates the terminology on branches that is used in this paper (Vos 1995a).

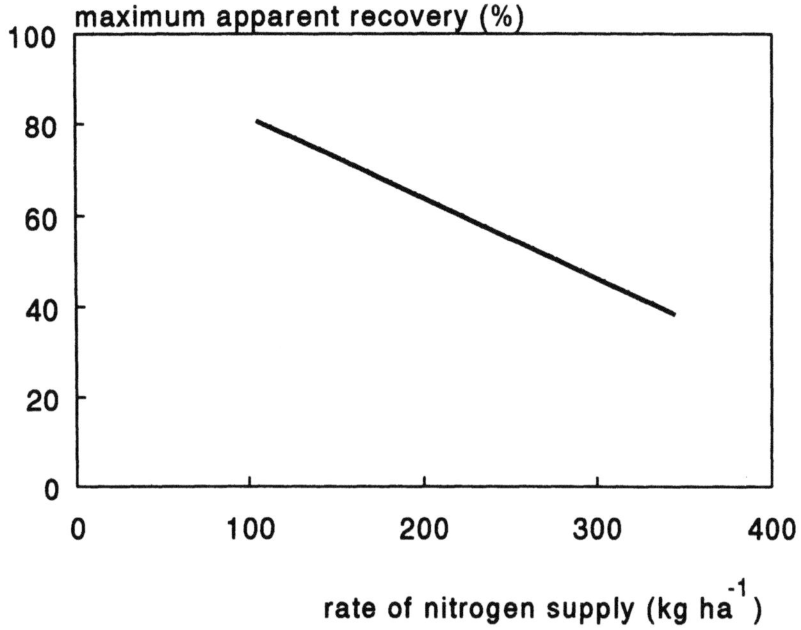

Figure 4. The maximum apparent recovery of nitrogen as a function of the rate of nitrogen supply; adapted from Neeteson et al. (1987).

Vos & Biemond (1992) found in a pot experiment that the areas of mature leaves and the ultimate order of apical branching that is reached were particularly sensitive to nitrogen. The data indicated a maximum in the response of leaf size to nitrogen above which leaf size is insensitive to further increase in nitrogen supply. The node number of the leaf was also a major determinant of leaf size. The duration of leaf expansion was fairly conservative, irrespective of nitrogen supply and leaf node number. Field observations corroborate the strong effect of nitrogen on apical branching (Figure 6; Millard and MacKerron 1986). The rate of appearance of leaves and branches was not affected by nitrogen, except of course at the point where nitrogen was becoming exhausted. The specific leaf area (SLA) increased with leaf age but was hardly affected by nitrogen supply. The absence of an effect of nitrogen on SLA was subsequently confirmed in field experiments (Vos 1995b). Vos and Biemond (1992) found that there was no difference in longevity of leaves at comparable nodes between a treatment with severely limiting nitrogen supply and a treatment where nitrogen limitation occurred first at advanced stages of growth; at still higher rates of nitrogen supply leaf longevity was somewhat longer. These observations suggest that leaf longevity is not extremely responsive to nitrogen limitation. However, Vos and Biemond made their observations on spaced plants.

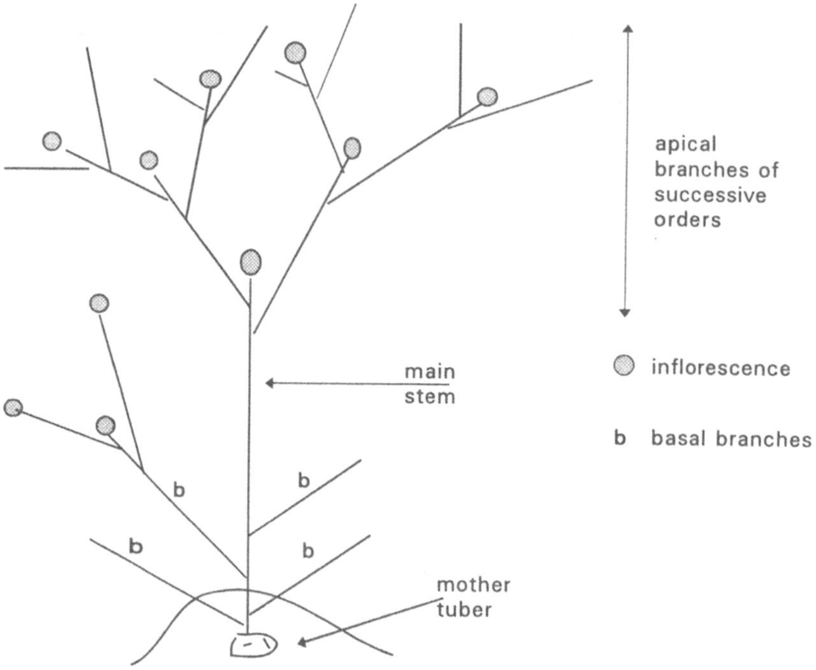

Figure 5. Schematic representation of a potato plant showing a main stem arising from the mother tuber, basal lateral branches and apical lateral branches. Calling the leaf subtending the flower 'leaf n', apical branches generally emerge from the axils of leaves 'n-1' and 'n-2'. The first order of apical branches consists of two parts, the second order of four parts, etc. (Vos 1995a).

Nitrogen, photosynthesis and radiation use efficiency

In crops leaf longevity could be greater in those raised under a comparatively low nitrogen regime than in well-nourished crops. This was postulated by Firman and Allen (1988) from measurements of the change in photosynthetic capacity (P_{max}) of leaves of particular numbers with leaf age, showing faster decline in P_{max} in leaves from high-N treatments than from low-N treatments. The explanation could be that larger leaves and a larger number of leaves (due to enhanced branching) induces 'shade effects' on leaves at lower positions in the canopy (cf. Olesinski et al. 1989) in high-N crops. 'Shade effects' mean enhanced remobilization of nitrogen from shaded leaves (and hence decline in P_{max}), perhaps due to decreased illumination of the lower leaves, but changes in the red/far red ratio (Casal et al. 1990) of the light received by the leaves deeper in the canopy could also play a role. Hirose and Werger (1987) postulated that the plant distributes its nitrogen in such a way as to optimize the photosynthetic productivity of the stand. Differences between low-N plants and high-N plants in the rate of remobilization of nitrogen from lower leaves can be understood from this point of view: it is not 'wise' for a plant to leave nitrogen in leaves that

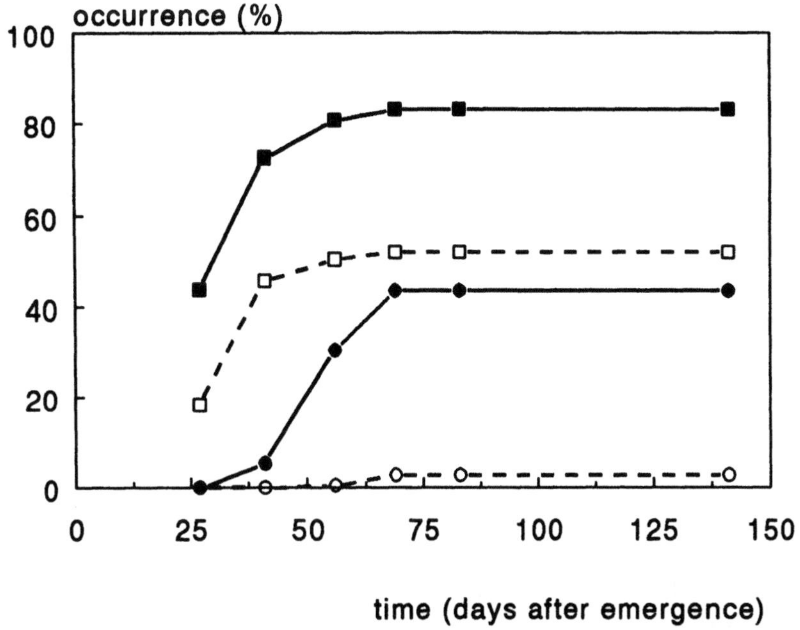

Figure 6. Changes with time in the occurrence of first order ('1o') (symbol: squares) and second order ('2o') (symbol: circles) apical branches in relation to nitrogen supply; open symbols 100 kg N ha^{-1}, closed symbols 300 kg N ha^{-1}; the percentage occurrence was calculated by: (total count of branch parts of a particular order m^{-2})/(the potential number of branch parts of that order m^{-2}) × 100; where the potential number is 2 per main stem for '1o' apical branches, and 4 for '2o' branches (Vos, unpublished; data from the same expt. as Figures 1 and 2).

only receive strongly attenuated light; that nitrogen can be employed better to produce leaves with a high P_{max} at the top of the plant. At our laboratory we made preliminary observations on the effects of shading the lower plant parts and found indications that shade causes faster remobilization of nitrogen from treated plant parts and reallocation to the top of the plant; however, more work needs to be done to examine this mechanism.

Implicit in the reasoning in this section is a relation between nitrogen concentration (N_{conc}) and P_{max}. Many authors describe a linear relation between P_{max} and N_{conc} indeed; see Vos and Oyarzún (1987) for potato and Van Keulen et al. (1989) for a compilation across species. Marshall and Vos (1991) found only marginal effects of nitrogen nutrition on the relation between P_{max} and N_{conc} of potato leaves. Although, physiologically speaking, we are not dealing with a fixed and causal relation between P_{max} and N_{conc}, for field conditions the relation between these variables appears sufficiently stable to justify its incorporation into models. Two inferences can be made from the discussion on P_{max} and N_{conc}:

(i) there is little or no specific effect of the nitrogen regime on the photosynthetic properties of leaves (Marshall and Vos 1991). In other words: differences in P_{max} between leaves from different nitrogen treatments can be explained largely from differences in N_{conc};

(ii) the (vertical) distributions in the canopy of both P_{max} and N_{conc} are congruent.

The ratio between the radiation intercepted and the dry matter produced over a particular period of time yields the radiation use efficiency (RUE; other terms are: light use efficiency, light conversion factor). Millard and Marshall (1986) observed no differences in RUE between nitrogen treatments, except for a somewhat smaller value in the absence of any nitrogen fertilizer. In our experiments we did not observe an effect of nitrogen on RUE either, while production differed substantially between treatments (Vos, unpublished).

Dry matter distribution

Biemond and Vos (1992) observed a lower harvest index the higher the rate of nitrogen supply for most of the period of tuber bulking, but the differences in harvest index had disapppeared in mature plants. Also Dyson and Watson (1971) reported delayed tuber bulking for increased rates of nitrogen supply. In field experiments the harvest index of approximately mature plants may decline with increase in nitrogen supply, but here relatively greater losses of haulm material from low-N treatments (which mature earlier) may play a role. Biemond and Vos reported no differences in the distribution of dry matter over stems and leaves, even though plant weights ultimately differed five-fold between extreme nitrogen treatments. Leaf/stem ratios appeared to decline with higher rates of N supply in field experiments (Vos, unpublished), but here differences in leaf longevity and rates of loss of dead leaf material may also play a role.

The effects of nitrogen on the ratios between shoot and root weight or between green leaf area and root length need to be known for a proper understanding of nitrogen effects, but there is little information in the literature on these topics. Based on the theory of a 'functional equilibrium' between shoots and roots (Brouwer 1983), one can speculate that the 'need' for high root length densities increases as nitrogen becomes more and more limiting. Relatively or absolutely lower root length densities at non-limiting rates of nitrogen supply could perhaps partly explain the decline in apparent nitrogen recovery with increase in nitrogen supply.

The relation between the accumulation of nitrogen and dry matter

The rate of uptake of nitrogen from the soil will depend on the availability of nitrogen in the soil and on the 'demand' by the crop. If, starting from a low concentration of nitrogen, more and more nitrogen is made available to the plant, the rate of uptake and the rate of growth will increase. The rate of uptake may even continue to increase beyond the level of nitrogen supply that 'saturates' the rate of growth, leading to increased nitrogen concentrations in the plant (sometimes called 'luxury consumption'), but there will be an upper limit above which there is no further increase in the rate of uptake in response to more abundant nitrogen supply, in other words a maximum concentration is reached. The best demonstrations of such reaction patterns are perhaps obtained from experiments in which experimental conditions during the exponential growth phase are manipulated such as to achieve 'steady state conditions', that is: a constant nutrient concentration in the plant and a constant shoot/root ratio. Such systems and their behaviour are described by Ingestad and co-workers (e.g. Ingestad and Ågren 1992).

An example of the relations between growth rates and nitrogen uptake under different nitrogen regimes for potted plants is provided by Biemond and Vos (1992; Figure 7). During the first 60 days after emergence the N3 treatment absorbed nitrogen at about the same rate as the N2 treatment (Figure 7b), although, more nitrogen was available for N3 plants (Figure 7a); this implies a maximum on the rate of uptake. The N3 plants absorbed the extra nitrogen they received beyond that given to N2 plants after 60 DAE (days after emergence), when nitrogen uptake in N2 plants levelled off. Yet, N2 and N3 plants maintained similar growth rates up to 100 DAE (Figure 7c); the slowing of N uptake in N2 plants had, apparently, no initial effect on productivity. In the N1 plants, also, the rate of nitrogen accumulation slowed down earlier in time than the rate of dry matter accumulation.

It is important to note that the average plant nitrogen concentration always declines with time (Figure 7d) and thus with increase in plant dry weight. This declining trend is attributed to a decrease in the fraction of the total amount of nitrogen in the plant which is associated with photosynthesis and a concomitant increase in the fraction of nitrogen tied up in structural tissue and storage constituents (Caloin and Yu 1984). Based on numerous experiments Greenwood and collaborators defined the 'critical nitrogen concentration'; this is the average concentration of nitrogen (N_c (%)) which is just sufficient for maximum growth. For potato Greenwood et al. (1985a) arrived at the following relation between N_c and the weight of the crop (W; tonnes ha-1; all plant parts, excluding fibrous roots):

$$N_c = 1.35 + 4.05\, e^{-0.26\, W} \tag{1a}$$

Greenwood et al. (1990) arrived at the following equation which is valid for C_3 crops in general and deviates numerically slightly from Eq. 1a:

$$N_c = 5.7\, W^{-0.5}. \tag{1b}$$

Greenwood et al. (1986) stated that the growth rate of the crop is approximately proportional to the average nitrogen concentration when this is less than the critical concentration (cf. Greenwood et al. 1991). Luxury consumption is recognized by Greenwood and collaborators, but their work offers no exact tools to

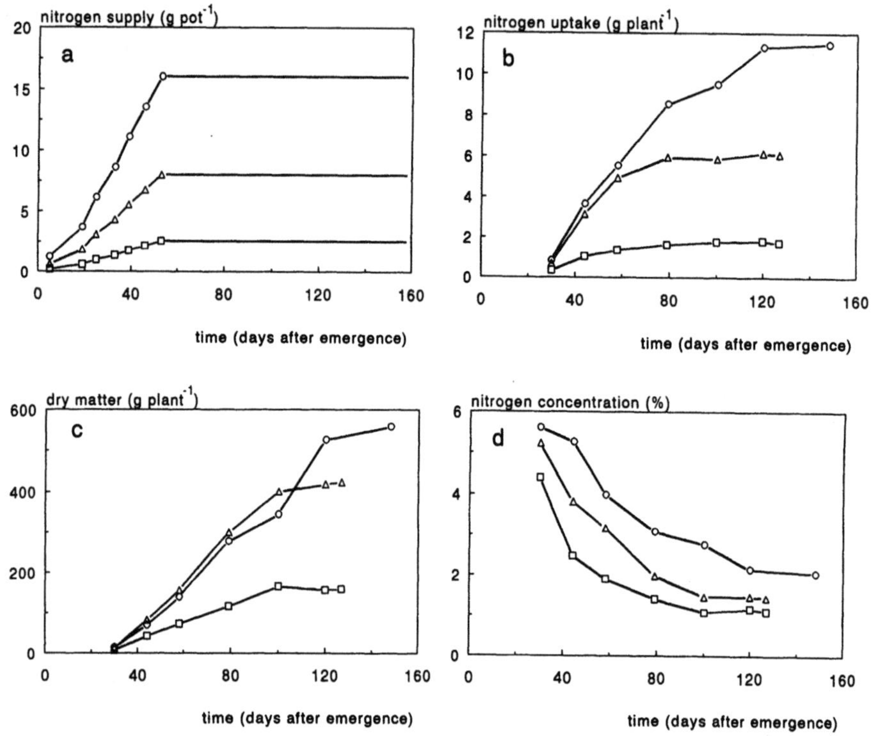

Figure 7. Effects of nitrogen supply on nitrogen uptake and dry matter accumulation of potato plants; (a) cumulative nitrogen supply, (b) nitrogen uptake with time, (c) dry matter accumulation, (d) average nitrogen concentration in plant dry matter; squares: N1 plants, total nitrogen supply 2.5 g pot^{-1}; triangles: N2 plants, 8 g pot^{-1}; circles: N3 plants, total nitrogen supply 16 g pot^{-1}. Data from a pot experiment described by Vos and Biemond (1992).

quantify the upper limit of that process other than the remark that the maximum concentration, N_m, is approximately 1.2 times N_c (Greenwood et al. 1985b).

Evidence from independent experiments corroborate that nitrogen concentrations, plotted as a function of crop dry weight, are closely scattered around the curve given by Eq. 1 (Figure 8), at least for crops for which (near) optimum rates of nitrogen supply can be assumed. It is noted that in young crops the nitrogen concentration tends to be higher than N_c (luxury consumption), whereas points tend to be lower than N_c near the end of the growing season.

For unfertilized crops Greenwood and Draycott (1989) found the following relation between the nitrogen concentration (N_o) and W:

$$N_o = 0.7673 + 2.3364\ e^{-0.26W}. \qquad (2)$$

Combining Eqs. 1a and 2 gives a relation between N_o and N_c such that for each value of W, N_o is equal to 0.58 N_c. The data from experiments independent of Greenwood's show that the nitrogen concentration of unfertilized crops was generally lower than given by Eq. 2, except for low values of crop weight (Figure

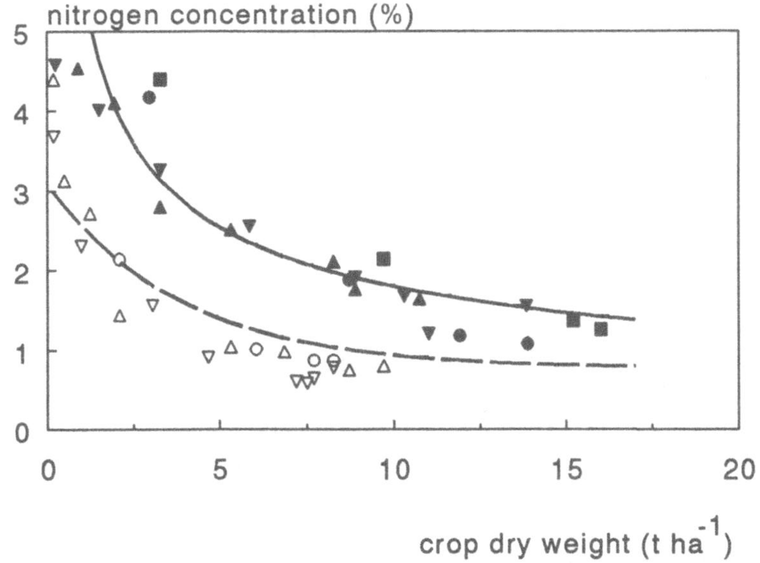

Figure 8. Data on the relation between the average nitrogen concentration and the total dry weight of the crop, obtained from sequential harvests throughout the growing period in several experiments. The full drawn line represents Eq. 1a, the broken line represents Eq. 2; symbols: full triangles upwards 240 kg N ha^{-1} 1984; full triangle downwards 240 kg N ha^{-1} 1985; open triangles upwards zero N 1984; open triangle downwards zero N 1985; full circles 180 kg N ha^{-1} 1992; full square 250 kg ha^{-1} 1992; open circles zero N 1992. Data from 1984 and 1985 are from MacKerron (unpublished), data from 1992 are from Vos (unpublished).

8). The minimum nitrogen requirement to produce a given crop mass is apparently lower than postulated by Greenwood et al.

Figures 3 and 7d show generally higher average nitrogen concentrations in the plant as nitrogen availability increases from limiting to (more than) sufficient to sustain maximum growth. However, it is generally found that the distribution of nitrogen over component plant parts is fairly conservative, nitrogen concentrations of all organs change more or less in proportion (Figure 4b of Greenwood et al. 1991; Biemond and Vos 1992).

Interactions between nitrogen and other growth factors

Other environmental factors such as temperature, day length, deficiencies of other nutrients, drought and stem density will modify the response to nitrogen. Temperature and daylength affect tuberization and thus the moment in the life cycle when tubers start to act as sinks for nitrogen. Most data discussed in this contribution are from experiments where care was taken that other nutrients were in abundant supply. In practice, also, nitrogen will generally be the most limiting nutrient, especially when soil fertility is maintained based on regular soil analysis. Stem density affected branching, leaf size and specific leaf area, but interactions with nitrogen supply were non-significant (Vos 1995b).

Elements of a (conceptual) model of potato growth, including nitrogen responses

If the objective of a model is to explain responses to different levels of nitrogen and different regimes of split applications, then the model should specifically deal with:

1. leaf growth and apical branching;
2. dynamics of nitrogen uptake in relation to nitrogen supply and growth rate.

SUCROS-type of models (Spitters et al. 1989) using daily radiation and temperature as input could be used as a basic model to simulate the general features of growth. The data on plant development discussed in this contribution are probably sufficiently detailed to design conceptual modules on plant development that can be plugged into a basic model upon parametrization. A difficulty is to assign numbers to the nitrogen concentration at which the full size of a leaf is less than its potential size and at which (apical) branching is reduced. There are no experimental data to verify whether Greenwood's critical nitrogen concentration – as related to crop mass – can be used to model nitrogen responses at this level of detail.

For modelling the rate of nitrogen uptake one could adopt as a guiding principle that the crop aims to maintaining N_c, and that an upper limit of uptake is reached when the average nitrogen concentration reaches 1.2 N_c. On a particular day the uptake rate by the crop would then be the lesser of what is available (to be calculated by a 'soil model') and the maximum uptake rate by the crop. Modelling uptake rates in relation to growth rates is particularly important when nitrogen is supplied in several doses. Preliminary results of experiments that are currently being conducted indicate that the pattern of uptake of nitrogen in time can be drastically influenced by the dose and the timing of split applications. Such changed patterns (in comparison to a single dose in spring) do not necessarily affect final yield, but if yield reductions occur because too large a fraction of the nitrogen is supplied late in the season, the yield reductions are not due to reduced uptake of nitrogen, but rather to a 'less efficient conversion of nitrogen into plant dry matter'. It is important to elucidate in more detail the regulation of nitrogen uptake and growth rate in response to split applications.

In Greenwood's approach the growth rate is reduced if the average nitrogen concentration, N_a, is below N_c, the reduction being approximately proportional to N_a/N_c. However, there is no direct causal link between the average nitrogen concentration of the crop and its growth rate. Therefore these should be decoupled in an explanatory model. The more elegant approach would be to modify a SUCROS-type of model in such a way that the rate of photosynthesis is dependent on the nitrogen status. Perhaps one could define a 'volumetric density of metabolically active nitrogen' in the canopy space. Maximum photosynthesis and continued growth of leaves and branches are ensured if that state variable is above a threshold value, while reductions would occur below that threshold. An alternative would be to use the dependence of P_{max} on the nitrogen concentration (Vos and Oyarzún 1987; Van Keulen et al. 1989; Marshall and Vos 1991) and model the nitrogen concentration of leaf layers. In both approaches the distribution of nitrogen over component plant parts needs to be modeled in sufficient detail. The latter seems necessary, anyway, if the model is to explain that the average nitrogen concentration generally increases with increase in rate of nitrogen supply while the relative distribution over plant parts (leaves, stems, tubers) remains fairly fixed.

Acknowledgements

Thanks are due to Dr. D.K.L. MacKerron of the Scottish Crop Research Institute for some of the data included in Figure 8.

References

Biemond H, Vos J (1992) Effects of nitrogen on the development and growth of the potato plant. 2. The partitioning of dry matter, nitrogen and nitrate. Annals of Botany 70:37–45.

Brouwer R (1983) Functional equilibrium: sense or nonsense? Netherlands Journal of Agricultural Science 31:335–348.

Burstall L, Harris P M (1983) The estimation of percentage light interception from leaf area index and percentage ground cover in potatoes. Journal of Agricultural Science, Cambridge 100:241–244.

Caloin M, Yu O (1984) Analysis of the time course change in nitrogen content in *Dactylus glomerata* L. using a model of plant growth. Annals of Botany 54:69–76.

Casal J J, Sanchez R A, Gibson D (1990) The significance of changes in the red/far-red ratio, associated with either neighbour plants or twilight, for tillering in *Lolium multiflorum* Lam. New Phytologist 116:565–572.

Dyson P W, Watson D J (1971) An analysis of the effects of nutrient supply on the growth of potato crops. Annals of Applied Biology 69:47–63.

Firman D M, Allen E J (1988) Field measurements of the photosynthetic rate of potatoes grown with different amounts of nitrogen fertilizer. Journal of Agricultural Science, Cambridge 111:85–90.

Greenwood D J, Draycott A (1989) Quantitative relationships for growth and N content of different vegetable crops grown with and without ample fertilizer-N on the same soil. Fertilizer Research 18:175–188.

Greenwood D J, Neeteson J J, Draycott A (1985a) Response of potatoes to N fertilizer: quantitative relations for components of growth. Plant and Soil 85:163–183.

Greenwood D J, Neeteson J J, Draycott A (1985b) Response of potatoes to N fertilizer: dynamic model. Plant and Soil 85:185–203.

Greenwood D J, Neeteson J J, Draycot A (19860 Quantitative relationships for the dependence of growth rate of arable crops on their nitrogen content, dry weight and aerial environment. Plant and Soil 91:281–301.

Greenwood D J, Lemaire G, Gosse G, Cruz P, Draycott A, Neeteson J J (1990) Decline in percentage N of C3 and C4 crops with increasing plant mass. Annals of Botany 66:425–436.

Greenwood D J, Gastal F, Lemaire G, Draycott A, Millard P, Neeteson J J (1991) Growth rate and %N of field grown crops: theory and experiments. Annals of Botany 67:181–190.

Hirose T, Werger M J A (1987) Maximizing daily canopy photosynthesis with respect to the leaf nitrogen allocation pattern in the canopy. Oecologia 72:520–526.

Ingestad T, Ågren G I (1992) Theories and methods on plant nutrition and growth. Physiologia Plantarum 84:177–184.

Marshall B, Vos J (1991) The relation between the nitrogen concentration and photosynthetic capacity of potato (*Solanum tuberosum* L.) leaves. Annals of Botany 68:33–39.

Millard P, MacKerron D K L (1986) The effects of nitrogen application on growth and nitrogen distribution within the potato canopy. Annals of Applied Biology 109:427–437.

Millard P, Marshall B (1986) Growth, nitrogen uptake and partitioning within the potato (*Solanum tuberosum* L.) crop, in relation to nitrogen application. Journal Agricultural Science, Cambridge 107:421–429.

Neeteson J J (1989) Evaluation of the performance of three advisory methods for nitrogen fertilization of suger beet and potatoes. Netherlands Journal of Agricultural Science 37:143–155.

Neeteson J J, Greenwood D J, Draycott A (1987) A Dynamic Model to Predict Yield and Optimum Nitrogen Fertilizer Application Rate for Potatoes. Proceedings 262. The Fertilizer Society, London.

Olesinski A A, Wolf S, Rudich J, Marani A (1989) Effect of leaf age and shading on photosynthesis in potatoes (*Solanum tuberosum*). Annals of Botany 64:643–650.

Spitters C J T, Van Keulen H, Van Kraalingen D W G (1989) A simple and universal crop growth simulator: SUCROS87. Pages 147–181 in Rabbinge R, Ward S A, Van Laar H H (Eds.) Simulation and Systems Management in Crop Protection. Simulation Monographs 32. Pudoc, Wageningen, The Netherlands.

Van Keulen H, Goudriaan J, Seligman J (1989) Modelling the effect of nitrogen on canopy development and crop growth. Pages 83–104 in Russel G, Marshall B, Jarvis P G (Eds.) Plant Canopies: Their Growth, Form and Function. SEB Seminar Series 31. Cambridge University Press, Cambridge, UK.

Vos J (1995a) The foliar development of the potato plant and modulations by environmental factors. In Kabat P, Van den Broek B J, Marshall B, Vos J, Van Keulen H (Eds.) Modelling and Parameterization of the Soil-Plant-Atmosphere System: A Comparison of Potato Growth Models. Wageningen Pers, Wageningen, The Netherlands (in press).

Vos J (1995b) The effects of nitrogen supply and stem density on plant attributes in potato (*Solanum tuberosum* L.) (submitted to Potato Research).

Vos J, Biemond H (1992) Effects of nitrogen on the development and growth of the potato plant. 1. Leaf appearance, expansion growth, life spans of leaves and stem branching. Annals of Botany 70:27–35.

Vos J, Marshall B (1994) Nitrogen and potato production: strategies to reduce nitrogen leaching. Pages 101–110 in Proceedings of the 12th Triennial Conference of the European Association for Potato Research (EAPR), Paris, France, July 18–23, 1993. INRA-France.

Vos J, Oyarzún P J (1987) Photosynthesis and stomatal conductance of potato leaves – effects of age, irradiance, and leaf water potential. Photosynthesis Research 11:253–264.

9. Modelling to optimize the use of both water and nitrogen by the potato crop

D.K.L. MacKERRON and G.J. LEWIS
Scottish Crop Research Institute, Invergowrie, Dundee, DD2 5DA, U.K.

Abstract. This work aims to extend an existing simulation model of water-constrained growth of potato, better to simulate the incidence of constraints in the water supply and to couple that development with a simulation of the requirement for and supply of nitrogen to the growing crop.

In the model, the appearance of roots and their extension growth are calculated using experimentally determined functions of temperature. A distributed root system is simulated by following the growth of all roots individually from their nodes of origin and by arbitrarily stopping growth in a proportion of them, linearly with depth. Water extraction is calculated from the rooted horizons within the soil profile, all of which are potentially allowed to contribute water equally. Water is abstracted in an exponential pattern from a cylinder of influence around each root and the difference between horizons in the calculated water abstraction is set only by the root length density and water content in each.

The method to incorporate this weather-driven model of potato growth with another that quantifies N supply and then to simulate the joint effects of limited water and nitrogen on growth has been described but has still to be implemented. Nitrogen is assumed to be taken up passively in the transpiration stream. An average fresh weight concentration of nitrogen over all the tissues is calculated and this value is used to assess whether the supply of N is sufficient for unrestricted growth. The principal merits of the method proposed are that growth and nitrogen uptake are not directly linked by a regression-type of relation determined *a priori* in other experiments and that the abstraction of water and nitrogen are not uniquely coupled as the sources of each in the soil are allowed to differ. It is likely that a successful model will include data from crop monitoring as an input, leading to corrective advice, if necessary.

Introduction

Models can be prepared to serve many purposes or operational functions but, always, they present an abstraction or sub-set of reality that is considered to be sufficient for the purpose in hand.

Detailed models are available of water supply (e.g. Feddes et al. 1978) or of nutrient supply (e.g. Johnsson et al. 1987). These are not usually of any specific crop, although the authors may subsequently apply them to particular crops (e.g. Belmans et al. 1982) and they are presented as a linked series of physical processes. Such models represent much of our scientific understanding of the processes involved but to set up such models requires information that is not generally available, and to run them may require even more information.

A.J. Haverkort and D.K.L. MacKerron (eds.), Potato Ecology and Modelling of Crops under Conditions Limiting Growth, 129–146.

Models that are to have practical or operational functions must be constrained to use only such data as is readily available (e.g. MacKerron and Waister 1985), even if specially obtained for their operation. Similarly, if such models are to be practicable their operational scale – both spatial and time-step – must be closely related to the intended function. That is, within two orders of magnitude or 2 levels of organization. Thus, models that are intended to inform our knowledge of a growing crop through the season, or to assist in the management of such a crop, will rarely be improved by operation at time-steps of less than a day and will, generally, treat the crop as one large homogeneous plant. For example, the leaves on every stem will be considered to be equivalent, developing in synchrony and all the roots will be considered to extend through and exploit the soil resources uniformly across the field. However, a system has been devised that enables parameters to be presented to models as distributions (Addiscott and Wagenet 1985) and it has been applied to at least one model of nutrient leaching (SLIM, Addiscott and Whitmore 1991).

Heterogeneity in time, which is called weather, is accommodated in daily time steps but heterogeneity in space, the result of parallel but differing conditions, can only be accommodated in separate runs of a model. Another distinction between models is in their degree of determinacy. The conventional models of crop growth are deterministic in that they provide a unique answer for a given set of inputs and yet real crops may present differing growth and yield with these same defined inputs. This condition may be referred to as uncertainty, and is an outcome of the model's simulation of only a part of the whole environment.

It is possible to construct or adapt models to take some account of such uncertainty (Gu 1994) but their output, while instructive to the scientist, policy maker, or market follower, depending on the nature of the uncertainty that is treated, is not helpful to the practical grower or his agronomic adviser. They wish to know, given a set of conditions, what they should do to optimize their chances of achieving a particular target. An example of this might be where a grower plans to irrigate his crop fully and then, in a dry year, finds that his supply of water has been restricted, perhaps by a licensing authority. Should he fully irrigate part of his crop and leave the rest? Should he irrigate all, fully until the water is done? Or should he eke out the water over the whole crop and season? He is asking, 'Can I modify my operations to return my system to profit or to maximize profit'. Applications of this nature require specific answers (perhaps with an error term added) so that action can be taken. The models to be considered here are such deterministic models.

There is a circular relation between growth and yield, the consumption of water, and the requirement for nitrogen so that a model that is to optimize the use of either or both of water and nitrogen must be based on an appropriate model of crop growth. It should have relatively simple inputs that are readily available for any field, or that can be made so and the model should be sufficiently robust to be 'transportable' across many crops and as many regions as possible. The model of potential growth of potato described by MacKerron and Waister (1985) and developed by Jefferies and Heilbronn (1991) to incorporate the effects of water constraints is a suitable such model and it is the one on which we will make our further developments. These models have been used successfully in comparisons with commercial crops (MacKerron 1987; Jefferies et al. 1991) and in providing a means to forecast regional and national yields of the potato crop in the UK (MacKerron 1992).

In this paper, the problem of optimizing water and nitrogen supply is set out in stages and at each stage we consider what has been done already, what is needed, and how the development should be tackled for that stage. Our model is still in the throes of parameterization and has not yet been tested but we believe that the consideration that we have had to give to the problems and the decisions that we have made on how to proceed should be of general interest.

Models of growth and water use

Evaporation from the soil surface can be treated quite simply but there are two options for estimating direct loss of water by the crop. One method, used by Jefferies and Heilbronn (1991) on potato and by Sinclair (1986) on soybean, is to use a value for water-use efficiency (W, the ratio of dry matter production to water consumption) to relate water use to radiation interception and increase in dry matter. That approach is rather unsatisfactory as it is known that W changes with crop development (Ehlers 1991; Silim et al. 1992), particularly where water supply is limiting, and it is not clear how W should be changed progressively. Further, in the Penman or Penman-Monteith models for calculating crop evaporation the aerodynamic term can exceed the radiation one. It is better, therefore, to have an independent method for estimation of the demand for water use by the crop and to use this as an input to the model. We will use the Penman-Monteith estimates of water use.

It is common to describe the accumulation of evaporative demand against rainfall and the reserves of soil water by calculation of a soil moisture deficit (SMD). A drawback to the simple estimation of SMD developed under a crop is that it makes no allowance for the effects of the distributions of water and roots in the soil. It is possible that an SMD of 25 mm developed in 30 cm of rooted soil presents a greater water-stress in the crop than an SMD of 50 mm developed in 100 cm of rooted soil. The model of Jefferies et al. (1991) used the fraction of the available water content (AWC) remaining in the soil as the descriptor of water-stress in order to allow for differences in rooted depth of soil and soils differing in their water-holding capacity. However, that model had an important limitation, that all the available water in the potential rooting volume could be extracted from the time of emergence, and it made no attempt to simulate the progressive exploration of the soil by roots. For a given soil, their AWC is entirely analogous to SMD. The effects of this limitation may not be evident where early water stress is not severe and the levels of water supply and extraction enable the crop to root to the full depth. However, it is necessary to simulate the differing distributions of roots and water in order to accommodate both early water-stress and the effects of re-wetting the soil surface. For example, when 25 mm of rain falls on a soil where an SMD of 75 mm had been developed the calculated SMD is still 50 mm and yet a significant proportion of the upper soil will have been re-charged with water and plant water-stress may be, temporarily, close to zero. Further, observations have shown (MacKerron and Peng 1989) that under conditions of limited water supply and high evaporation, roots may dry out the soil in the region of root extension so that further root growth is prevented even though there is moist soil below the dry zone and at a depth that the roots could otherwise reach.

For all of these reasons it is necessary to extend the model of potato growth to simulate a distributed root system and to calculate water and nutrient extraction from individual horizons of the soil profile.

Models of soil water availability

Soil characteristics vary with depth in the soil, often discontinuously, and it is necessary to treat the description of the soil by dividing its profile into discrete horizons each characterised by its own levels of soil-water parameters. Models of soil water usually require a range of parameters for each soil and horizon. These include the soil moisture volume fraction (Θ) at 'field capacity' and at the limit to which the crop can extract water – conventionally -1.5 MPa, together with the relations between pressure head or soil water potential (ψ) and Θ, and between Θ and hydraulic conductivity (e.g. Belmans et al. 1982). The model, SLIM, of Addiscott and Whitmore (1991) divides a 2 m depth of soil into forty slices of 5 cm each. The level of subdivision of the soil influences the size of any model that describes it and the time that a computer programme takes to run. There is little point in operating at finer levels of resolution than there is information available about the soil and we intend that our model should have a general advisory application. We, therefore, treat the potato ridge as two zones, an outer shell (2 cm) that dries rapidly, and an inner one, to a depth of 25 cm below the top of the ridge. Below that the soil is divided into 10 cm horizons; currently eight of them to 105 cm. It is convenient to attempt to characterise the nutrient supply within the same soil horizons. We shall, also, limit the information required to the plant-available water between 'field capacity' and -1.5 MPa. This should be an input variable but where the information is lacking we follow Sinclair (1986) and Jefferies and Heilbronn (1991) in using a volumetric fraction of 0.13 for amount of plant available water in each horizon (Ratliff et al. 1983). Figure 1 presents a schematic diagram of the structure of our model for water abstraction.

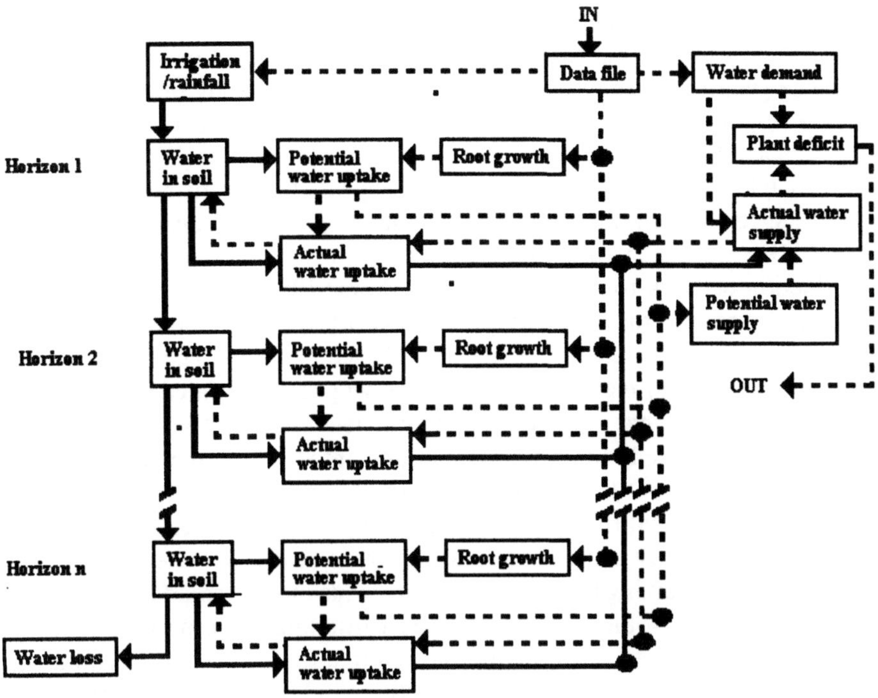

Figure 1. Schematic diagram of the sub-model for extraction of soil water. 'Data file' is an input file containing rainfall, PE, root length in each soil layer and at each time interval. Plant (water) deficit is the difference between the plant's requirement for water in one day and that which it gets from the soil layers. The value is used as a constraint in the growth sub-model. Pecked lines represent flow of information, solid lines the flow of water.

Models of root growth

Several authors present data on the configuration of potato root systems at harvest (e.g. Stalham. 1989; Mackie-Dawson et al. 1990), but these reports give no indication of the changes in the distribution of roots during the season in response to developmental changes, weather, and edaphic factors. A few authors have reported the development of the root system at only a few stages through the season, determined by sequential sampling (Lesczynski and Tanner 1976; Vos and Groenwold 1986). Parker et al. (1989) used minirhizotrons to observe rooting depth on as many as 11 occasions in a growing season. In their experiments, potato roots extended continuously until the end of July, more than 100 days after planting.

In the potato, each stem is rooted independently of other stems originating from the same seed-tuber so that each stem is, in effect, an independent plant. Peng (1990) studied roots grown on single-stemmed plants and reported considerable genotypic variation in the vigour and pattern of rooting that was, also, influenced by soil moisture. There was a tendency for the proliferation of lateral roots to be less in dry soil than in wet. In most cases the main roots did not normally branch but extended, undivided, to the depth of the root profile. In some genotypes there was a limited amount of branching of main roots.

The important feature of a root system is its absorptive surface area. Often the total length of root is used as a surrogate for surface area (Hillel and Talpaz 1976; Taylor and Klepper 1978) and this appears to be satisfactory for those plants, such as potato, that do not have tapering roots (Hamza and Aylmore 1992). One approach to simulating root growth is to partition dry matter production between root and shoot (e.g. Dewar 1993) and then to generate a length of root from the available root dry matter. We do not consider this to be a useful approach for a working model for several reasons. The first is that the estimate of total dry matter fixed by the growing crop is related to the amount of solar radiation that is intercepted by the leaf canopy. This total available energy is 'converted' in most simulations into dry matter using a light conversion coefficient (LCC) that was derived empirically in experiments that did not measure total root mass. It follows that the LCC does not contain an allowance for the dry matter fixed that was translocated into the roots in those experiments. Presuming that an appropriate value of LCC could be found there would remain the problems of speculative partitioning between shoot, root, and tubers (MacKerron and Peng, 1989, showed that dry mass of roots could vary between 3% and 17% of the total), converting total root dry matter into length specific root weight and length are not constants (Allmaras et al. 1975; Jefferies 1993), and of distributing it in the soil. An approach that avoids those problems is to simulate root length directly from estimates of numbers of roots initiated and of root extension rates (Hackett and Rose 1972) and death rates (Hillel and Talpaz 1976). This is the approach that we have taken.

The distribution of root length in the soil (root length density, RLD) is not uniform, generally declining from a maximum to a small value at maximum rooting depth. Feddes and Rijtema (1972) used an exponential function to describe the change in RLD with depth, although they found evidence of a change to a linear pattern with prolonged drought. Van Bavel and Ahmed (1976) used a linear distribution.

In potato, short, fine, lateral roots form adventitiously on the main roots. The published accounts of root distribution in potato do not, generally, distinguish between main root and lateral roots and, from the data given by Vos and Groenwold (1986) on the frequency distribution of root diameters, it would appear that most of the roots measured were lateral roots. This feature of the published data makes it particularly difficult to use to calibrate the simulation of the growth of main roots.

Root growth and root growth rate are continuously variable but it is convenient to partition the growth made, once simulated, into discrete soil horizons so as to permit estimation of the uptake of both water and nitrogen, each as a function of supply (what is available) and of uptake capacity which will be dependent on RLD within the horizon. The interactions between soil environmental conditions and root growth are harder to quantify.

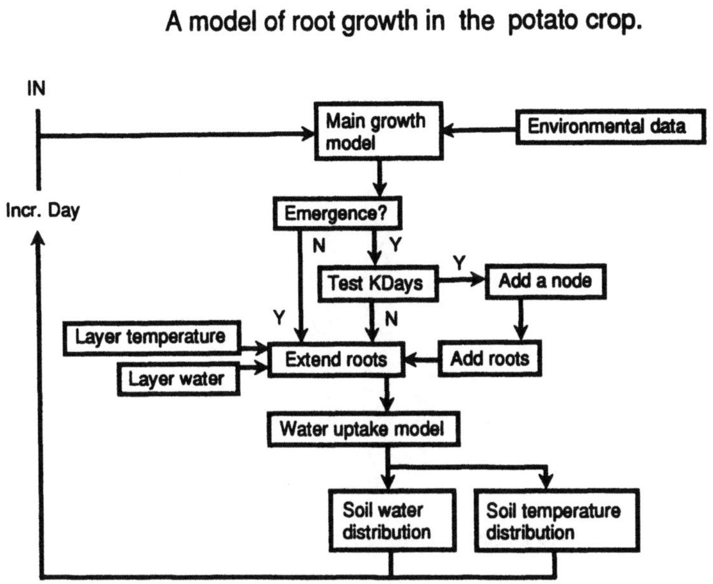

Figure 2. Schematic diagram of the sub-model for root initiation and extension.

Root growth sub-program

The root-growth sub-model that we have developed runs once for each simulated day. There are two central data structures that are updated and contain the state of the root system and soil profile.

1. The state of the root system is stored as a dynamic, linked list of records. Each record contains the node of origin and current length of the root and the soil layer in which its growing tip is currently located. This data structure is more complex than is strictly necessary at present but allows for extension of the programme to include such features as root death.
2. The state of the soil profile is stored as an array of records. Each record holds information on the state of an identified horizon of the soil, generally a 10 cm layer. The record contains the current temperature, water content, number of roots present, and root length. Other details such as mechanical resistance and nitrogen concentration can easily be added.

The state of the root system is updated in each simulated day (Figure 2), followed by that of the soil-water profile (Figure 1).

Updating the state of the root system

The growth and development of the root system is modeled as the result of two concurrent processes, an increase in numbers of roots and an increase in the length of each root. The two processes are simulated separately. Thermal time drives sprout growth to emergence (MacKerron and Waister 1985) and,

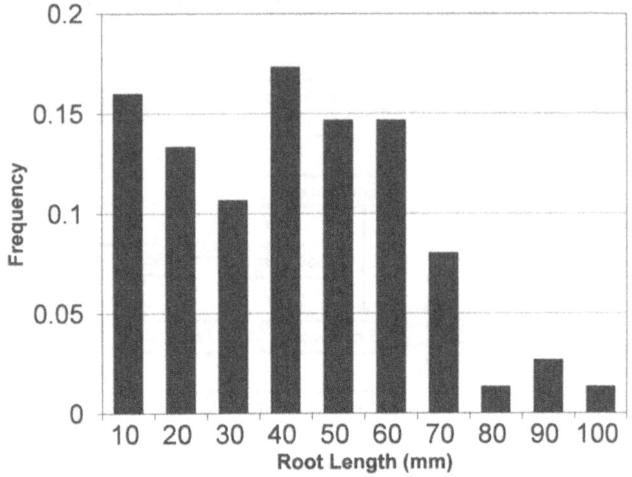

Figure 3. Observed frequency distribution of roots grown for 18 days at 10 °C.

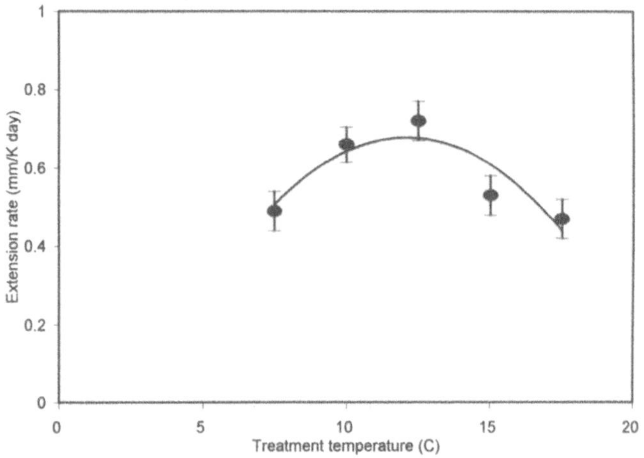

Figure 4. Extension rates of potato roots as a function of root temperature.

together with planting depth, defines the number of underground nodes on the sprout. Thermal time then determines the production of a single batch of roots at each node and a fixed number of roots is added to a linked list of root records. The initial lengths of the roots are not uniform but are distributed to simulate the size distribution of real roots at a node shortly after their formation (Lewis, unpublished data; Figure 3). Each simulated root extends daily by an amount that is determined by the temperature of the soil in the layer in which its tip is located at the start of the growth interval. The temperature relations used have been determined experimentally (Lewis, unpublished data; Figure 4). All the roots respond to soil temperature and moisture content in the same way. The new location of each root tip is then determined.

Two limitations are imposed on root extension. One is designed to produce a profile of RLD that declines linearly with depth from a maximum in the ridge (Vos and Groenwold 1986; Parker et al. 1989) to a limiting potential root depth. At present that is achieved by stopping the growth of individual roots at intervals. We do not yet have a soundly-based function to decelerate root extension with decreasing soil water content, but root extension is prevented below a moisture volume fraction of 0.15 (Sharp and Davies 1985; Wraith and Baker 1991).

Updating the state of the soil profile

The state of the soil profile is updated so that the numbers and lengths of roots in each soil layer are determined. The temperature in each layer for the following day is determined using a simple linear interpolation between the 10 cm soil temperature for that day and the temperature at 100 cm depth (assumed constant). The model then proceeds to calculate the soil-water uptake for the next day (Figure 1).

Water extraction by roots

The movement of water through the soil-plant-air system is, properly, a function of the gradient in water potential. However, although several models use the gradient in ψ to drive water movement, and the value of ψ_{soil} and the hydraulic conductivity of the soil can be calculated from soil moisture content, the value of ψ in the plant is not readily calculable as it is influenced by many factors besides the supply of water from the soil (Taylor and Klepper 1978). An approach to the problem of quantifying water extraction from the soil is to calculate what is available in a day from each horizon in the soil and to allow extraction to a limit set by PE or the total daily available water, whichever is the lesser, and to recognize that the drying power of the air is so great that potential evaporation will be met even at the expense of desiccating the plant, unless the stomata in the leaves close. The crop is considered to be water-stressed when the evaporative demand exceeds the daily supply.

The absorption pathway for water comprises a soil cylinder, a soil-root interface, epidermal and cortical layers, an endodermis, and stelar tissue to the lumen of the xylem. Taylor and Klepper (1978) suggested values around 5 mm for the length of the radial pathway for water flux from bulk soil to xylem vessels for a 'typical crop plant' with a rooting density of 1 cm/cm^3 and a root diameter of 0.5 mm. Hamza and Aylmore (1992) used high-resolution, computer-aided tomography to observe the draw-down of water by roots of lupin from a cylinder of radius 15 mm, and showed that water was abstracted in an exponential pattern from a cylinder of radius 10–12 mm over an 8 h period of transpiration when soil water potentials went as low as -0.14 MPa. (A water potential of $-$ 0.14 MPa is well below the point at which an irrigated crop would be re-irrigated and, translated through the bulk soil, would represent a significant SMD in many agricultural soils.) This can conveniently be summarized as being equivalent to the removal of 1/3 of the available water in the root's zone of influence in one day. We, therefore, simulate the amount of water available on a given day (AVW_h), in a given horizon (h) as a function of the number of roots (N_{roots}) in that horizon (strictly, the total root length), the radius of influence of the roots (R_r), and the total available water in the horizon (PWU_h) thus:

$$AVW_h = 1/3 \times PWU_h \times N_{roots,h} \times R_r \times R_r\ \pi/G$$

(where G = ground area per stem and $N_{roots,h} \times R_r \times R_r\ \pi/G$ is limited to $\leq$ 1). If the total available water from all rooted horizons is ΣAVW_h Then the actual water use (AWU_h) from each horizon is:

$$AWU_h = PE \times AVW_h/\Sigma AVW_h \qquad \text{where } PE \leq \Sigma AVW_h$$

and

$$AWU_h = AVW_h \qquad \text{where } PE > \Sigma AVW_h$$

The presence and function of the fine lateral roots can be simulated by adjusting the size of the zone of influence (R_r) around the main roots.

No allowance is made, at present, for axial resistance to water flow in the roots so water can be drawn from each horizon simultaneously and is drawn from each at rates that depend solely on the amount of root present in the horizon and its water content. A non-zero value for the axial, root resistance would have a marked effect on the pattern of water abstraction from the soil, reducing the amount that is drawn initially from the lower layers (Taylor and Klepper 1978). This approach allows evaporative demand to drive water extraction and avoids the need to estimate water potentials in shoot, root, and soil. It, also, allows soil water content in the rooted zone to limit transpiration to less than PE. Further it has the property that the roots are not able totally to deplete the soil water in only a few drying cycles. If the whole cross-sectional area of a horizon were rooted – ($N_{roots,h} \times R_r \times R_r\ \pi$) = (area per stem) – and PE were greater than ΣAVW_h then it would still take 6 days to remove 91% of available water.

The difference between PE and AWU is called the 'plant water deficit' and is used as the term to constrain leaf expansion, LCC, and nitrogen uptake. It is necessary to assume that the water is uniformly distributed throughout the horizon and that it is uniformly re-distributed each night so that transpiration can continue the next day.

Estimating the requirement for and supply of nitrogen

The requirements of the potato crop for nitrogen are directly related to the growth that will be made and so, indirectly, to the expected length of the growing season. The supply of nitrogen from mineralized organic matter is not generally sufficient to meet the requirements of the crop and so the soil-supplied N is supplemented by application of fertilizer. The requirements for N-fertilizer differ greatly from field-to-field in ways that are difficult to predict such that the economic optimum level of application has been found in the Netherlands to range between 0 and 400 kg/ha (Neeteson and Wadman 1987). The best single estimator of the N available from the soil (and so, by difference, of the fertilizer requirements) is the amount of mineral N in the top 60cm of soil in spring (Greenwood et al. 1986), but even this accounts for only 30% of the variation in optimum application rate (Neeteson and Zwetsloot 1989) and, by inference, the actual supply of soil-N. In fact, the relations reported by Neeteson and Zwetsloot (1989) may be misleading in that they did not measure uptake from individual layers of the soil, but estimated the relative importance of the three layers 0–30 cm, 30–60 cm and 60–100 cm by non-linear regression analysis. These authors pointed out, themselves, that in their equation relating total N uptake to the N available as fertilizer and as soil mineral N at 0–30 and 30–60 cm, ($N_t = N_f + 0.67N_{m30} + 0.33N_{m60}$) the coefficients are based on non-linear regression and not at all to do with measured abstraction. Since there was a high correlation between N_{m30} and N_{m60} the importance of the lower layer could not properly be assessed.

Much of the N in the soil is bound up in the organic matter fraction and it is microbial activity that causes the breakdown of some of that organic matter and the mineralization of the N which is then available for uptake by the crop (Greenwood 1986). The average rate of mineralization calculated over a season may vary from < 0.4 kg/ha/day to > 2.0 kg/ha/day and between seasons (Neeteson et al. 1987). That is a range of 5 times and the short-term rates must differ more widely. Not only may rates of mineralization fluctuate according to environmental conditions but it has been shown that there is an underlying decline in available N in the soil down to a minimum value from around mid-summer (Greenwood et al. 1985a). There have been reports that the action of potato roots may stimulate the mineralization of soil-N but Griffiths and Robinson (1992) argued that the effect was so small that it was unlikely to influence the N-supply to the plant to an appreciable extent.

Efforts to estimate the optimum requirements for N-fertilizer hinge on forecasting the rates at which the organic N will be mineralized, on the fate of applied fertilizer-N, on calculating the changing requirement for N by the crop, and on estimating its changing ability to absorb the N that is available (Greenwood 1982; Greenwood et al. 1985b; Neeteson et al. 1987; Neeteson 1989). It is evident that quantifying the dynamics of N availability is a particularly difficult task.

It is important to appreciate that the requirement for N is dependent upon the growth of the crop. There is, therefore, a circular relation coupling growth and N-requirement. A low supply of N leads to poor growth which, in turn, leads to low demand for N. On the other hand, where there is an ample supply of N, as in early growing-season, the potato crop can take up excess or luxury levels of nitrogen that are later re-distributed within the plant (Millard and MacKerron 1986). This results in the well-known decline in [NO_3^-] in petiole sap.

Although Greenwood et al. (1985a; 1990) showed exponentially declining [N]

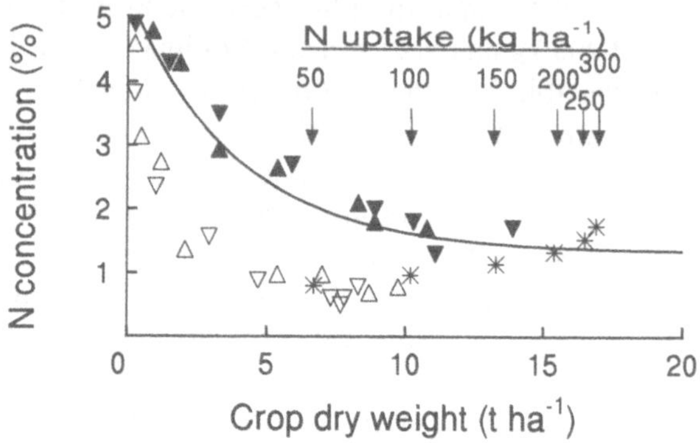

Figure 5. The relation between the average nitrogen concentration (dry weight basis) and crop dry weight re-drawn after Vos and Marshall (1993, their Figure 1) with the addition of calculated asymptotic values (*) of N concentration at indicated final uptake of N (arrows) calculated from relations describing both nitrogen concentration and crop dry weight as functions of N uptake (Vos and Marshall 1993; Figure 2). Data points are from Mackerron (unpublished); closed triangles 240 kg N applied ha-1, open triangles 0 kg N applied ha-1, right triangles 1984, inverted triangles 1985.

in the dry weight of crops, including potato, the results to which equations were fitted were from crops that were well grown, with 'sufficient N to permit near maximum growth' (their Figures 3 and 4 in 1990). It is possible to have crops with lower [N] and lower DWt than fit the curve that they published (e.g. their Figure 2 in 1985a, and Figure 1b in 1990). Vos and Marshall (1993) pointed out that the published equations described the response of optimally-grown crops, and did not describe the results of fertilizer dose-response experiments. They presented schematized relations between total N taken up and both total final dry matter production and average [N]. Those relations can be used to derive values for the asymptotic [N] of data such as those shown by Greenwood et al. (1990) for N-limited growth (Figure 5) that themselves show a near-linear relation between final [N] and final crop dry weight. It also shows how very insensitive the final [N] is to the total dry matter production.

There may be three principal criticisms of the approach estimating nitrogen requirements from the [N] in the dry matter. One, the relations are derived from crops given single applications of N at the start of the growing season and, two, the high concentrations associated with the lower levels of production are obtained with an early end to growth and represent 'luxury' uptake of N (Millard and MacKerron 1986). The third criticism is that the relation represents merely a 'dilution' of the functional parts of the plant by the structural and storage parts. In contrast, [N] as determining continued plant growth is not in the structural material or in storage but in the symplast. MacKerron et al. (1990; 1993) showed that [N] in the tissue fresh weight or in the plant water-content is conservative (almost constant) through all but the earliest part of the growing season, suggesting that growth appears to be set, relatively early in the season, in a pattern related to the availability of N. Recent data (MacKerron and Young, unpublished) suggest that when allowance is made for the nitrogen contribution to the crop from the seed tuber the relation between N uptake and fresh weight or water content is linear from emergence onwards.

Wadman et al. (1990) examined the success of soil-testing in providing sound fertilizer recommendations for potato and sugar beet. They found that the recommendations were 'rather crude' and could not be considered to be field-specific without information on rates of nitrogen mineralization. It may be that the way forward is to combine a model simulating mineralization and leaching (e.g. Addiscott et al. 1991) with a growth model for the early part of the growth season, and to include [N] measurements from the crop as an input from around mid-season (MacKerron et al. 1993).

Incorporating nitrogen into the model

Addiscott and Whitmore (1987) calculated soil mineral N and crop N using a model in which both crop uptake of N and dry matter production were driven by thermal time based on soil temperature. In that model, potential uptake of N was matched against the available N and any shortfall was translated into a reduction of thermal time, so checking both processes. This method did have the merit of keeping growth of dry matter in line with N-uptake, but it did not allow [N] that was lower than ideal. Nor was growth driven by the variable that actually influences it – solar radiation.

The success of models to simulate the amounts of mineral N found in the soil and the uptake of N has been mixed. Addiscott et al. (1991) found that their model, SLIM, was reasonably successful in simulating the amounts of mineral N found in the soil in early spring and the changes from applying small supplements of fertilizer in February. However, the model was not as successful in simulating the effects of larger, later applications of N, mainly because the mineral N found in the soil subsequently did not reflect these applications. Wheatley and Ritz (1994) reported this feature of the dynamics of nitrate in soil under potato. They reported that applied, mineral N is incorporated rapidly in the microbial biomass and is later released in a manner that is not readily predictable.

Recently, Porter (1993) combined the model, AFRCWHEAT1, with a soil water and leaching model, SLIM (Addiscott and Whitmore 1991), to simulate wheat growth incorporating responses to both water and nitrogen supply. In that combined model, AFRCWHEAT2, water and nitrogen are the only factors to limit crop growth rate to below the potential set by solar radiation and temperature. In the model, the uptake of water and nitrogen are each limited by their availability in the rooting zone and by the ability of roots to absorb them. The model adequately met several of its validation criteria but it consistently overestimated the decline in shoot nitrogen concentration (dry matter).

We consider that the necessary balance between input and calculation to keep the simulation both realistic and usable will require the combination of information on available soil-N at planting and actual N-uptake, both as inputs to our model. An initial condition of the model must be the amount of soil-N that can be incorporated into the crop. This soil- and fertilizer-derived N is assumed to be freely available to the crop during the initial stages of growth. The total available N is assumed to be in solution in the available soil water and to be taken up passively in the transpiration stream. In order to calculate the fresh weight concentration, $[N]_{FWt}$, it is necessary to estimate total fresh weight from the modeled dry matter production.

From around the time of tuber initiation, $[N]_{FWt}$ can be measured and used as an input to the model. We expect to compare the measured values of $[N]_{FWt}$ with 'standard' values to modify growth of the crop or to flag corrective action.

The growth parameters that will be influenced by $[N]_{FWt}$ are: the maximum leaf size of each leaf, branching (limiting numbers of leaves), and leaf area duration (Vos and Biemond 1992). The value of the light conversion coefficient will not be changed except at very low levels of [N] (MacKerron, unpublished data).

The sequence of relations that operate are:

N taken up in transpiration stream = > N_{up}
HI (calculated from thermal time) = > HI_{Fresh} (MacKerron and Heilbronn 1985)
HI_{Fresh} × Modeled total dry matter = > Total fresh weight (FWt)
N_{up}/FWt = > $[N]_{FWt}$
where HI and HI_{Fresh} are harvest indices on dry and fresh weight bases, respectively.

MacKerron et al. (1990; 1993) have shown conservative values of $[N]_{FWt}$ for crops not limited for N at any time in the growing season, $[N]_{240}$, and limited only at the end of the season, $[N]_{160}$ (subscript indicates level of applied N in kg/ha). These provide reference points for comparison. Jefferies and Heilbronn (1991) calculated the progress of canopy expansion by simulating leaf appearance and then leaf expansion to a maximum size, characteristic of variety and temperature. Maximum leaf size and rate of expansion were both reduced by water stress. Maximum leaf size will be reduced in direct proportion to the ratio of $[N]_{FWt}:[N]_{160}$. Thus:

If $[N]_{FWt} \geq [N]_{160}$ leaf growth is standard by temperature and water supply.
If $[N]_{FWt} < [N]_{160}$ leaf growth is standard × $([N]_{FWt}/[N]_{160})$.

The same comparison can be used as an aid to deciding on adequate fertilization. Thus:

If $[N]_{160} < [N]_{FWt} < [N]_{240}$ then no further N-fertilizer is required.
If $[N]_{FWt} < [N]_{160}$ then recommend a supplementary application of N
Re-determine $[N]_{FWt}$ after another week.

The measured $[N]_{FWt}$ has the possibility of being used to trigger the simulation of senescence.

Discussion

The supply of adequate amounts of water and nitrogen to a crop are among the most critical aspects determining its growth and yield for both strongly influence the development of the leaf canopy, hence setting the ability of the crop to intercept light and so fix dry matter. Nitrogen is also required in the formation of proteins in leaves and in the storage organs, tubers where it can comprise 8.5% of the dry matter.

The supply of water can be arranged by irrigation but the supply of N is variable and is influenced by the several factors that have been considered in the previous sections.

The conventional approach to arranging adequate N-nutrition in the latter part of the 20th century has been to apply 'enough' nitrogen fertilizer at planting or early in the season to be sure that the crop will not run out of N before the planned senescence or haulm destruction. Rule-of-thumb methods have provided rough estimates of the likely N-supply from the soil based upon soil-type, previous cropping history, and, only sometimes, the anticipated duration of the crop.

Environmental concerns rather than economic considerations, have created pressures to limit the application of fertilizers, especially N, to all crops, including potato. Economic considerations do play a small part in the decision on application rates but, in general, the relatively low cost of N-fertilizer has meant that growers will prefer to be generous with N-applications rather than risk being short.

Further, where organic fertilizers are used, growers will still use applications of mineral N rather than rely on the uncertain quantities that will become available from the manure.

There is a growing awareness that providing the 'correct' amount of N-fertilizer can help to produce a crop that is in the right condition for its market at harvest. Excessive applications of N can result in crops with heavy foliage growth and delayed maturity causing difficulties with haulm destruction, pre-harvest, and poor delayed skin-set on the tubers. Tubers destined for processing may not be at the correct stage of maturity with stable, low levels of reducing sugars.

For all these reasons, it is important to be able to estimate the requirement of a potato crop for nitrogen and, more particularly, for applied nitrogen. The literature is full of examples of experiments done to derive optimum application rates for N, that have provided the answers retrospectively and that have shown the optimum to vary with yield, soil, and year. That is quite unhelpful to the grower. Increasingly interest is being developed in either model-based systems or crop monitoring, or both to estimate N-requirements.

There are several aims in this work. One is to extend an existing simulation model of water-constrained growth of potato, better to simulate the incidence of constraints in the water supply. This is done by calculating root growth to simulate a distributed root system and by calculating water extraction from several horizons within the soil profile. The first approach has been potentially to allow all horizons to contribute water equally. The difference between horizons is set only by the root length density and water content in each. If tests against independent data show that water from the deeper levels is extracted less readily than at the shallower ones then the presence of axial resistances could be simulated by scaling the notional zone of influence around each root.

The outstanding problem is to couple this weather-driven model of potato growth that considered N not to be limiting with another that quantifies N supply and then to simulate the joint effects of limited water and nitrogen on growth. A method has been described that has still to be tested. The principal merits of the method proposed are that growth and nitrogen uptake are not directly linked by a regression-type of relation determined *a priori* in other experiments and that the abstraction of water and nitrogen are not uniquely coupled as the sources of each in the soil are allowed to differ. Rather, nitrogen uptake is determined by the supply of nitrogen in the soil, the transpiration rate, and the source of the transpired water. Given the difficulties in estimating the rates of mineralization of N in soil, it is likely that a successful model will include data from crop monitoring as an input, leading to corrective advice, if necessary.

Acknowledgements

We gratefully acknowledge the support of the Scottish Office Agriculture and Fisheries Department in this work. G.J. Lewis is supported by a grant from the Potato Marketing Board of Great Britain.

References

Addiscott T M, Bailey N J, Bland G J, Whitmore A P (1991) Simulation of nitrogen in soil and winter wheat crops: a management model that makes the best use of limited information. Fertilizer Research 27:305–312.

Addiscott T M, Wagenet R J (1985) A simple method for combining soil properties that show variability. Soil Science Society of America Journal 49:1365–1369.

Addiscott T M, Whitmore A P (1987) Computer simulation of changes in soil mineral nitrogen and crop nitrogen during autumn, winter and spring. Journal of Agricultural Science, Cambridge 109:141–157.

Addiscott T M, Whitmore A P (1991) Simulation of solute leaching in soils of differing permeabilities. Soil Use and Management 7:94–102.

Allmaras R R, Nelson W W, Voorhees W B (1975) Soybean and corn rooting in southwestern Minnesota: II. Root distribution and related water inflow. Soil Science Society of America Proceedings 39:771–777.

Belmans C, Dekker L W, Bouma J (1982) Obtaining soil physical field data for simulating soil moisture regimes and associated potato growth. Agricultural Water Management 5:319–333.

Dewar R C (1993) A root-shoot partitioning model based on carbon-nitrogen-water interactions and Münch phloem flow. Functional Ecology 7:356–368.

Ehlers W (1991) Leaf area and transpiration efficiency during different growth stages in oats. Journal of Agricultural Science, Cambridge 116:183–190.

Feddes R A, Kowalik P J, Zaradny H (1978) Simulation of Field Water Use and Crop Yield. Simulation Monographs. Pudoc, Wageningen, The Netherlands.

Feddes R A, Rijtema P E (1972) Water withdrawal by plant roots. Journal of Hydrology 17:33.

Greenwood D J (1982) Modelling of crop response to nitrogen fertilizer. Philosophical Transactions of the Royal Society, London B B296:351–362.

Greenwood D J (1986) Prediction of nitrogen and fertilizer needs of arable crops. Advances in Plant Nutrition 2:1–61.

Greenwood D J, Lemaire G, Gosse G, Cruz P, Draycott A, Neeteson J J (1990) Decline in percentage N of C3 and C4 crops with increasing plant mass. Annals of Botany 66:425–436.

Greenwood D J, Neeteson J J, Draycott A (1985a) Response of potatoes to N fertilizer: Quantitative relations for components of growth. Plant and Soil 85:163–183.

Greenwood D J, Neeteson J J, Draycott A (1985b) Response of potatoes to N fertilizer: Dynamic model. Plant and Soil 85:185–203.

Greenwood D J, Neeteson J J, Draycott A (1986) Quantitative relationships for the dependence of growth rate of arable crops on their nitrogen content, dry weight and aerial environment. Plant and Soil 91:281–301.

Griffiths B S, Robinson D (1992) Root-induced nitrogen mineralisation: A nitrogen balance model. Plant and Soil 139:253–263.

Gu Y, Crawford J W, Peiris D R, Jefferies R A (1994) An approach for treating the uncertainties in the impact of climate change. Environmental Pollution 83:87–93.

Hackett C, Rose D A (1972) A model of the extension and branching of a seminal root of barley, and its use in studying relations between root dimensions. Australian Journal of Biological Science 25:669–679.

Hamza M A, Aylmore L A G (1992) Soil solute concentration and water-uptake by single lupin and radish plant-roots. 1. Water extraction and solute accumulation. Plant and Soil 145:187–196.

Hillel D, Talpaz H (1976) Simulation of root-growth and its effect on the pattern of soil water uptake by a non-uniform root system. Soil Science 121:307.

Jefferies R A (1993) Cultivar responses to water stress in potato: effects of shoot and roots. New Phytologist 123:491–498.

Jefferies R A, Heilbronn T D (1991) Water-stress as a constraint on growth in the potato crop. I. Model development. Agricultural and Forest Meteorology 53:185–196.

Jefferies R A, Heilbronn T D, MacKerron D K L (1991) Water-stress as a constraint on growth in the potato crop. II. Validation of the model. Agricultural and Forest Meteorology 53:197–205.

Johnsson H, Bergstrom L, Jansson P E, Paustian K (1987) Simulated nitrogen dynamics and losses in a layered agricultural soil. Agricultural Ecosystems and Environment 18:333–356.

Lesczynski D B, Tanner C B (1976) Seasonal variation of root distribution of irrigated field-grown Russet Burbank potato. American Potato Journal 53:69–78.

MacKerron D K L (1987) A weather-driven model of the potential yield in potato and its comparison with achieved yields. Acta Horticulturae 214:85–94.

MacKerron D K L, 1992. Agrometeorological Aspects of Forecasting Yield of Potato within the European Community: Commission of the European Communities, Luxembourg, 249pp.

MacKerron D K L, Davies H V, Marshall B, Millard P (1990) Optimum nitrogen supply and yield – The crop as an indicator. Pages 135–136 in Abstracts of the 11th Triennial Conference of the European Association for Potato Research.

MacKerron D K L, Heilbronn T D, 1985. A method for estimating harvest indices for use in surveys of potato crops. Potato Research 28:279–282.

MacKerron D K L, Peng Z Y (1989) Genotypic comparisons of potato root growth and yield in response to drought. Pages 199–206 in Aspects of Applied Biology 22, Roots and the Soil Environment. Association of Applied Biologists, Wellesbourne, UK.

MacKerron D K L, Waister P D (1985) A simple model of potato growth and yield. Part I. Model development and sensitivity analysis. Agricultural and Forest Meteorology 34:241–252.

MacKerron D K L, Young M W, Davies H V (1993) A method to optimize N-application in relation to soil supply of N, and yield of potato. Plant and Soil 155:139–144.

Mackie-Dawson L A, Millard P, Robinson B (1990) Nutrient uptake by potato crops growing on two soils with contrasting physical properties. Plant and Soil 125:159–168.

Millard P, MacKerron D K L (1986) The effects of nitrogen application on growth and nitrogen distribution within the potato canopy. Annals of Applied Biology 109:427–437.

Neeteson J J (1989) Effect of reduced fertilizer nitrogen application rates on yield and nitrogen recovery of sugar beet and potatoes. Netherlands Journal of Agricultural Science 37:227–236.

Neeteson J J, Greenwood D J, Draycott A (1987) A Dynamic Model to Predict Yield and Optimum Nitrogen Fertiliser Application Rate for Potatoes. The Fertiliser Society of London, London, UK.

Neeteson J J, Wadman W P (1987) Assessment of economically optimum application rates of fertilizer N on the basis of response curves. Fertilizer Research 18:37–52.

Neeteson J J, Zwetsloot H J C (1989) An analysis of the response of sugar beet and potatoes to fertilizer nitrogen and soil mineral nitrogen. Netherlands Journal of Agricultural Science 37:129–141.

Parker C J, Carr M K V, Jarvis N J, Evans M T B, Lee V H (1989) Effects of subsoil loosening and irrigation on soil physical properties, root distribution and water uptake of potatoes. Soil and Tillage Research 13:267–285.

Peng Z Y, 1990. MSc Thesis, University of Dundee. The physiological basis of genotypic differences in drought tolerance in potatoes. 141 pp.

Porter J R (1993) AFRCWHEAT2: A model of the growth and development of wheat incorporating responses to water and nitrogen. European Journal of Agronomy 2:69–82.

Ratliff L F, Ritchie J T, Cassell D K (1983) Field-measured limits of soil water availability as related to laboratory-measured properties. Soil Science Society of America Journal 47:770–775.

Sharp R E, Davies W J (1985) Root growth and water uptake by maize plants in drying soil. Journal of Experimental Botany 36:1441–1456.

Silim S N, Hebblethwaite P D, Jones C (1992) Irrigation and water use in leafless peas (*Pisum sativum*). Journal of Agricultural Science, Cambridge 119:211–222.

Sinclair T R (1986) Water and nitrogen limitations in soybean grain production. I. Model development. Field Crops Research 15:125–141.

Stalham M A (1989) Growth and Water Use of the Potato Variety Record on Contrasting Sites. Ph.D. Thesis, University of Cambridge, Cambridge, UK.

Taylor H M, Klepper B (1978) The role of rooting characteristics in the supply of water to plants. Advances in Agronomy 30:99–128.

Van Bavel C H M, Ahmed J (1976) Dynamic simulation of water depletion in the root zone. Ecological Modelling 2:189.

Vos J, Groenwold J (1986) Root growth of potato crops on a marine-clay soil. Plant and Soil 94:17–33.

Vos J, Biemond H (1992) Effects of nitrogen on the development and growth of the potato plant. 1. Leaf appearance, expansion growth, life spans of leaves and stem branching. Annals of Botany 70:27–35.

Vos J, Marshall B (1993) Nitrogen and potato production: Strategies to reduce nitrate leaching. Pages 101–110 in Proceedings 12th Triennial Conference of the EAPR, Paris. EAPR, Paris, France.

Wadman W P, Neeteson J J, Zwetsloot H J C, 1990. Development of nitrogen fertilizer recommendations for potatoes and sugar beet on the basis of soil testing. Kali-Briefe 20:171–180.

Wheatley R E, Ritz K (1995) The dynamics of mineral nitrogen in soils supporting potato crops. Biology And Fertility Of Soils 19:36–40.

Wraith J M, Baker, J M (1991) High-resolution measurement of root water uptake using automated time-domain reflectometry. Soil Science Society of America Journal 55:928–932.

10. Simulation of nitrogen uptake from soil and partitioning in potato plants: model description and sensitivity analysis

SVETLANA FISHMAN[1] and B. BAR-YOSEF[2]

[1] *Department of Statistics and Operations Research and* [2] *Institute of Soils and Water, Agricultural Research Organization, the Volcani Center, Bet Dagan 50250, Israel*

Abstract. A model is presented describing nitrogen transport and biochemical transformations in soil, the uptake of N by the potato crop and its partitioning between the plant organs. The model enhances an earlier model of the potato crop, which described the grows of leaves, stems, roots and tubers as affected by meteorological conditions assuming of optimal nitrogen status in the plants. Coupling these two submodels enabled a simulation of the influence of N fertilization (amount of N applied, timing of fertilization and the kind of the fertilizer) on the yield of tubers and on nitrate concentration in the tubers. A detailed sensitivity analysis was performed, which identified the most critical parameters in the model for its calibration under different environmental and managerial conditions.

Introduction

An appropriate supply of nitrogen to potatoes is essential for high tuber yields. Excessive N application may lead to accumulation of nitrates in the tubers and to pollution of ground water. Since high concentrations of nitrates in food and water are toxic to mammals (Barker et al. 1971; Wilson 1949; Wright and Davison 1964) and under-fertilization may jeopardize the profitability of potato production, the application of nitrogen must be very carefully managed.

A potato crop model (ISPOTA) was developed by Fishman et al. (1984; 1985; 1994) based on observations of potatoes grown in arid zones. That model predicts the growth of leaves, stems, roots and tubers during the season, assuming that the nitrogen status in the plant is optimal. The objective of the present study was to improve ISPOTA by adding to it a sub-model describing the transport of nitrogen and water in the soil and their uptake by roots, N transformations in soil, partitioning of N species among plant organs, and the dependence of tuber yield on the N and water status of the plant. Previous models that described the N nutrition of plants either emphasized the behaviour of N in the soil and its absorption by roots without due consideration of the accumulation of different N species in plant organs (Hadas et al. 1979; Rao et al. 1981; Greenwood et al. 1985a,b), or dealt with N-transport in plants without accounting for soil-N processes (Pate 1980). In the model described here both approaches are integrated, to give a comprehensive description of the effects of N and water management on yield and tuber quality.

A.J. Haverkort and D.K.L. MacKerron (eds.), Potato Ecology and Modelling of Crops under Conditions Limiting Growth, 147–166.

The model

Overview

The model considers the reactions of NH_4^+ and NO_3^- in soil and the uptake by roots. For transport and uptake considerations, the soil profile is divided into layers. Uptake rates are determined by the Michaelis-Menten flux multiplied by the root weight in each soil layer. Accumulation of NO_3^- in roots inhibits its uptake from the soil. Absorbed NO_3^- is reduced by nitrate reductase in leaves and roots. The nitrogen is distributed among plant organs according to a sink concept with restricted supply, which is consistent with the dry matter partitioning approach in ISPOTA (Fishman et al. 1984; 1985). The plant water status is determined by transpiration and water uptake by roots. The leaf water potential and N concentration affect the rate of photosynthesis, thus coupling the simulated soil and uptake processes with the ISPOTA crop model. The plant N concentration also affects the dry matter partitioning in the plant and canopy senescence.

The nitrate and ammonium ions in the soil profile are assumed to move downwards and upwards by convection only. The ion concentrations and water potential at the root surface are calculated from the simulated concentrations in the bulk soil in each layer and each time step by using a steady-state transport approximation with nonlinear flux at the root surface.

The schematic description of the model is presented in Figure 1. The shaded blocks show the main components of the ISPOTA crop model, while the unshaded blocks represent the new components added to it.

Governing equations for water in the soil and plant systems

The rate of water uptake (w) is controlled by the difference between the water potentials of the root (H_R) and of the adjacent soil (H_a)

$$w = p\,(H_a - H_R) \qquad (1)$$

where p is the conductivity of the soil-plant pathway (Hillel et al. 1976). The resistance to water movement inside the plants is assumed to be negligible compared with the resistances at the root and leaf surfaces. This assumption allows the plant water content, P_W, (presented here as a decimal fraction) and water potential to be treated as average entities over the whole plant. Transpiration is assumed to be equal to pan evaporation attenuated in response to a decrease in P_W by an empirical water stress factor, f_t . According to Van Loon (1981), the reduction in transpiration in potato plants starts when $P_W <$ 0.85 (denoted as $P_{W_{opt}}$); f_t approaches zero when $P_W < 0.70$ (denoted as $P_{W_{min}}$). Assuming a linear decrease in transpiration between the $P_{W_{opt}}$ and $P_{W_{min}}$, f_t is expressed as:

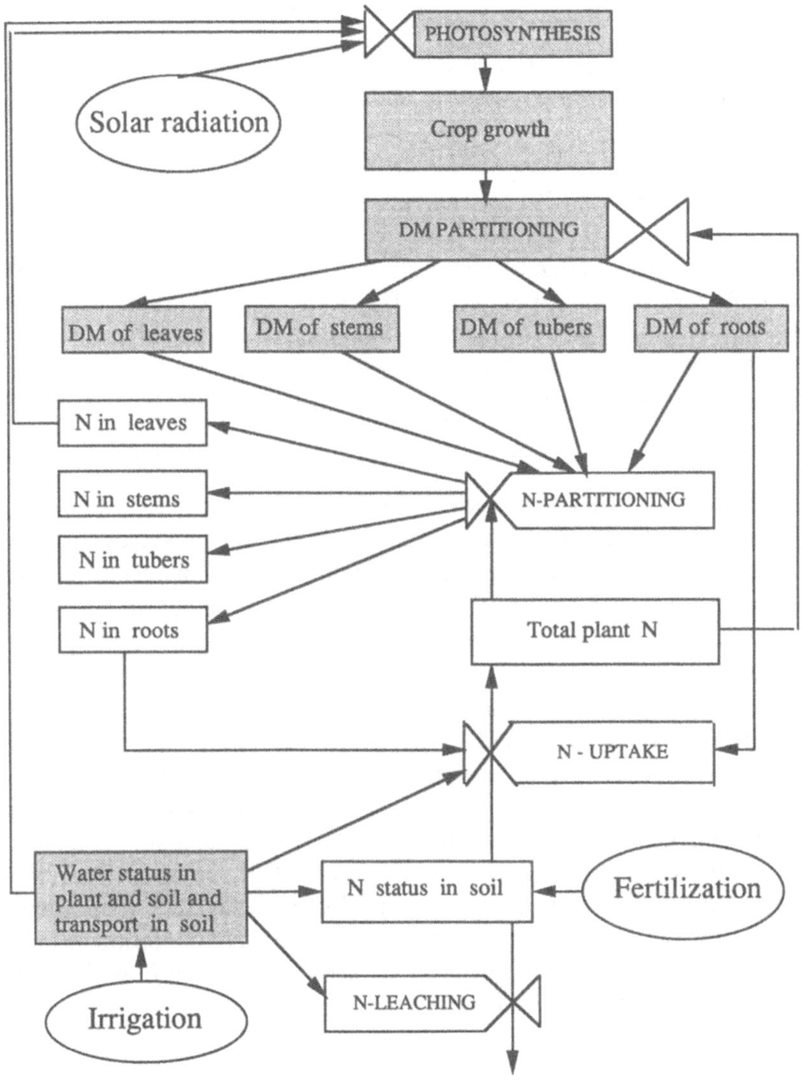

Figure 1. Schematic description of the nitrogen model and its coupling with the plant growth model. Grey blocks refer to the plant and water submodel; white blocks represent the nitrogen submodel.

$$f_t = \begin{cases} 1 & ; \text{ if } P_W \geqq P_{Wopt} \\ (P_w - P_{Wmin})/(P_{Wopt} - P_{Wmin}); & \text{ if } P_{Wmin} \leqq P_W < P_{Wopt} \\ 0 & ; \text{ if } P_W < P_{Wmin} \end{cases} \quad (2)$$

Eq. 2 is consistent with the 'message' controlled switch-on mechanism suggested by Johnson et al. (1991).

The relation between P_W and plant water potential, H_L, is based on a function obtained for tomato leaves (Bar-Yosef et al. 1980):

$$H_L = -6.0\,(P_{Wmax} - P_W)/(P_W - 0.65\,P_{Wmax}) \tag{3}$$

Here P_{Wmax} is the saturated plant water content.

The soil water retention curve (Eq. 4) was derived for the Gilat loess soil:

$$H_i = 23.01\,(100\,\theta_i - 5.89)^{-1.92} \tag{4}$$

where H_i and θ_i ($\theta_i > 0.059$) are the bulk soil matric potential and volumetric soil water content in the i-th soil layer, respectively.

The daily water loss due to evaporation from the soil, E, is estimated as the mean daily water evaporation from a free water surface multiplied by a soil coefficient which depends on the water potential in the top layer of soil. The water loss E is divided between the soil compartments, attenuating exponentially with soil depth according to Van Keulen (1975):

$$E_i = E\,(\theta_i - \theta_A)\exp(-K\,Z_i)/\Sigma_j[(\theta_j - \theta_A)\exp(-K\,Z_j)] \tag{5}$$

where E_i is the loss of water in layer i, due to evaporation, K is an extinction coefficient, Z_i is the depth of soil layer i and θ_A is the soil moisture content at air dryness, and dummy subscript j is used to denote summation over all soil layers.

Variations in θ due to rainfall or irrigation are described in two steps: (i) infiltration during water application; and (ii) redistribution among the soil layers afterwards. During step (i) water enters the soil surface at a steady flux, thereby increasing soil water content in the i-th soil layer, θ_i, up to saturation, θ_S, layer by layer. When a soil layer reaches θ_S, added water moves down instantly to the next layer. Once the water application terminates, step (ii), which is time dependent, takes place. Here θ_i decreases exponentially with time to θ at field capacity (Rao et al. 1981). The characteristic time for the redistribution process is t_r. Following Gardner (1960), the water flow towards the roots was approximated at each time step by the steady-state water transport equation. Solving this equation, subject to the boundary condition at the root surface (Eq. 1), and defining H_b as H in the bulk soil and b as the midway distance between adjacent roots, gives

$$H_a - H_R = (H_b - H_R)/(1 - a\,p\ln(a/b)/K_W) \tag{6}$$

The root conductivity, p, is defined per unit root surface area. To calculate flow per unit root length, the parameter $P = 2\pi a p$, replaces p, a being the root radius. The soil hydraulic conductivity, K_W, is calculated at each time interval and soil layer according to Gardner (1965):

$$K_W = \alpha/(1 + \beta\,H_b^{\gamma}) \tag{7}$$

where α, β and γ are empirical soil parameters.

Governing equations for nitrogen behaviour in soil and its uptake by plants

The total organic nitrogen (Q_{TN}) contains a mineralizable fraction (Q_{NOR}) defined as

$$Q_{NOR} = K_{OR}\, Q_{TN} \tag{8}$$

where K_{OR} is a soil constant.

The ammonium adsorption by the soil, A_{NH4}, was approximated by the Freundlich equation

$$A_{NH4} = K_A\, C_{NH4}{}^{n} \tag{9}$$

where K_A is an affinity constant, n is assumed to be 1, and C_{NH4} is the concentration of $NH_4^+ - N$ in soil solution. It is assumed that NO_3^- adsorption is negligible. The total amount of $NH_4^+ - N$ per cm^3 of soil is given by

$$Q_{NH4} = \varrho A_{NH4} + \theta\, C_{NH4} \tag{10}$$

where ϱ is the soil bulk density.

The mineralization, nitrification and denitrification of the soil nitrogen are described as follows:

$$\begin{aligned} (dQ_{NOR,i}/dt)_c &= -k_M\, g_{MW,i}\, Q_{NOR,i} \\ (dC_{NH4,i}/dt)_c &= (k_M\, g_{MW,i}\, Q_{NOR,i} - \theta_i\, k_N\, g_{NW,i}\, C_{NH4,i})\, (\theta_i + \varrho\, K_A)^{-1} \\ (dC_{NO3,i}/dt)_c &= (k_N\, g_{NW,i}\, C_{NH4,i} - k_{DN}\, g_{DW,i}\, C_{NO3,i}) \end{aligned} \tag{11}$$

where k_M, k_N and k_{DN} are temperature-dependent rate constants for the mineralization, nitrification and denitrification processes, respectively, discussed by Hadas et al. (1979). The factors $g_{MW,i}$, $g_{NW,i}$ and $g_{DW,i}$ describe the water effect on the respective processes (Hadas et al. 1979; Rao et al. 1981). The subscripts i and c relate to the i-th soil layer and to changes due to chemical and microbiological transformation of nitrogen in soil, respectively.

The rate of change in nitrogen concentrations due to convective flow (subscript t) between soil layers is described as follows:

$$\begin{aligned} (dC_{NH4,i}/dt)_t &= [(J_i\, C_{NH4,i-1} - J_{i+1}\, C_{NH4,i})/\Delta x - (d\theta_i/dt)\, C_{NH4}]\, (\theta_i + \varrho\, K_A)^{-1} \\ (dC_{NO3,i}/dt)_t &= [(J_i\, C_{NO3,i-1} - J_{i+1}\, C_{NO3,i})/\Delta x - (d\theta_i/dt)\, C_{NO3,i}]\, (\theta_i)^{-1} \end{aligned} \tag{12}$$

where J_i is the water influx from the $(i-1)$-th to the i-th soil layer and Δx is the width of the layer. The last terms in Eq. 12 express the changes in ion concentrations due to the changes in θ.

Nitrogen uptake from the i-th soil layer (subscript u) is calculated as the product of the active root dry weight in layer i, $W_{R,i}$, and the uptake rate per unit root weight, F_{NO3} and F_{NH4}.

$$\begin{aligned} (dC_{NH4,i}/dt)_u &= F_{NH4}\, W_{R,i}\, (\theta_i + \varrho\, K_A)^{-1} \\ (dC_{NO3,i}/dt)_u &= F_{NO3}\, W_{R,i}\, \theta_i{}^{-1} \end{aligned} \tag{13}$$

The flux functions are approximated by the competitive Michaelis-Menten equation:

$$F_{NO3} = F_{NO3max}\, C_{NO3}^{s}/(K_{NO3M}\,(1 + K_I\, C_{NO3}^{R}) + C_{NO3}^{s})$$
$$F_{NH4} = F_{NH4max}\, C_{NH4}^{s}/(K_{NH4M} + C_{NH4}^{s}) \qquad (14)$$

where F_{NO3max} and F_{NH4max} are maximal values of F_{NO3} and F_{NH4}, C_{NO3}^{s} and C_{NH4}^{s} are ion concentrations at the root surface, K_{NO3M} and K_{NH4M} are the Michaelis-Menten constants, K_I is the equilibrium constant for the competitive ion interaction with the transporting unit and C_{NO3}^{R} is the nitrate concentration inside the roots. For NH_4^+ ions, $K_I = 0$ is assumed.

Nitrogen losses due to plant senescence are taken into account by assuming that part of the nitrogen from aging plant organs is lost and the rest is redistributed among the younger tissues.

Total changes in nitrogen concentration in the bulk soil solution account for microbial transformations (subscript c, Eq. 11), the vertical transport (subscript t, Eq. 12) and uptake by roots (subscript u, Eq. 13)

$$dC_{NH4,i}/dt = (dC_{NH4,i}/dt)_c + (dC_{NH4,i}/dt)_t + (dC_{NH4,i}/dt)_u$$
$$dC_{NO3,i}/dt = (dC_{NO3,i}/dt)_c + (dC_{NO3,i}/dt)_t + (dC_{NO3,i}/dt)_u \qquad (15)$$

In order to estimate the ion concentrations at the root surface, it was assumed that a quasi-steady-state concentration profile perpendicular to the root axis was established at each time step. This profile may be calculated analytically taking into account both convective and diffusive flow of ions towards the root sink. Let a stand for the root radius, r for radial distance from the root axis and $C(r)$ for the ion concentration at distance r. The steady-state radial flow of ions per unit root length, J_R, under these conditions equals to

$$J_R = 2\pi r\,(\theta + \varrho\, K_A)\, D\, dC(r)/dr + 2\pi a\, w\, C(r) \qquad (16)$$

D is the effective diffusion coeficient of the ions in soil. It was shown (Olsen and Kemper 1968) to be an exponential function of volumetric water content:

$$D = D_0 \exp(b_0\theta)$$

where D_0 is the value of D in water and b_0 is a soil parameter.

Olsen and Kemper (1968) described the solution of the differential Eq. 16 subject to boundary condition where the concentration $C(r)$ is maintained at a distance b from the axis of the root. They assumed D to be constant with r. The profile of soil water content in the vicinity of root surfaces was studied by Gardner (1960). For three soils studied experimentally, the suction at the roots does not exceed the soil suction appreciably in a wide interval of water concentrations. This means that during each time step and in each soil layer in our model, the diffusion coefficient may be considered as independent of the radial distance but is to be recalculated for each time step because of the temporal changes in the moisture content. If $F(a)$ is the flux of ions at the root surface, the solution of Eq. 16 subject to the boundary condition $C(b) = \text{const}$ (b being the halfway distance between adjacent roots) is

$$C(r) = F(a)/2\pi a\, w - [F(a)/2\pi a\, w - C(b)]\,(b/r)^{u} \qquad (17)$$

where $u = a\, w/(\theta + \varrho\, K_A)\, D$. Substitution $F(a)$ in Eq. 17 by Eq. 14, alows calculation of the ion concentration at the root surface, $C(a)$, from Eq. 18:

$$C(a)^2 + 2\eta\, C(a) - K_M\, C(b)\, (b/a)^u = 0 \tag{18}$$

where

$$\eta = 0.5\, \{K_M - (F_{max}/2\pi a\, w)\, [1 - (b/a)^u] - C(b)\, (b/a)^u\} \tag{19}$$

Here $C(a)$ plays the role of $C_{NH4}{}^s$ or $C_{NO3}{}^s$; $K_M = K_{NH4M}$ in the case of ammonium ions and $K_M = K_{NO3M}(1 + K_I\, C_{NO3}{}^R)$ in the case of nitrate; F_{max} equals F_{NH4max} or F_{NO3max}, respectively, and the dimension of the flux $2\pi a\, w$ must equate with this of F_{max}. Eq. 18 has the following solution

$$C(a) = -\eta + [\eta^2 + K_M\, C(b)\, (b/a)^u]^{0.5} \tag{20}$$

Only the positive value of the square root function is valid for $C(a) \geq 0$.

Nitrogen partitioning in the plant

Nitrogen is translocated in plants at velocities of about 1 cm/min (Pate 1971; 1980). The time step over which water and N transport are simulated is one hour, therefore, the N partitioning can be considered as instantaneous within each time step (Fishman 1992).

Total N in plant is the sum of the nitrogen located in roots (subscript R), stems (S), leaves (L) and tubers (T):

$$N = \Sigma_j\, N_j \qquad (j = R, S, L, T) \tag{21}$$

Defining nitrogen concentration in the dry matter of each organ as $p_j = N_j/W_j$ and combining it with Eq. 21 yields

$$N_i = N\, \zeta_i\, W_i/\Sigma_j \zeta_j\, W_j \qquad (i, j = R, S, L, T) \tag{22}$$

where ζ_j is defined arbitrarily as

$$\zeta_j = p_j/p_T \tag{23}$$

which means that for tuber $\zeta_T = 1$. The parameters ζ_R, ζ_S and ζ_L must be estimated independently. As a first approximation, the parameters s_j are assumed to be constant during the growing season.

Nitrates taken up by roots at a given time step, U_{NO3}, enter the plant nitrate pool, n. The nitrate reduction rate is described by Eq. 24, where k_r and k_l are the rate constants for the leaves and roots, respectively:

$$dn/dt = U_{NO3} - k_r\, n_R - k_l\, n_L \tag{24}$$

Defining a partitioning function $\phi_i = n_i/n$, with i = R, S, L, T, one obtains

$$dn/dt = U_{NO3} - (k_r\, \phi_R + k_l\, \phi_L)\, n \tag{25}$$

Defining $q_j = n_j/N_j$, and $\zeta_j = q_j/q_T$, the partitioning may be presented as

$$n_i/n = \eta_i\, N_i / \Sigma_j\, \eta_j\, N_j \tag{26}$$

Assuming ϕ_j to be constant during the day, Eq. 25 may be integrated to yield

$$n = n_0 + [U_{NO3}/k_r\,(\phi_R + \phi_L) - n]\,\{1 - \exp[-k_r\,(\phi_R + \phi_L)\,(t - t_0)]\} \tag{27}$$

Effect of nitrogen on plant growth and development

Nitrogen concentration in leaves is strongly correlated with photosynthetic activity (Van Keulen et al. 1989). For C_3 plants the net assimilation was reported to approach zero at a nitrogen concentration of 3.8 g-N/kg DM and to be linearly related to nitrogen concentration over a wide range of concentrations (Van Keulen et al. 1989), until saturation is obtained (Findenegg 1990). Evans (1989) reported that assimilation by C_3 plants approached zero at about 8.4 g-N/kg. Based on this information, the effect of reduced N concentration in the leaves $(N_L - n_L)/W_L$ on the maximal photosynthesis is expressed as

$$F_{NPHOT} = 1 - \exp\{-\alpha_N \times [(N_L - n_L)/W_L - \omega_{min}]\} \tag{28}$$

where ω_{min} is the threshold leaf-N concentration at which assimilation becomes zero, and α_N is the slope of the curve when N concentration approaches ω_{min}. Note that $0 \leq F_{NPHOT} \leq 1$.

To account for the effect of nitrogen application on the tuber yield observed by Millard (1986), the dry matter partition parameters for leaves (b_L) and stems (b_S) used by Fishman et al. (1984) were replaced by the functions:

$$\begin{aligned} b_L &= 0.457\,\{1 - \exp[-1.587\,(N/W/0.012 + 0.525)]\} \\ b_S &= 0.838\,\{1 - \exp[-1.587\,(N/W/0.012 + 1.343)]\} \end{aligned} \tag{29}$$

The aging initiation parameter (T_a, days) was replaced by the function

$$T_a = 45\,\{1 - \exp[-0.811\,(N/W/0.012)]\} \tag{30}$$

Here, N/W is the whole-plant nitrogen concentration. Calculation of b_L, b_S and T_a with N/W = 0.012 gives $b_L = 0.42$, $b_S = 0.82$ and $T_a = 25$, the parameter values used in ISPOTA.

To perform a computer simulation using this model, the following data must be input:

(i) Daily minimum and maximum air temperatures, daily pan evaporation, precipitation events, and solar radiation during the growing season.
(ii) Soil chemical and physical characteristics (Table 1) and latitude.
(iii) Agrotechnical details: planting, emergence and tuber initiation dates; plant density, and N fertilization rates, dates and type (NH_4^+ or NO_3^--N).
(iv) Additional parameters, which appear in the governing equations (Table 1).

Table 1. Parameters of the model

Eq. no.	Parameter	Process described by Eq.	Value	Dimension
1	$P = 2\pi a\ p$	Water uptake rate per cm of root	0.023	cm^3/cm-root/bar/d
	$P = 2\pi a\ p$	Water uptake rate per g of root	120	cm^3/g-root/bar/d
2	P_{Wopt}	Control of water stress	0.85	dimensionless
2	P_{Wmin}	Control of water stress	0.70	dimensionless
3	P_{Wmax}	Water retention in plant	0.95	dimensionless
5	K	Extinction coefficient in soil	5.0	1/m
7	α	Soil capillary conductivity	12	cm/d
7	β	Soil capillary conductivity	$2.4\ 10^4$	bar
7	γ	Soil capillary conductivity	1.7	dimensionless
8	K_{OR}	Organic N mineralization	0.2	dimensionless
9	K_A	Ammonium adsorption	15.0	cm^3/g-soil
10	ϱ	Soil bulk density	1.25	g/cm^3
11	k_{DN}	Denitrification	$6\ 10^{-4}$	1/d
14	F_{NO3max}	Nitrate uptake	0.01	g-N/g-root/d
14	K_{NO3M}	Nitrate uptake	$3.64\ 10^{-6}$	g-N/cm^3
14	K_I	Nitrate uptake	10^4	cm^3/g-N
14	F_{NH4max}	Ammonium uptake	0.1	g-N/g-root/d
14	K_{NH4M}	Ammonium uptake	$3.64\ 10^{-7}$	g-N/cm^3
16	D_0	Ion diffusion in soil	0.216	cm^2/d
16	b_0	Ion diffusion in soil	10	dimensionless
22	ζ_R	Nitrogen partitioning in plant	1.0	dimensionless
22	ζ_S	Nitrogen partitioning in plant	1.0	dimensionless
22	ζ_L	Nitrogen partitioning in plant	1.3	dimensionless
24	k_r	Nitrate reduction in plant	0.35	1/h
24	k_l	Nitrate reduction in plant	0.7	1/h
26	η_R	Nitrate partitioning in plant	7.0	dimensionless
26	η_S	Nitrate partitioning in plant	7.0	dimensionless
26	η_L	Nitrate partitioning in plant	7.0	dimensionless
28	α_N	Nitrogen effect on photosynthesis	200	dimensionless
28	ω_{min}	Nitrogen effect on photosynthesis	0.003	dimensionless

Results and discussion

Reference conditions

The reference conditions describe a spring potato crop which was planted on 9 February (day of year, DOY = 40) and was burnt-down at DOY = 165. The meteorological data for the 1982 season in southern Israel were used. Irrigation was applied weekly: 28 mm per irrigation in February and March, 35 mm in April, 42 mm in May and 49 mm per irrigation in June. The soil profile comprised 10 layers, each 10 cm thick. No rain occurred during the growing season.

Initial NH_4^+- and NO_3^--N concentrations in the soil solution of the top two soil layers were 1.3 mg N/l and 250 mg N/l, respectively. The concentrations in

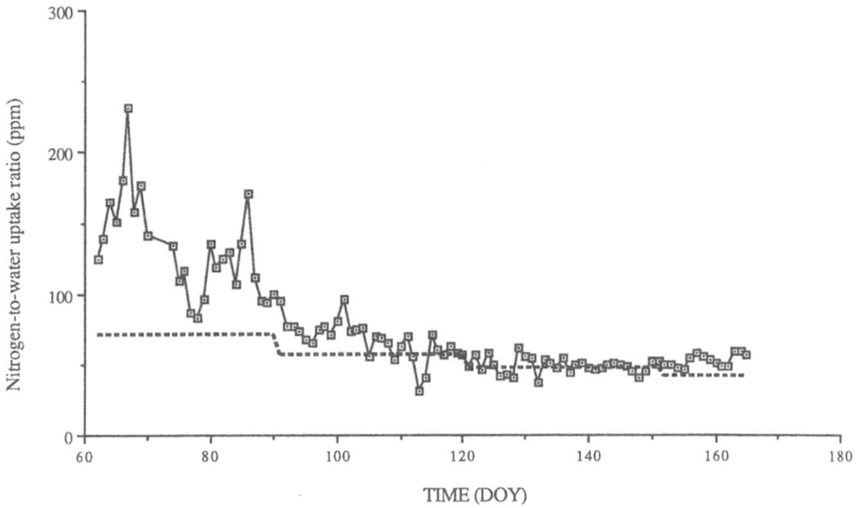

Figure 2. Simulated nitrogen to water uptake ratios by potato plants (mg N/l) during the season. The dotted line shows the N concentration in the irrigation water. Reference conditions.

deeper soil layers were assumed to be zero. Initial organic N concentration in all soil layers was 750 mg N/kg soil. Supplementing N-fertilization started after emergence (DOY = 62) as NH_4NO_3 and was applied in the water at each irrigation, at a rate of 20 kg/ha per irrigation. Ferilization was ended 15 days prior to canopy burn-down.

Simulation examples under reference conditions

The simulated tuber yield under the reference conditions was 61.5 t/ha. The computed maximal LAI was 5.3, obtained 38 days after emergence. Both results are characteristic of field grown potatoes in southern Israel, which suggests that the model and its reference parameters are reasonable.

The simulated ratio of total nitrogen to total water uptake is presented as a function of time in Figure 2. The general trend shows that the ratio declined with crop age, indicating a decline in the efficiency of N-uptake relative to water uptake efficiency over time. During March, C_{NT} was twice as high as N concentration in the irrigation water, C_N. Between April and June C_{NT} accurately matched C_N, thus maintaining a fixed N concentration in the soil solution between successive fertigations. The anomalous peaks and troughs (Figure 2) reflect daily climatic effects on N and water uptake rates. The total simulated N uptake at harvest (NH_4^+ and NO_3^-) was 198 kg/ha, taking into account the N losses due to senescence of plant organs.

The simulated total-plant N concentration is presented as a function of time in Figure 3. The concentration dropped from 55 gN/kg to 10 gN/kg in about two weeks, followed by slow variations during the rest of the season to give a mean

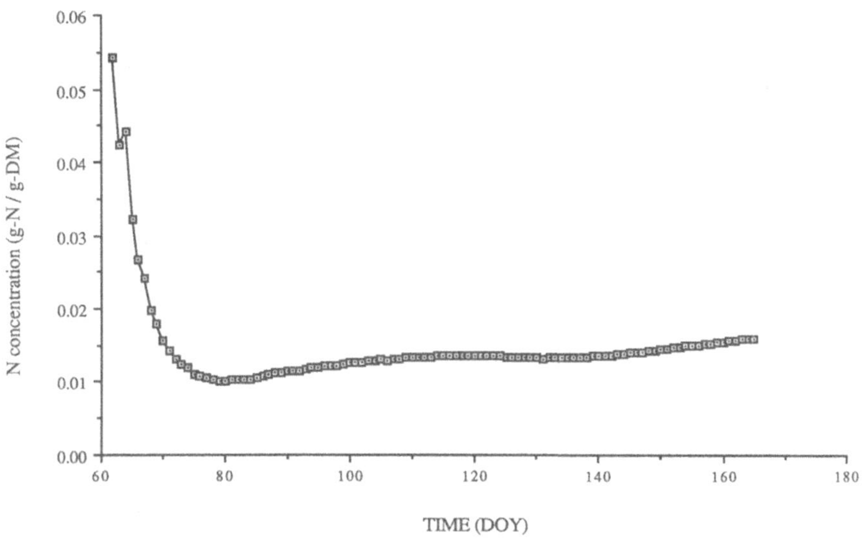

Figure 3. Simulated total plant N concentration (g N/g DM) as a function of time under reference conditions.

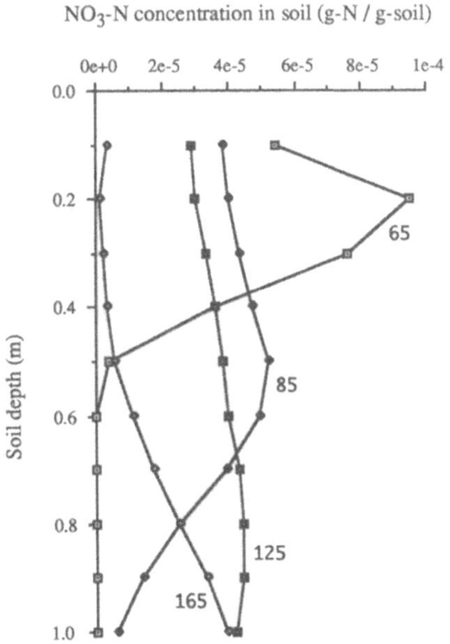

Figure 4. Simulated NO_3-N concentration as a function of depth and time under reference conditions.

concentration of about 16 gN/kg. This behavior is not in full agreement with the data of Greenwood et al. (1985a) who reported a progressive decline of the N concentration throughout the growth of the plant.

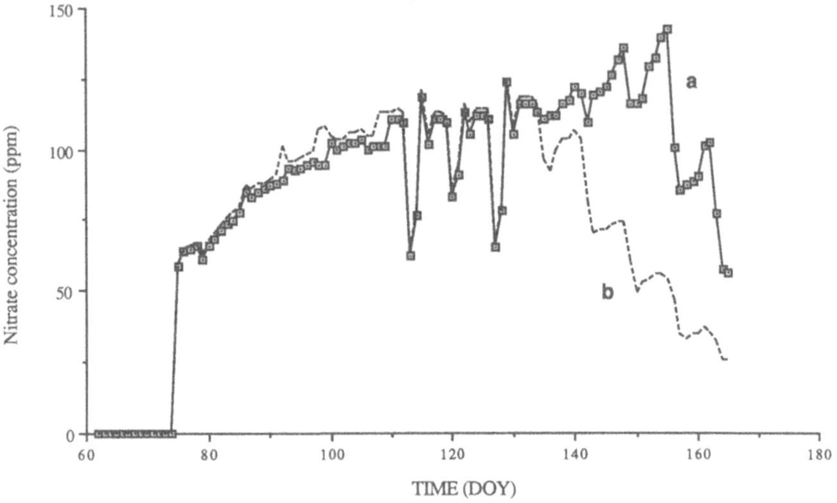

Figure 5. Simulated nitrate concentration in tubers (mg NO_3/kg DM) as a function of time. (a) Reference conditions; (b) N applications ended 35 days before harvest (instead of 15 days in the reference).

The simulated NO_3 distribution in the soil profile is presented as a function of time in Figure 4. The maximum in NO_3-concentration shifted to deeper soil layers with time, indicating leaching beyond the effective root zone. The simulated depth of the root system changed with time as follows: 10 cm at DOY = 65, 20 cm at DOY = 70, 30 cm at DOY = 75 and 40 cm at DOY = 95; after DOY = 115 the total active root weight declined because of senescence.

One of the major model objectives was to simulate nitrate accumulation in the tubers. The simulated nitrate concentration in tubers shows a gradual increase over time and a decrease after N fertilizations ended (Figure 5). The simulated nitrate concentration in leaves as a function of time (not presented) were similar to data published by Biemond and Vos (1992). It is noted that according to the model assumptions, nitrate concentration in leaves is proportional to the concentration in tubers, as was observed by Carter and Bosma (1974). The simulated fluctuations in nitrate concentration in tubers on DOY 113–114 and 126 stemmed from extreme evaporation on those days – 14 and 11 mm/d, respectively.

Simulated response to fertilization rates

The simulated response in tuber yield to N fertilization is presented in Figure 6. In this case, the amount of N supplied via the water at each irrigation was changed, without varying the fertilization scheduling. The simulated response is in qualitative agreement with observations on potato yield response to N reported by Millard (1986) and Millard and Marshall (1986). The simulated slight decline in tuber yield at large fertilizer doses (Figure 6) stemmed from an

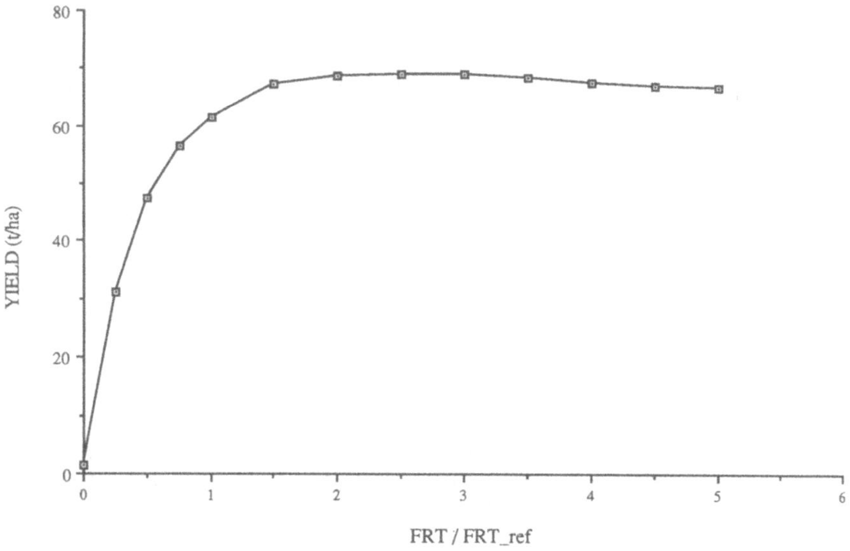

Figure 6. **Simulation of tuber yield response to seasonal N fertilizer (FRT) rates. The reference rate (FRT—ref) was 200 kg N/ha.**

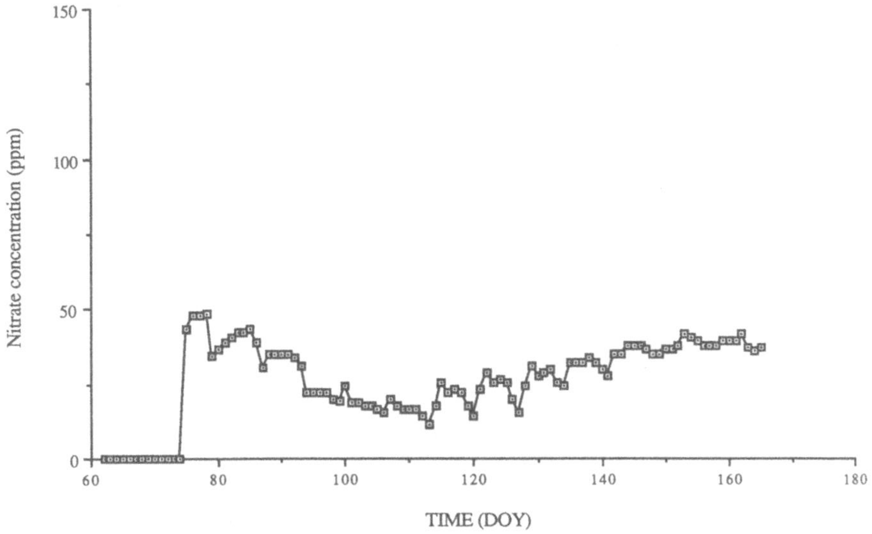

Figure 7. **Simulated nitrate concentration in tubers (mg NO_3/kg DM) as a function of time. Reference conditions, except that all N was applied as NH_4-N.**

increase in the fraction of total dry matter allocated to leaves at high N concentrations in the plant as described by Eq. 29. Further decrease in yield at enhanced fertilization rates is expected, owing to build up of salinity. Currently, the model does not take into account the salinity effect on water uptake, and hence nitrate mass flow to roots.

To reduce nitrate leaching in soil and accumulation in tubers, farmers used to stop fertilization at a certain time before harvest. To simulate this practice, it was assumed that N applications ended 35 days before harvest (instead of 15 days in the reference). The total N application was unchanged. The yield changed slightly in this case, but the decrease in nitrate concentration in tubers started earlier, resulting in a lower nitrate level at harvest than in the reference (Figure 5, curve b). Another way to reduce nitrate accumulation in tubers is to supply the N as a NH_4 fertilizer. Simulation of this effect reveals a significant reduction in nitrate content in tubers relative to the reference (compare Figures 7 and 5, respectively).

Sensitivity analysis

The criteria for testing the sensitivity of the model to its parameters were chosen to be tuber yield and nitrate concentration in tubers (NCT) at harvest. All the results are relative to those obtained under the reference conditions and parameters.

Parameters related to N uptake

The response of yield and NCT to the parameters related to N uptake by roots is shown in Figure 8. Decreasing parameter F_{NH4max} to 0.2 of the reference

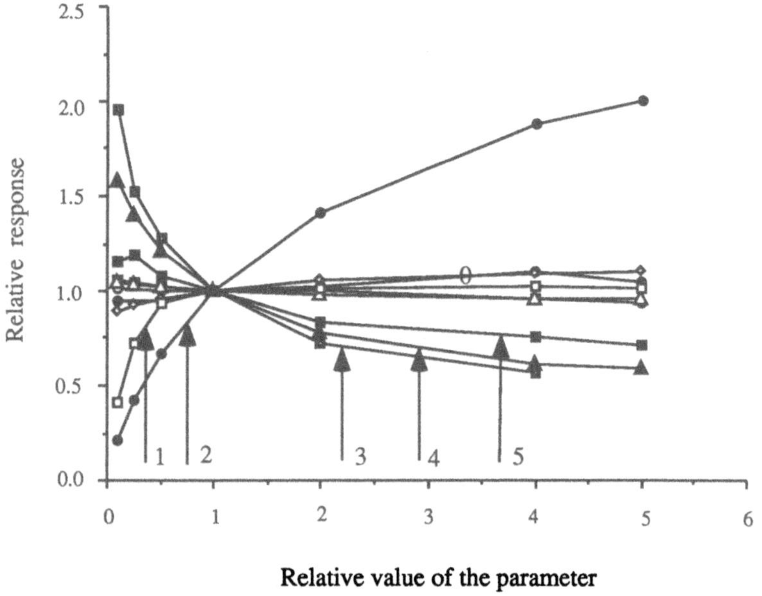

Figure 8. Analysis of the model sensitivity to parameters of N uptake by roots. The strongest effects are: response of the tuber yield to changes in F_{NH4max} (no. 1) and response of the nitrate concentration in tubers (NCT) to F_{NO3max} (no. 2), K_{NO3M} (no. 3), K_I (no. 4) and F_{NH4max} (no. 5).

reduced yield to 0.4 of its reference value (curve no. 1). This particular sensitivity of the yield to F_{NH4max} is expected as the NH_4 concentrations in the soil solution considerably exceeded K_{NH4M} and under such conditions the flux approaches F_{NH4max}. The tuber yield was only slightly affected by the parameters K_{NH4M}, F_{NO3max}, K_{NO3M} and K_I (non-numbered curves). Increasing the values of these parameters 25-fold (from 0.2 to 5) resulted in a 20% change in yield.

These parameters had a stronger effect on the NCT than on tuber yield. The greatest sensitivity was to F_{NO3max}: when it was increased 25-fold, NCT increased by a factor of 10 (curve no. 2). As expected, variations in F_{NO3max} and K_{NO3M} (curve no. 3) had opposite effects; the magnitude of the effect of these parameters was similar. Parameter K_I which is responsive for inhibition of nitrate uptake from soil, affects NCT in the same direction (curve no. 4) as K_{NO3M}. The sensitivity in NCT to K_{NH4M} was relatively small (the response enters in the group of non-numbered curves). Increasing F_{NH4max} from (0.2 × reference) to (5 × reference) caused a decline in NCT from (1.2 × reference) to (0.8 × reference) (curve no. 5).

Parameters related to N functions in plant

Nitrogen concentration in leaves affects photosynthesis rate via two parameters: α_N and ω_{min} (Eq. 28). Increasing the reference ω_{min} to (5 × reference) caused a decline in yield to (0.65 × reference) (curve no. 1, Figure 9). Decreasing α_N from the reference value to (0.2 × reference) decreased tuber yield to (0.58 × reference) (curve no. 3, Figure 9). The effect of ω_{min} and α_N on NCT (curves no. 2 and 4, Figure 9, respectively) was qualitatively similar, but was smaller than on the tuber yield.

The effects of the rate constant of nitrate reduction in leaves, k_l, and the parameters determining nitrate partitioning among the plant organs, η_L, η_R and η_S, are presented in Figure 10. When changing k_l tenfold, tubers yield changed by merely 10%. Decreasing k_l to a tenth of its reference value, caused a fivefold increase in NCT, while increasing k_l by the same factor caused a sevenfold reduction NCT (curve no. 1, Figure 10). The nitrate reduction rate constant in roots, k_r, had a much smaller effect on yield and on NCT than k_l and is not presented here. The decrease of the nitrate partitioning coefficients, η_L and η_R, from the reference value to (0.1 × reference) increased NCT. In the case of η_L, NCT increased fivefold (curve no. 2, Figure 10), while in the case of η_R, NCT increased by only 50% (curve no. 3, Figure 10). The yield was only slightly affected by variations in these parameters, as represented by the group of non-numbered curves in Figure 10.

Parameters related to plant water status

The water stress coefficients, P_{Wopt} and P_{Wmin}, and the water retention parameter, P_{Wmax}, cannot be changed arbitrarily, because their values are limited to: $0 < P_{Wmin} < P_{Wopt} < 1$ and $0 < P_{Wmax} < 1$. Increase of P_{Wopt} and P_{Wmin} menas diminishing the range of plant water content over which the plant

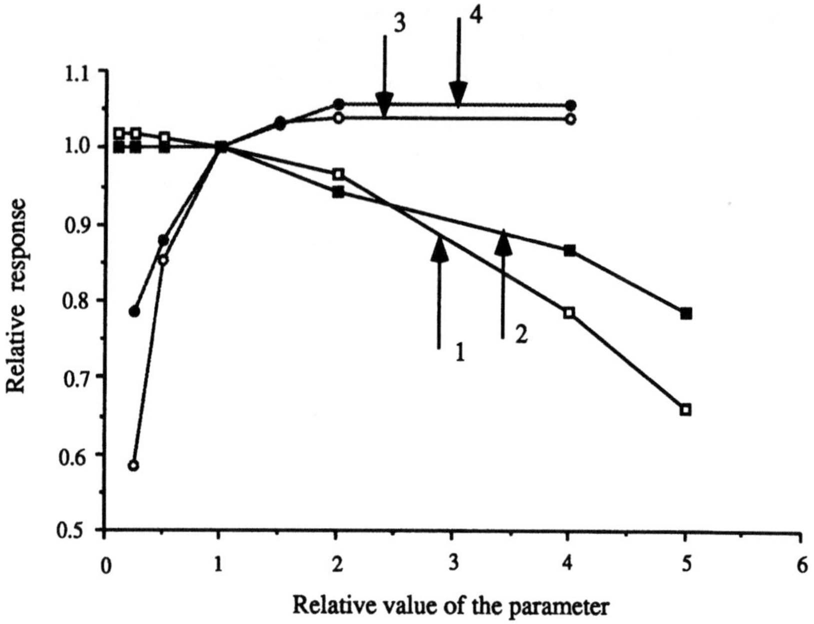

Figure 9. Analysis of the model sensitivity to parameters of photosynthesis rate: response of the tuber yield to changes in ω_{min} (no. 1) and α_N (no. 3); response of nitrate concentration in tubers (NCT) to changes in ω_{min} (no. 2) and α_N (no. 4).

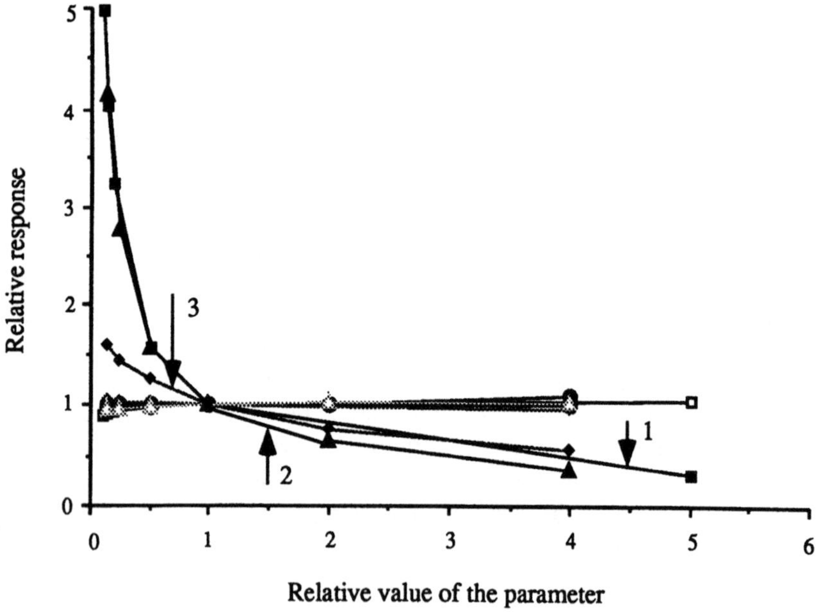

Figure 10. Analysis of the model sensitivity to parameters determining nitrate partitioning among the plant organs and nitrate reduction rate in leaves. The strongest effects are: response of the nitrate concentration in tubers (NCT) to changes in k_l (no. 1), ζ_L (no. 2) and ζ_R (no. 3).

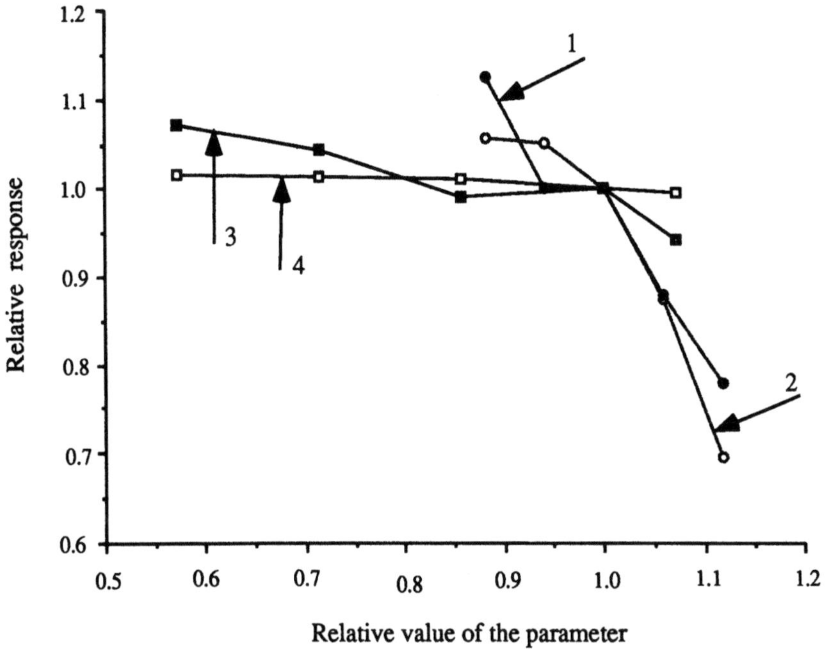

Figure 11. Analysis of the model sensitivity to parameters determining the effect of plant water content on transpiration: response of nitrate concentration in tubers (NCT) to changes in P_{opt} (no. 1) and P_{Wmin} (no. 3); response of the tuber yield to changes in P_{opt} (no. 2) and P_{Wmin} (no. 4).

operates without water-stress. Therefore, the increase of both P_{Wopt} and P_{Wmin} leads to reduction in yield (curves no. 2 and 4 in Figure 11, respectively). Perturbation of these parameters influences the water uptake by roots and affects NCT. Curves no. 1 and 3 in Figure 11 show the influence of P_{Wopt} and P_{Wmin} on NCT.

Perturbation of P_{Wmax} affects the water potential in the plant, which influences the water uptake by roots and hence yield and NCT. Changing P_{Wmax} by ± 15% caused a change of about ± 25% in yield and in NCT (not represented in figures).

Parameters related to soil processes
Evaluation of the sensitivity to the parameters describing water and N transport in soil shows that the greatest sensitivity was to p, the coefficient of root permeability to water: when it declined to 25% of the reference value, the yield decreased by approximately 30% and NCT by about 60%. Changing the nitrification rate in soil by a factor of four, slightly changed tuber yield and led to a change of about 20% in NCT. The 4-fold change in the time constant which characterizes the water redistribution after precipitation also affects the yield and NCT in the range of 20% relative to the reference.

Conclusions

The model presented here satisfactorily simulated tuber yield, dry matter production and total N and nitrate accumulation in various plant organs as functions of time, under reference conditions characteristic of southern Israel. It also described the changes in the distribution of nitrate, ammonium and roots in the soil in response to N fertilization rate, application scheduling and type of N fertilizer. The satisfactory agreement of the simulation results with practical experience supports the basic assumptions of the model and partly confirms some phenomenological functions that are used in the model to describe certain processes for which the mechanisms are insufficiently understood.

Sensitivity analysis of the model showed that nitrate concentration in tubers (NCT) was more sensitive than tuber yield to a unit change in most model parameters. The parameters which affected NCT most strongly were the leaf nitrate reduction rate constant and the coefficient of nitrate partitioning among plant organs. The next most important parameters were the Michaelis-Menten nitrate uptake coefficients, which changed NCT by a factor of 4 as they increased from (0.2 × reference) to (5 × reference).

The parameters which exerted most influence on tuber yield were the slope of the curve which relates leaf N percentage to net photosynthesis rate, and the threshold leaf N percentage below which the photosynthesis rate approaches zero. Next in importance was the parameter which defines the leaf water content below which transpiration starts to decline.

Soil-related parameters were less effective than plant-related parameters in determining tuber yield and NCT. This conclusion depends to some extent on the chosen reference conditions. Under less abundant N and water supply, the role of soil factors might be more critical.

Acknowledgements

The authors gratefully acknowledge the valuable suggestions of Dr. D. MacKerron.

References

Barker A V, Peck N H, MacDonald G E (1971) Nitrate accumulation in vegetables. I. Spinach grown in upland soils. Agronomy Journal 63:126–129.

Bar-Yosef B, Sagiv B, Eliah E (1980) Fertilization and Irrigation of Winter Tomatoes Grown in a Glasshouse in the Besor Area. Report no. 775, Division of Scientific Publications, The Volcani Center, Bet Dagan, Israel.

Biemond H, Vos J (1992) Effects of nitrogen on the development and growth of the potato plant. 2. The partitioning of dry matter, nitrogen and nitrate. Annals of Botany 70:37–45.

Carter J N, Bosma S M (1974) Effect of fertilizer and irrigation on nitrate-nitrogen and total nitrogen in potato tubers. Agronomy Journal 66:263–266.

Evans J R (1989) Photosynthesis and nitrogen relationships in leaves of C_3 plants. Oecologia 78:9–19.

Findenegg G R (1990) Effect of varied shoot/root ratio on growth of maize (*Zea mays*) under nitrogen-limited conditions: Growth experiment and model calculations. Pages 21–27 in Van Beusichem ML (Ed.) Plant Nutrition – Physiology and Applications. Kluwer Academic Publishers, Dordrecht, The Netherlands.

Fishman S, Talpaz H, Dinar M, Levy M, Arazi Y, Rozman Y, Varshavsky S. (1984). A phenomenological model of dry matter partitioning among plant organs for simulation of potato growth. Agricultural Systems 14: 159–169.

Fishman S, Talpaz H, Winograd R, Dinar M, Arazi Y, Roseman Y, Varshavski S (1985) A model for simulation of potato growth on the plant community level. Agricultural Systems 18:115–128.

Fishman S (1992) A phenomenological model of nitrogen uptake and partitioning in plants for use in plant growth simulation. Acta Horticulturae 313:65–71.

Fishman S, Talpaz H, Roseman Y, Varshavski S, Arazi Y (1995) Simulation of potato crop growth: ISPOTA – Model design and parametrization in Kabat P, Van den Broek B, Marshall B, Vos J, Van Keulen H (Eds.) Modelling and Parametrization of the Soil-Plant-Atmosphere System. A Comparison of Potato Growth Models. Wageningen Press, Wageningen, The Netherlands (in print).

Gardner W R (1960) Dynamic aspects of water availability to plants. Soil Science 89:63–73.

Gardner W R (1965) Dynamic aspects of soil-water availability to plants. Annual Review of Plant Physiology 16:323–342.

Greenwood D J, Neeteson J J, Draycott A. (1985a) Response of potatoes to N fertilizer: Quantitative relations for components of growth. Plant and Soil 85:163–183.

Greenwood D J, Neeteson J J, Draycott A (1985b) Response of potatoes to N fertilizer: Dynamic model. Plant and Soil 85:185–203.

Hadas A, Bar-Yosef B, Kafkafi U, Yaniv J (1979) A model to determine N, P, K fertilization recommendation under field conditions. Pamphlet No. 217. Division of Scientific Publications, the Volcani Center, Bet Dagan, Israel.

Hillel D, Talpaz H, Van Keulen H (1976) A macroscopic-scale model of water uptake by nonuniform root system and of water and salt movement in the soil profile. Soil Science 121:242–255.

Johnson I R, Melkonian J J, Thornley J H M, Riha S J (1991) A model of water flow through plants incorporating shoot/root 'message' control of stomatal conductance. Plant, Cell and Environment 14:531–544.

Millard P (1986) The nitrogen content of potato (*Solanum tuberosum* L.) tubers in relation to nitrogen application – the effect on amino acid composition and yields. Journal of the Science of Food and Agriculture 37:107–114.

Millard P, Marshall B (1986) Growth, nitrogen uptake and partitioning within the potato (*Solanum tuberosum* L.) crop, in relation to nitrogen application. Journal of Agricultural Science, Cambridge 107:421–429.

Olsen S R, Kemper W D (1968) Movement of nutrients to plant roots. Advances in Agronomy 20:91–151.

Pate J S (1971) Movement of nitrogenous solutes in plants. Pages 165–187 in Nitrogen-15 in Soil-Plant Studies. Proceedings, 1–5 Dec. 1969. IAEA–P1–341/13. International Atomic Energy Agency, Vienna, Austria.

Pate J S 1980) Transport and partitioning of nitrogenous solutes. Annual Review of Plant Physiology 31:313–340.

Rao P S C, Davidson J M, Jessup R E (1981) Simulation of nitrogen behaviour in the root zone of cropped land areas receiving organic wastes. Pages 81–95 in Frissel M J, Van Veen J A (Eds.) Simulation of Nitrogen Behaviour of Soil-Plant Systems. Papers of a Workshop on Models for the Behaviour of Nitrogen in Soil and Uptake by Plant. Comparison Between Different Approaches, Wageningen, Jan. 28–Feb. 1, 1980. Pudoc, Wageningen, The Netherlands.

Van Keulen H (1975) Simulation of Water Use and Herbage Growth in Arid Regions. Simulation Series Monographs, Pudoc, Wageningen, The Netherlands.

Van Keulen H, Goudriaan J, Seligman N G (1989) 5. Modelling the effects of nitrogen on canopy development and crop growth. Pages 83–104 in Russell G, Marshall B, Jarvis P G (Eds.) Plant Canopies: Their Growth, Form and Function. SEB Seminar Series, 31. Cambridge University Press, Cambridge, UK.

Van Loon C D (1981) The effect of water stress on potato growth, development and yield. American Potato Journal 58:51–68.

Wilson J K (1949) Nitrate in foods and its relation to health. Agronomy Journal 41:20–22.

Wright M J, Davison K L (1964) Nitrate accumulation in crops and nitrate poisoning in animals. Advances in Agronomy 16:197–247.

11. Crop physiological responses to infection by potato cyst nematode (*Globodera* spp.)

A.J. HAVERKORT[1] and D.L. TRUDGILL[2]

[1] *Research Institute for Agrobiology and Soil Fertility (AB-DLO), P.O. Box 14 6700 AA Wageningen, the Netherlands*

[2] *Scottish Crop Research Institute (SCRI), Invergowrie, Dundee, DD2 5DA, U.K.*

Abstract. Fresh potato tuber yields are determined by the total amount of photosynthetically active radiation intercepted by the green foliage of the crop, its conversion efficiency into dry matter, the proportion of dry matter allocated to the tubers and the tuber dry matter concentration. Reduced light interception at the beginning and at the end of the growing season, best explain yield losses following infection by potato cyst nematode (*Globodera* spp.). Retarded canopy closure during canopy establishment is due to a reduced photosynthetic rate (caused by reduced nutrient uptake rates, hormonal signalling and disrupted plant water relations), increased dry matter allocation to the roots, the formation of fewer stems and the leaves being smaller and thicker. In the second half of the growing season crops infected by potato cyst nematode senesce earlier because the reduced assimilation rate leads to reduced formation of new leaves and the reduced water and nutrient uptake is aggravated because rooting depth is less whereas water and nutrients are depleted in the rooted zone leading to increased leaf shedding. Moreover, tuber fresh yields are reduced because infection by potato cyst nematode reduces the number of tubers and increases the tuber dry matter concentration.

The short-term and long-term effects of potato cyst nematode infection on crop functioning are illustrated and their implications for plant breeding for tolerance , cultural practices and modelling are discussed.

Introduction

Crop performance may be expressed in terms of factors that set the upper limits to yield (e.g. temperature and solar radiation), that restrict yield (e.g. lack of water and nutrients) and reduce yield (e.g. pests and diseases). Those stress factors restricting and reducing yield may be abiotic or biotic in origin and several such factors may operate in the course of the growing season. For instance, emergence of potato sprouts may be hampered by unfavourable soil conditions (cold or wet) or soil-borne pathogens (e.g. *Rhizoctonia solani* and nematodes) or seed-borne (viruses) ones. Between emergence and canopy development the same stress factors affect crop growth but more hazards (night frosts or drought) are added. The canopy is subjected gradually to air borne diseases such as late blight caused by *Phytophthora infestans* and aphid-transmitted viruses. Towards the end of the growth cycle depletion of water and minerals as well as certain diseases such as potato wilting caused by *Verticillium*

A.J. Haverkort and D.K.L. MacKerron (eds.), Potato Ecology and Modelling of Crops under Conditions Limiting Growth, 167–184.

dahliae may become increasingly important. The objective of crop management and disease control is to reduce the chance of growth being restricted by unfavourable factors.

Crop reactions to nematode attack may be partitioned into those that are short- and long-term. Short-term reactions include effects on plant and cell water relations, stomatal conductance, photosynthesis, respiration, nutrient uptake and chemical composition. To measure these effects techniques such as the pressure chamber, porometry, infrared gas analysis and isotope labeling have been applied. Longer-term effects are often expressed as growth retardation or earlier senescence and by wilting, yellowing and leaf shedding. They may be determined by crop growth analysis, measurement of intercepted radiation and radiation-use efficiency, dry matter partioning and specific leaf area. Other longer-term effects may be detected through special measurements such as crop reflectance, transpiration efficiency, stable isotope fractionation and crop mineral content.

The objectives of this chapter are to review the short-term and long-term effects of potato cyst nematodes on crop growth, development and productivity. We discuss the relevance of the knowledge of the mechanism of damage by potato cyst nematodes to modelling crop development and growth, to breeding strategies and to cultural practices. Attention is paid to the interaction of the effects of nematode infection with abiotic soil factors, especially the availability of water, nutrients and the soil pH.

Short-term effects

From measurements (Fatemy et al. 1985; Haverkort et al. 1991a) it is clear that potato cyst nematodes affect water relations in potato plants. The moisture condition of a plant cell is described by its water potential (Ψ) comprising an osmotic component (Ψo) and a pressure component (Ψpi): $\Psi = \Psi o + \Psi pi$, expressed in MPa. When plant or cell moisture content decreases because transpiration exceeds water transport through the roots (due to a high evaporative demand, a dry soil or root damage) Ψ, and Ψpi decrease as well as the relative water content of the cell or the tissue (Turner 1988). Leaf water potential may be observed with a pressure chamber or psychrometrically (Brown and Van Haveren 1972) and the osmotic potential with the aid of an osmometer. Vos and Oyarzún (1987) observing potatoes growing in moist soil found typical values of Ψ and Ψo of -0.5 and -0.8 Mpa respectively. Drought coupled with infection of roots by potato cyst nematodes can decrease leaf water potentials from -0.6 MPa to about -1.1 MPa (Haverkort et al. 1991a). Fatemy et al. (1985) found that nematode infection increased stomatal resistance, reduced photosynthetic rates and reduced water use of the plants. Exogenously applied abscissic acid temporarily increased these effects (for 2 days with cv. Pentland Dell and for 6 days with the tolerant cv. Cara); apparently the plants grow accustomed to increased abscissic acid levels and the

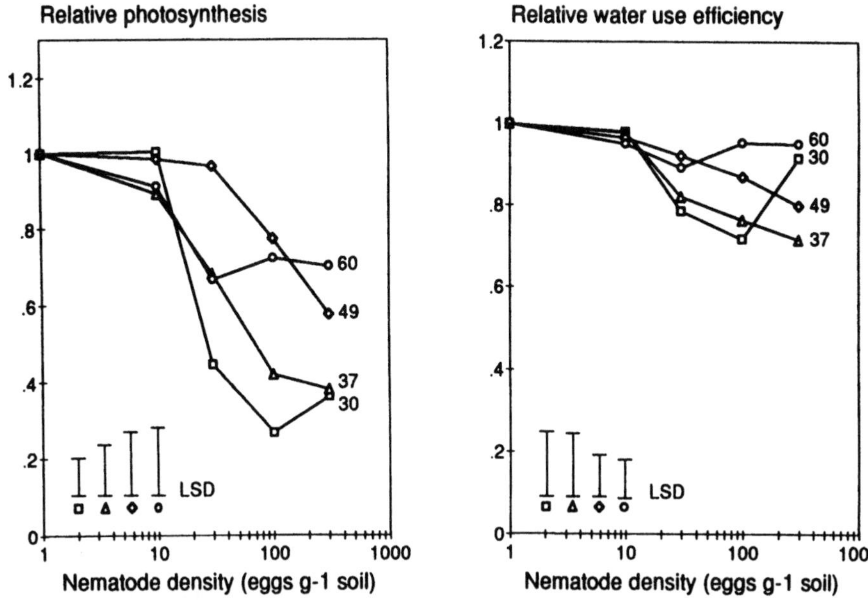

Figure 1. Photosynthetic rate and water use efficiency at 30 (□), 37 (Δ), 49 (◇) and 60 (○) days after planting at various nematode densities (from Schans and Arntzen 1991).

effect disappears. Water stress also increased abscisic acid levels so the initial effect of drought on stomatal regulation may be through hormonal effects.

Schans and Arntzen (1991) showed that infection by potato cyst nematodes decreased rates of photosynthesis of potato leaves. This effect was progressively more severe at higher nematode densities but at each nematode density, the effects appeared to be lesser with increasing age of the plant (Figure 1). The authors suggested that the gradual relative increase of photosynthesis is due to two factors. With time the attack decreases due to depletion of the inoculum, leading to less hormonal signalling as less infected root tips are present, and, with time the leaf area increases, leading to a possible dilution of the signal in the leaves, resulting in a reduced stomatal response. The authors did not record levels of ABA or other hormones. A simultaneous effect of potato cyst nematode attack on photosynthesis and on water-use efficiency (Figure 1) shows that not are only stomatal processes involved (as then the water-use efficiency increases) but that photochemical or biochemical processes are also implicated. Reduced uptake of nutrients is a major component as will be discussed later. Schans and Arntzen (1991) conclude that their observation of a decreasing effect of potato cyst nematodes on the water-use efficiency of individual leaves agrees with Evans' (1982) finding that whole plant water-use efficiency was decreased up to 32 days after planting but increased from 32 days onward. The reported relationships between gas-exchange and water-use efficiency from literature, however, have not yet proven unambiguously that the two mechanisms (signalling and root damage leading to disrupted water and nutrient uptake) operate. The methodology used and the results obtained are conflicting still and the processes involved need further elucidation.

Nematode attack (Haverkort et al. 1991a; Schans 1993), like water – and heat – stress can reduce the photosynthetic rate indirectly by closure of the stomata (prompt effect) or directly by a reduction of the photosynthetic capacity of the leaves (after a few days). As a side-effect of stomatal closure the assimilation/transpiration ratio increases as does the proportion of stable isotope 13C within the plant (Farquhar et al. 1982). No consensus exists on the primary site of the nematode effect on photosynthesis and whether photoreactions in the thylakoid membranes or biochemical reactions of the Calvin cycle are the most affected. In vivo fluorescence signals give information on the light use efficiency and on the rate of electron flow in the thylakoids. Combined with gas exchange measurements, analysis of fluorescence signals provides information on rate limitation of processes related to resistance to flow in the gas phase and to internal photosynthetic processes, but these approaches have not been applied to effects of potato cyst nematodes on potato.

Table 1. Effects of potato cyst nematode infection (15 living eggs per g soil) and drought (no water applied from emergence) on plant water relations (container experiment, 42 days after planting, Haverkort et al. 1991a) Data, within a row, with different letters differ at the 1% confidence level

Observation	Treatment		
	Control	Infected	Droughted
Leaf water potential (10^4 Pa)	6.9a	10.8b	11.8b
Diffusion resistance (s cm^{-1})	1.36a	1.51a	5.11b
Transpiration (ug H_2O cm^{-2} s^{-1})	11.0a	10.21a	4.09b
Water use efficiency (g kg^{-1})	7.34a	6.43b	9.19c
Leaf dry matter concentration (%)	8.5a	10.5b	12.3c
Stem dry matter concentration (%)	6.5a	8.0b	8.0b
Tuber dry matter concentration (%)	15.0a	15.2a	21.7b

The effects of infection by potato cyst nematode (*G. pallida*) and drought on the water relations of the potato plant are shown in Table 1. Nematodes and drought both reduced leaf water potentials as measured with a pressure chamber and increased dry matter concentrations of all plant parts. The effects of nematodes on stomatal diffusion resistance and transpiration were negligible whereas drought led to a strong increase in stomatal resistance and consequently to a strongly reduced transpiration rate (Haverkort et al. 1991a). Drought increased the water-use efficiency but, as was seen in Figure 1, potato cyst nematodes decreased the water use efficiency showing that, in contrast to drought, nematodes reduced photosynthesis more than transpiration.

During the later stages of growth, the water use efficiency of infected plants increased strongly, even above that of uninfected plants. Similar results were obtained by Evans et al. (1975): later in the season, the efficiency with which water is taken is taken up is impaired, intolerant cultivars suffering most leading to water stress and premature senesence. Haverkort et al. (1991a) concluded that at least two yield reducing mechanisms exist of which the relative importance varies with time. Firstly, reduced apparent assimilation rates which are not wholly related to a change in the water balance caused by the initial attack by the cyst nematodes. Secondly a reduced dry matter accumulation resulting from a decrease of water and/or nutrient uptake. Effects of drought and cyst nematodes (as observed in field experiments which will be discussed later) were often additive. Where they were not additive this was because infected plants used less water than uninfected plants leading to less water stress.

Long-term effects

Effects on morphology

The most convenient means of observing differences in crop development caused by environmental factors is to make comparative assesments of the proportion of the ground covered by the canopy as this is strongly correlated with the percentage of light interception. It is also highly highly correlated with the leaf area index (below 3), and with the proportion of the incoming infra-red radiation reflected by the canopy i.e. leaves are more reflective than the soil (Haverkort et al. 1991b). This is illustrated in Figure 2 where the proportion of infra-red reflectance of potato crops affected by drought (unirrigated), by potato cyst nematodes (unfumigated) and by night frost (7 weeks after planting) are shown. Ground cover and hence infra-red reflectance were reduced more strongly by nematodes than by drought. Night frost did not decrease the proportion of the ground covered by green leaves (data not shown) but it did reduce the canopy reflectance.

The effect of nematode attack and an early drought period on some morphological characteristics of potato plants grown in containers under a rain shelter is shown in Table 2. At the end of the dry period the leaf area of the infected plants was less than half that of the control 3181 cm^2 per plant instead of 8253 cm^2 per plant, explained because thicker (SLA decreased from 301 to 263 cm^2 g^{-1}) and smaller 72 instead of 150 cm^2 per leaf) leaves were formed. The effect of drought on these morphological characteristics were similar. The plants subjected to either or both stress factors remained shorter, had lower tuber yields and lower shoot/root ratios than the control plants. Four weeks after rewatering the droughted plants, droughted and undroughted plants had similar leaf areas and weights although the plants that had been droughted before still had thicker leaves and lower leaf area ratios. The unstressed plants were senescing at 70 days after planting whereas the plants which were subjected

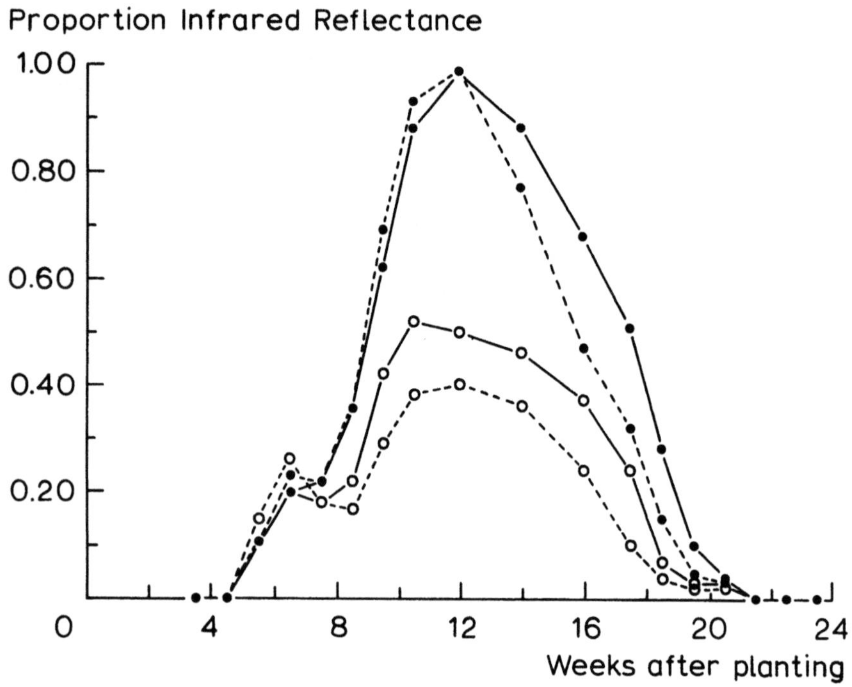

Figure 2. Proportion of infrared reflectance of potato crops subjected to drought and potato cyst nematodes. — irrigated, --- unirrigated, ● nematodes controlled, o nematodes not controlled (after Haverkort and Schapendonk 1994).

Table 2. The main effects of potato cyst nematode and drought on plant characteristics of potato cv. Mentor, 43 days after planting. Crop characteristics differed significantly $P < 0.05$ between the control and nematode or drought treatments (Haverkort et al. 1991a)

Plant characteristic	Potato cyst nematode (eggs per g soil)			
	0 moist	18.5 moist	0 dry	18.5 dry
Leaf, dry (g/plant)	27.4	12.1	20.2	10.4
Leaf area (cm^2/plant)	8253	3181	4661	2258
Leaf size (cm^2/leaf)	150	72	108	58
SLA (cm^2 g^{-1} leaf dry weight)	301	263	228	218
LAR (cm^2 g^{-1} plant dryweight)	93.1	89.0	74.2	73.1
Number of stems	5.4	4.5	5.2	4.5
Stem length (cm)	52	30	34	24
Spec. stem weight (mg cm^{-1})	62	46	62	52
Tuber dry weight (g/plant)	42	15	27	13
Number of tubers per plant	30	16	28	16
Shoot/root ratio	17.3	8.6	15.1	7.9

to an early drought stress continued to grow longer, probably because they had not yet depleted soil nutrients.

In Table 1 it was shown that similarities exist in the water relations of droughted crops and those infected by potato cyst nematodes. The same holds for the influence of the two stress factors on plant morphology. The number of tubers, however, is more strongly reduced by potato cyst nematodes (from 30 and 28 to 16 tubers per plant) than by drought The difference is only partly explained by a reduction in numbers of stems due to cyst nematode infection (from 5.4 per plant to 4.5 per plant). The shoot/root ratio in this experiment was only marginally affected by drought but strongly decreased in the presence of potato cyst nematodes, indicating that a major shift in relative dry matter allocation occurred in favour of the roots.

The recently constructed Wageningen Rhizolab (Smit et al. 1994) allows a frequent and non-destructive observation of root growth of plants subjected to various soil stress factors including nematodes and drought. Recent studies using that facility (Haverkort et al., 1994), to compare uninfected potato plants with plants infected with *Globodera pallida*, (40 eggs per g soil) showed increased root growth in the top 30 cm of soil but reduced root growth in the subsoil (Figure 3). Potato cyst nematodes initially caused the formation of less roots in the topsoil but root formation (branching) continued until 120 days after planting, whereas in the control, root formation in the topsoil stopped after 50 days after planting to allow exploitation of the subsoil.

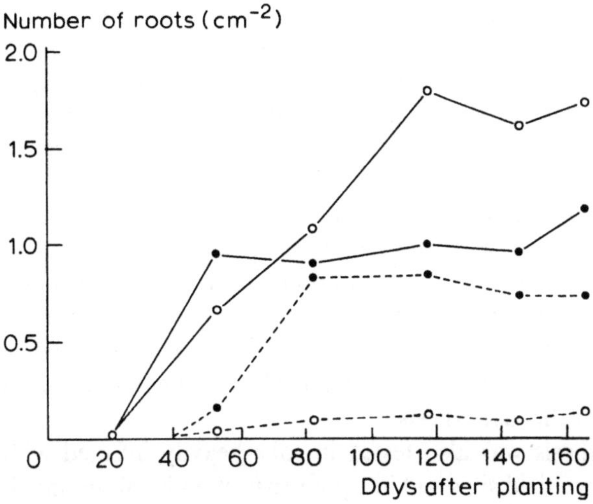

Figure 3. Mean number of roots observed per cm^2 surface of minirhizotrons with time at 0 (●) and 40 (o) living juveniles of *Globodera pallida* per g soil at 0–30 (—) and 30–100 cm (---) soil depth (Haverkort et al., 1994).

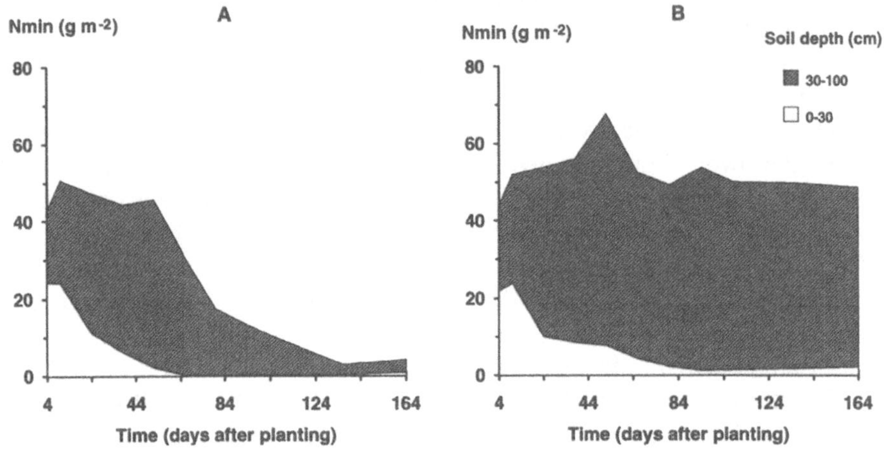

Figure 4. Total soil mineral nitrogen in the 0–30 cm and the 30–100 cm soil layers in the experiment in the wageningen Rhizotron. (A) no potato cyst nematodes present, (B) initial population of 40 living juveniles per g soil.

Effects on nutrient uptake

The total amount of soluble nitrogen (nitrate and ammonia) was determined every two weeks in the topsoil from 0–30 cm and in the subsoil from 30–100 cm soil depth (same experiment as shown in Figure 3). Without cyst nematodes, the soil profile towards crop senescence was depleted of mineral nitrogen to a depth of at least 1 m whereas, following high initial population densities there was no change in nitrogen concentration in the subsoil between 30 and 100 cm (Figure 4). The spatial heterogenity of roots (Figure 3) and nitrogen (Figure 4) in the soil (as there were hardly any roots in the subsoil so the plants could not exploit the nitrogen in that layer), caused by potato cyst nematodes apparently is an important mechanism of damage, especially in the second part of the growing season and may contribute to the earlier senescence which is observed in infected fields.

The effects of potato cyst nematode on the growth of potato were analysed in a long-term trial involving irrigation, fumigation and resistant cultivars (Trudgill et al. 1975). Based on their observations the authors hypothesized that reduced rates of top growth due to potato cyst nematodes were a consequence of root damage which decreased rates of nutrient uptake and led to chronic deficiency of the least available nutrient. The findings shown in Figures 3 and 4 are in line with this hypothesis.

This hypothesis was also tested in soils heavily infected with potato cyst nematode at two field sites and in a pot experiment by observing the interaction between rates of fertilizer and a non-fumigant nematicide. Percentage ground cover and yield of potato plants infected with potato cyst nematodes were affected much more by changes in fertilizer inputs than by nematicide (Trudgill

Table 3. Effects of nematicide and additional fertilizer on cv. Pentland Crown (pot trial 14 weeks after planting, weights in g per plant, nutrient concentrations in haulm dry matter, additional fertilizer was 2 × standard dose of N, or 2 × K or 3 × P). Weights in g per plant

Observation	With nematicide fertilizer increased				Without nematicide fertilizer increased			
	Standard	N	P	K	Standard	N	P	K
Haulm (g)	298	448*	306	251	98	265*	208*	160*
Root (g)	60	84*	77	34	11	64*	56*	22
Tuber (g)	363	372	363	357	100	158	168	142
% N	2.5	3.4*	2.5	2.4	3.3	3.5	3.3	3.3
% P	0.16	0.22*	0.26*	0.15	0.18	0.20	0.27*	0.20
% K	3.5	3.3	3.4	5.2*	2.6	3.1	2.9	4.5*

* statistically significant different ($P < 0.01$) within a nematode treatment.

1987). In one field trial the relatively tolerant cv. Cara was much less responsive to rates of either nematicide or fertilizer than the intolerant cv. Pentland Dell. Increasing fertilizer rates from 0.5 to 1.5 t per ha increased tuber yields of the unprotected cv. Pentland Dell from 0.13 to 0.82 kg per plant, compared with 1.30 to 1.60 kg per plant for those protected with a nematicide. In a pot experiment to vary P, K and N independently, Trudgill (1980) showed that nematode damage interacted strongly with the availability of N and of P. Doubling the amount of N and P more than doubled top weight in the absence of nematicide. Amounts of K had a smaller but still significant effect (Table 3). Greater effects were observed in the pot experiment than in one of the field trials possibly because of adsorption of the P in the heavy clay soil at that site.

Yield formation

As indicated before, the yield of a crop (Y) is determined by the amount of photosynthetically active solar radiation it intercepts (R), the conversion efficiency of radiation to dry matter (E), the proportion of the total dry matter in the harvested parts (H) and its dry matter concentration (D). In a formula: $Y = R\ E\ H/D$.

The value of each parameter in this simple model of crop growth is determined by a number of underlying processes. R depends on the (duration of) leaf expansion rate and light extinction by the canopy, E depends on photosynthesis and respiration rates and H depends on dry matter distribution processes related to crop development. In a series of field trials in the north east of the Netherlands in which four cultivars were either irrigated or not irrigated and either fumigated or not fumigated (leading to averages of 5 or 44 eggs per g soil) periodic harvests were made to determine the effect of drought and cyst nematodes on components of yield. The effects on infrared reflection by cv. Mentor were shown in Figure 2. Both intercepted radiation and radiation use efficiency were negatively affected by increased levels of potato cyst nematodes and drought.

Table 4. Relative values of yield components in unirrigated and unfumigated plots (mean data 1989 and 1990, values of irrigated and fumigated plots = 100) (Haverkort et al. 1992)

Year	Cultivar	Irrigated	Fumigated	Yield =	R ×	E ×	H /	D
88–90	Darwina	+	–	48	61	90	94	105
	Désirée	+	–	52	71	86	93	102
	Elles	+	–	73	86	95	102	107
	Mentor	+	–	49	57	92	100	101
89–90	Darwina	–	+	55	62	99	94	105
	Désirée	–	+	77	88	99	94	105
	Elles	–	+	80	93	90	95	101
	Mentor	–	+	73	87	97	97	111

Table 4 shows the values of intercepted radiation, radiation-use efficiency, harvest index and dry matter concentration of the unirrigated and unfumigated plots relative to those of the irrigated and fumigated controls of the four cultivars (including cv. Mentor) tested.

Apparently cyst nematodes and drought act similarly on the components of yield: yields are mainly reduced because crops intercept less solar radiation. Other factors reducing fresh tuber yields are lower efficiencies whereas reduced harvest indices and increased dry matter concentration also play a role. The cultivar least tolerant of nematode damage (Darwina) was, also least tolerant of drought and cv. Elles was most tolerant of both stress factors. Trudgill et al. (1990) found a value of 1.05 g MJ^{-1} for tolerant Cara and a value of 1.25 g MJ^{-1} for intolerant Pentland Dell which maintained ground cover for a longer period and hence intercepted more solar radiation. This phenomenon of a lower radiation use efficiency associated with a higher value of intercepted radiation is also presented in Table 4 where the unfumigated drought tolerant cv. Elles had a 5% lower E-value and a 14% lower R-value than the fumigated treatment whereas the cv. Mentor (less tolerant of potato cyst nematodes) had a 43% lower R-value when not fumigated. With drought this phenomenon is even more apparent: cv. Darwina lost only 1% of its efficiency but 38% of its intercepted radiation when not irrigated. Cultivars that are able to maintain a well developed canopy apparently are not able to keep it functioning at maximum rates as, during stress, photosynthetic rates decline whereas respiration does not decline. Beside long-term visible effects of stress factors on crop morphology and measurable effects on intercepted radiation and radiation use efficiency, stresses affect the transpiration efficiency, 13C-discrimination and mineral uptake. Haverkort et al. (1991a; Table 1) reported transpiration efficiencies of 7.34 g kg^{-1} in plants that were not subjected to stress factors but a value of 6.43 g kg^{-1} in potato plants subjected to cyst nematodes and 9.19 g kg^{-1} in plants subjected to drought.

Different cultivars react differently to potato cyst nematode infection in the development of ground cover as was shown by Mulder (1994). Comparison of

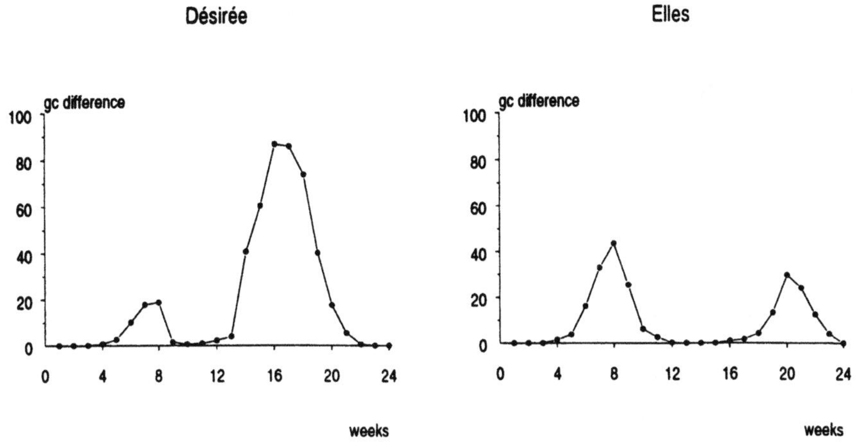

Figure 5. Difference in percentage ground cover (gc difference) between lightly and heavily infected crops of Désirée and Elles (Mulder 1994).

ground cover values of healthy crops and infected crops shows (Figure 5) that ground cover development of cv. Désirée was much less affected early in the growing season than in the later stages of crop growth, whereas with cv. Elles (tolerant cultivar), the reverse was observed.

Potato cyst nematodes and drought both decrease 13C-discrimination indicating that nematodes disrupt plant water relations (Haverkort and Valkenburg 1992). Drought and potato cyst nematodes, however, are not identical in their effects as are shown by the reduced water use efficiency following nematode infection and increased water-use efficiency following drought. Apparently, the negative effect of potato cyst nematode infection on plant processes such as photosynthesis and dry matter allocation to the roots is greater than the increase in water use efficiency. Another dissimilarity between the effects of drought and potato cyst nematodes is that drought decreases Ca-uptake whereas potato cyst nematode infection increases it (Fatemy and Evans 1986). Calcium is taken up by the root tips and is transported within the plant with the transpiration stream; a process that is enhanced where synchytia break the endodermis allowing apoplastic movement. As is shown in Figure 3, the ramification of the roots induced by potato cyst nematodes leads to many more root tips in the topsoil and hence contributes to the increased calcium uptake.

Tolerance of potato cyst nematodes

Resistance to potato cyst nematodes is expressed as the ratio of final to initial nematode population density. Resistance is of importance in reducing the nematode population density but does not necessarily reflect the degree of damage which is to be expected. Tolerance of potato cyst nematodes in one cultivar, is defined as a smaller reduction in yield at high nematode densities

Table 5. Relations between soil pH, tuber yields, dry matter and initial density of *Globodera pallida* (Haverkort et al. 1993)

pH_{KCl}	Tuber dry yield (kg m^{-2})	Initial density (eggs g^{-1} soil)
4.5	1.09	17.5
5.0	1.04	13.8
5.5	1.03	14.4
6.0	0.92	11.3
6.5	0.75	8.8
$LSD_{0.05}$	0.11	8.3

than in another cultivar. It is a relative measure. Evans and Haydock (1990) reviewed some of the (curvi-)linear relations between the initial nematode density and associated tuber yield losses conceived by authors such as Brown, Oostenbrink and Seinhorst. Although such relations may be valid when expected damage has to be forecast, they convey no information as to the mechanism of damage occurred. Evans and Haydock (1990) in their review, attributed differences in tolerance between cultivars to differences in nematode hatching, invasion and colonisation of roots, root vigour, form of the root system, local reaction of roots to nematode attack, physiological responses of plants to nematode attack, and to interactions with other organisms. They found no evidence that tolerance is consistently greater to one species of potato cyst nematode (*Globodera rostochiensis*) than the other (*Globodera pallida*). Damage due to potato cyst nematodes has been reported to be increased by biotic soil-borne factors such as *Rhizoctonia solani*, *Verticillium dahliae* and *Verticillium albo-atrum*. Abiotic soil-related factors, however, may be of equal or greater importance than biotic factors. Mulder (1994) compared relative damage due to potato cyst nematodes with that caused by soil type, soil fertility, soil water availability and soil pH. Higher soil pH-values appeared to be especially important and Haverkort et al. (1993) found a strong interaction between yield losses and pH in a light, sandy soil in the north-east of the Netherlands. Through annually liming, plots with pH levels of 4.5 to 6.5 were obtained in a reclaimed peat soil with about 22% organic matter. Potato yields (Table 5) in these plots decreased from 45 t per ha at pH 4.5 to 33 t per ha at pH 6.5, whereas the nematode population densities decreased from about 18 to 9 juveniles per g soil. In a container experiment with the same soil a strong interaction was observed between soil pH and the presence of potato cyst nematodes. Tuber yields were about 11% lower at pH 6.5 than at 4.5 in the absence of nematodes, but 44% lower when an initial population of 27 juveniles per g soil was present. Plants grew more vigourously at pH 4.5 even though initial nematode densities were higher than at pH 6.5; the more vigourous (root) growth at the lower pH values probably contributed to higher multiplication rates that were observed.

When modelling the effect of potato cyst nematodes on crop growth, development and tuber yield, such interactions should be taken into account in

Table 6. Effect of inoculation with 10 000 eggs per small pot on root length of cultivars differing in tolerance (Trudgill and Cotes 1983)

Cultivar	Nematodes	Root length (cm/pot)		
		Main	Lateral	Total
Maris Piper	not inoculated	138	384	522
(tolerant)	inoculated	275	490	766*
Epicura	not inoculated	118	298	416
(intermediate)	inoculated	87	307	397
Maris Anchor	not inoculated	179	973	1153
(intolerant)	inoculated	219	260*	480

* significantly different ($P < 0.05$) per cultivar and root class.

order to accurately estimate model parameter values of uninfected and infected crops. Breeding new cultivars with resistance to potato cyst nematodes and capable of high yields in infested soils needs screening techniques for tolerance. Evans and Haydock (1990) reviewed several direct and indirect methods of screening for tolerance in the field and in pots. Trudgill and Cotes (1983) showed that an infected tolerant cultivar had a greater total root length in pots than an uninfected intolerant cultivar (Table 6).

Arntzen (1993) found no relation between tolerance and maturity class of potato as indicated in the Netherlands Recommended list of Varieties. However, Haverkort et al. (1992) and Trudgill et al. (1990) concluded that culivars which develop abundant foliage and intercept more light (generally late-maturing cultivars) usually suffer less damage from potato cyst nematode infection than cultivars which develop less foliage and hence keep the ground covered for a shorter period (early maturing cultivars). Arntzen also found that, in the greenhouse, the reduction of the leaf area due to potato cyst nematode infection was mainly because the plants formed smaller leaves rather than because they formed fewer leaves. He reported a good correlation between reduction in total plant weight in pots in the greenhouse and yield losses in the field. The depression of growth in pots was mainly associated with the rate of hatching of the nematodes and the differences in growth of the roots, which became apparent within a few days in vitro.

Screening techniques in pots may be useful to determine the degree of tolerance of potato cyst nematodes in potato genotypes. They are not useful for crop ecological purposes with a view to mechnistic modelling of the development and growth of field grown crops. In pots the inoculum and nutrients may be depleted sooner than in the field, and fluctuations in water availability are likely to occur within a shorter time span. In pots it is often observed (Fasan and Haverkort 1991) that infected plants, after an initial depression of growth continue to grow for a longer period than uninfected plants once the inoculum is depleted as they continue to benefit from nutrients present within the soil while all the nutrients have already been taken up by the healthy plants.

Discussion

Three major processes determine crop growth and their contribution to reduced crop growth following potato cyst nematode infection is of greatest importance during two distinct phases in the growing season.

In the presence of potato cyst nematodes, the three following effects are found:

1. reduced assimilation rates,
2. dry matter allocation unfavourable to the leaves, and
3. morphological changes unfavourable to leaf expansion.

Considering the interception of light and its use, the phases of crop growth that are most crucial to yield formation are:

1. canopy establishment from emergence to canopy closure, and
2. crop senescence at the end of the season when light interception is reduced from 100% to 0%.

The three major effects on crop physiological processes operate in both those phases and the mechanism of damage to the crop is explained as follows. For maximum yields, the time between planting and emergence should be as short as possible. The time between emergence and 100% light interception should also be as short as possible then the period of 100% light interception should be as long as possible. The time between planting and emergence is not affected by potato cyst nematode (Haverkort et al. 1992) but the initial stage of crop growth between emergence and 100% light interception (canopy establishment) and the final stage of crop growth when ground cover declines from 100 to 0% (canopy senescence), are both affected.

As was shown, potato cyst nematodes lead to a direct reduction of photosynthesis. This reduction may be mediated initially by a hormonal signal such as abscissic acid. Water uptake, by infected crops is also reduced leading to symptoms of drought in the plant (stomatal closure, lowered leaf water potential and reduced transpiration). However, it is evident that rates of photosynthesis are affected more than those of transpiration. Reduced mineral uptake (nitrogen and phosphorus) may also lead to a decrease in the efficiency of the photosynthetic apparatus (rubisco) thus contributing to reduced radiation use efficiency. Later canopy closure following potato cyst nematode infection may be due mainly to disrupted water relations, whereas in the second part of the growing season assimilation is reduced because roots of infected crops penetrate less deeply into the soil and in the second half of the growing season the topsoil is becoming depleted of water and nutrients. The resulting reduction in assimilation rates leads to a reduction in the rate of appearance of new leaves (beside leaf shedding contributing to reduced radiation interception) and to reduced rates of tuber bulking due to the combined effect of reduced radiation interception and lower radiation-use efficiency. A reduction in the number of tubers formed following infection with potato cyst nematode may also contribute to the reduction of assimilation rates as fewer sinks are formed which, through feed-back mechanisms, render the source less effective.

Potato cyst nematode damage is associated with a relative increase in dry matter partitioning to the roots, as is shown by decreased shoot/root ratio and increased rooting in the Wageningen Rhizolab. Dry matter allocation favouring the roots decreases the amount available to be invested in the leaves thereby reducing the amounts of intercepted radiation and decreasing crop growth rates. The timing of tuber initiation is hardly altered by potato cyst nematodes which means that tuber initiation in infested crops occurs with smaller plants than in healthy crops, again leading to a pattern of dry matter allocation unfavourable to the leaves contributing to the later closure of the canopy and to an earlier onset of senescence.

To have a maximum effect on ground cover and radiation interception, the amount of dry matter which is allocated to the leaves should be spread over the soil surface as widely as possible. This means that a larger specific leaf area (the formation of thinner leaves) a higher leaf water content (leading to larger leaves) and more and longer stems (allowing a better spread of the leaves over the soil surface with less overlap) are the major factors contributing to light interception with a given amount of dry matter partitioned to the plant tops. Potato cyst nematodes, however, unfavourably influence these four factors. As was shown in the previous paragraphs potato cyst nematodes damaged plants have smaller specific leaf areas, lower leaf water contents, less and shorter stems resulting from disrupted water relations and decreased uptake rates of N, P an K by the damaged roots.

Finally, marketable yields and crop profitability, do not only depend on the amount of dry matter partitioned to the tubers but also to tuber size distribution. Infected crops yield smaller tubers because of reduced crop growth rates, and, because tuber water content is lower.

The relative importance of three processes in their influence on tuber yield varies. Which underlying process influences assimilation (signalling, lack of water or nutrients), partitioning (favouring the roots or the reproductive organs) and morphology (leaf thickness and stem length), depends on many factors. Between emergence and canopy senescence, lack of nitrogen is likely to become increasingly limiting for optimal photosynthesis and may gradually replace the immediate signalling effect which is likely to play a major role in reducing photosynthesis during the early stages of crop growth. Dry matter partitioning favouring the roots initially withdraws assimilates which in uninfested conditions would contribute to leaf growth. In the second half of the growing season dry matter allocation to the tubers probably plays a more significant role: tuber initiation in smaller plants leads to an earlier allocation of virtually all assimilates that are produced to the tubers and consequently leads to earlier senescence as no assimilates are left for the formation of new leaves and/or maintenance of the old ones. Thicker leaves initially lead to later closure of the canopy but the earlier reduction in light interception following potato cyst nematode infection may rather be due to the reduced number of leaves that are formed.

The relative importance of the three major processes to yield reduction following potato cyst nematode infection, vary with cultivar (degree of tolerance and physiological processes on which this tolerance is based), cultural practices (soil type, pH, soil moisture and nutrient availability throughout the season) and environmental factors that the grower cannot alter, such as temperature and radiation. The many interactions between plant growth processes and environmental and genetic factors cannot possibly be quantified within a single comprehensive simulation model of crop growth and development. Modelling, therefore will focus on part of the interactions such as, for instance, root growth and nitrate uptake as influenced by cultivar, soil type, and moisture availability. At a very high level of integration such as shown in the fresh yield formula based on radiation interception, radiation use efficiency and harvest index, models may be used in simple crop guidance systems.

Acknowledgements

D.L. Trudgill acknowledges the financial support of the Scottish Office of the Agriculture and Fisheries Department

References

Arntzen F K (1993) Some Aspects of Resistance to and Tolerance of Potato Cyst Nematodes in Potato. Ph.D. Thesis, Wageningen, The Netherlands.

Brown R W & Haveren B.V. van (1972) Psychrometry in water relations research. Utah Experimental Station, Logan, 45 pp.

Evans K (1982) Water use, calcium uptake and tolerance of cyst-nematode attack in potatoes. Potato Research 25:71–88.

Evans K and Haydock P P J (1990) A review of tolerance by potato plants of cyst nematode attack with consideration of what factors may confer tolerance and methods of assaying and improving it in crops. Annals of Applied Biology 117:703–740.

Evans K, Parkinson P L, Trudgill D L (1975) Effects of potato cyst nematodes on potato plants III. Effects on the water relations and growth of a resistant and a susceptible variety. Nematologica 21:273–280.

Farquhar G D, O'Leary M H, Berry J A (1982) On the relationship between carbon isotope discrimination and intercellular carbondioxide concentration in leaves. Australian Journal of Plant Physiology 9:121–137.

Fasan T, Haverkort A J (1991) The influence of cyst nematodes and drought on potato growth. 1. Effects on plant growth under semi-controlled conditions. Netherlands Journal of Plant Pathology 97:151–161.

Fatemy F, Evans K (1986) Effects of globodera Rostochiensis and water stress on shot and root growth and nutrient uptake of potatoes. Revue de Nematologie 9:181–184.

Fatemy F, Trindler P K E, Wingfield J N, Evans K (1985) Effects of Globodera rostochiensis, water stress and exogenous abscissic acid on stomatal function and water use of Cara and Pentland dell potato plants. Revue de Nematologie 8:249–255.

Haverkort A J, Boerma M, Velema R, Van de Waart M (1992) The influence of drought and cyst nematodes on potato growth. 4. Effects on crop growth under field conditions of four cultivars differing in tolerance. Netherlands Journal of Plant Pathology 98:179–191.

Haverkort A J, Fasan T, Van de Waart M (1991a) The influence of cyst nematodes and drought on potato growth. 2. Effects on plant water relations under semi-controlled conditions. Netherlands Journal of Plant Pathology 97:162–170.

Haverkort A J, Groenwold J & Waart M van de (1994) The influence of cyst nematodes and drought on potato growth. 5. Effects on root distribution and nitrogen depletion in the soil profile. European Journal of Plant Pathology 100:381–394.

Haverkort A J, Mulder A & Waart, M van de (1993) The effect of soil pH on yield losses caused by the potato cyst nematode *Globodera pallida.* Potato research 36:219–226.

Haverkort A J, Schapendonk A H C M (1994) Crop reactions to environmental stress reactions. In Struik P C (ed) Plant production on the threshold of a new century. Kluwer Academic Publishers: 339–347.

Haverkort A J, Uenk D, Veroude H, Van de Waart M (1991b) Relationships between ground cover, intercepted solar radiation, leaf area index and infrared reflectance of potato crops. Potato Research 34:113–121.

Haverkort A J, Valkenburg G W (1992) The influence of cyst nematodes and drought on potato growth. 3. Effects on carbon isotope fractionation. Netherlands Journal of Plant Pathology 98:12–20.

Mulder A (1994) Tolerance of the Potato to Stress Associated with Potato Cyst Nematodes, Drought, and pH. An Ecophysiological Approach. Thesis, Wageningen, The Netherlands.

Schans J (1993) Population Dynamics of Potato Cyst Nematodes and Associated Damage to Potato. Ph.D. Thesis, Wageningen, The Netherlands.

Schans J, Arntzen F K (1991) Photosynthesis, transpiration, and plant growth characters of different cultivars at various densities of *Globodera pallida.* Netherlands Journal of Plant Pathology 97:297–310.

Smit A L, Groenwold J, Vos J (1994) The Wageningen Rhizolab – a facility to study soil-root-atmosphere interactions in crops. II. Methods of root observations. Plant and Soil 161:289–298.

Trudgill D L (1987) Effects of rates of nematicide and of fertilizers on the growth and yields of cultivars of potat which differe in their tolerance of damage by potato cyst nematode (*Globodera rostochiensis* and *G. pallida.* Plant and Soil 104:235–243.

Trugill D L (1980) Effects of Globodera rostochiensis and fertilizers on the mineral nutrient content and yield of potato plants. Nematologica 26:234–254.

Trudgill D L, Cotes L M (1983) Tolerance of potato cyst nematodes (*Globodera rostochiensis* and *G. pallida*) in relation to the growth and efficiency of the root system. Annals of Applied Biology 102:385–397.

Trudgill D L, Evans K, Parrot D M (1975) Effects of potato cyst nematodes on potato plants. Nematologica 21:183–191.

Trudgill D L, Marshall B, Phillips M (1990) A field study of the relationship between pre-planting density of *Globodera pallida* and the growth and yield of two potato cultivars differing in tolerance. Annals of Applied Biology 117:107–118.

Turner N C (1988) Measurement of the plant water status by the pressure chamber technique. Irrigation Research 9:289–308.

Vos J, Oyarzún P J (1987) Photosynthesis and stomatal conductance of potato leaves. – Effects of leaf age, irradiance and leaf water potential. Photosynthesis Research 11:253–264.

12. Modelling the interaction between potato crops and cyst nematodes

M. VAN OIJEN[1]*, F.J. DE RUIJTER[1] and R.J.F. VAN HAREN[2]

[1] *Research Institute for Agrobiology and Soil Fertility (AB-DLO), P.O. Box 14, 6700 AA Wageningen, The Netherlands*

[2] *Research Institute for Plant Protection (IPO-DLO), P.O. Box 9060, 6700 GW Wageningen, The Netherlands*

* *Present address: Wageningen Agricultural University, Dept. of Theoretical Production Ecology, P.O. Box 430, 6700 AK Wageningen, The Netherlands*

Abstract. Simulation modelling of the effects of cyst nematodes on potato crop growth is still in its infancy. The first attempts to simulate nematode population dynamics together with host growth were done in the 1970s. For many years the models focussed on the rate of growth of the nematode population, while using oversimplified, descriptive functions for host growth and its response to nematodes. Further development was hampered by lack of information on the physiological damage mechanisms involved.

However, recent years have witnessed the development of sophisticated combination models that treat both pest and host in a more realistic manner. It is becoming increasingly clear that cyst nematodes affect both the carbon and nutrient metabolism of their host. Possible ways to incorporate the various damage mechanisms into simulation models are discussed.

Introduction

The use of statistical regression models, linking crop yield to pre-planting densities of nematodes, has been reviewed elsewhere (Elston et al. 1991). In this chapter we will restrict ourselves to simulation models. In such models the physiological processes involved in the interaction of crops with nematode populations in soil and roots are simulated dynamically. We will focus mainly on the role of the models in accounting for nematode effects on host growth.

The earliest simulation models were developed in the 1970s. In a model by Ferris (1976; 1978) of *Meloidogyne arenaria* in grapevine (*Vitis vinifera*), host growth rate is temperature-driven but decreases in inverse proportion to the fraction of the root system infected by the nematodes. In contrast to this simple approach to host growth, the model included much detail of the nematode life cycle, and its dependence on various environmental conditions. This approach to plant-nematode simulation modelling, involving much complexity for the nematode but only very simple descriptive functions for host growth, remained the common approach during the following years (McSorley et al. 1982). Jones (Jones and Perry 1978; Jones et al. 1978; Jones and Kempton 1980) produced such a model for cyst nematodes, including *Globodera rostochiensis*, the golden potato cyst nematode. One of the new features of Jones' model study was the incorporation of nematode genetics, which allowed application of the model to

A.J. Haverkort and D.K.L. MacKerron (eds.), Potato Ecology and Modelling of Crops under Conditions Limiting Growth, 185–196.

the problem of selection for virulence. The study also showed how simulation models could be used to study the effect of nematicides, resistant cultivars in various rotation schemes, and competition between nematode species on nematode populations.

For some years there remained a gap between modelers of nematode population dynamics and crop growth modelers. At a conference in 1982, nematologists expressed the wish to bridge the gap: 'if we wish to build valid plant models, we had better bring in some agronomists and work closely with them' ...'they had models of nutrition, photosynthesis and growth and carbon allocation and no one uses the models' (Freckman 1982, pp. 200–201). The need to couple pests to crop growth simulations was reaffirmed by Boote et al. (1983), who indicated how a soybean crop growth model might easily be adapted for that purpose.

However, simulation models in which a very detailed nematode submodel was coupled to an oversimplified crop model remained the trend (e.g. Bird et al. 1985; Caswell et al. 1986; Schmidt 1992; Van Haren et al. 1994). An exception was the model of Ward et al. (1985), who incorporated the dynamics of potato cyst nematodes into the sophisticated crop growth model SUCROS (Van Keulen et al. 1982). Unfortunately, they did not take full advantage of their crop growth simulation in that they restricted the effects of the nematodes to altered assimilate partitioning, while neglecting effects on photosynthesis and leaf senescence. SUCROS was later used by Den Toom (1990) to examine hypotheses about relations between populations of the ectoparasitic nematode

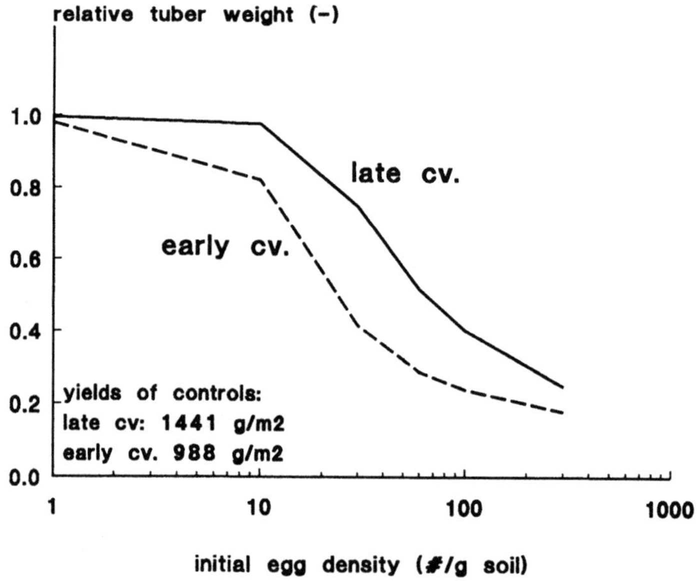

Figure 1. Simulated final tuber yields at different levels of soil nematode density, for an early and a late maturing cultivar. (Source: Schans, 1993).

Tylenchorynchus dubius and damage to *Lolium perenne*. Den Toom's model study showed that *T. dubius* probably impaired the growth of its host by obstructing water uptake.

Wallace (1987) reviewed the literature on the effects of nematodes on photosynthesis and strongly advocated incorporating the effects into existing crop growth models. This was finally done by Schans (1993) who simulated the population dynamics of potato cyst nematodes in great detail in his model. Again, SUCROS was used to simulate host growth. Schans (1991) had earlier concluded from experiments that invasion of roots by second stage juvenile potato cyst nematodes triggers an hormonal signal from root to shoot, causing stomatal closure and impairing photosynthesis. He incorporated this damage mechanism in his model by decreasing photosynthetic rate proportionately to the ratio of root penetration rate to leaf area index. By using this ratio he accounted both for stimulation of hormone production in penetrated roots and for dilution of the hormonal signal in the leaf tissue. Schans used his model to explain several observations on nematode-related damage in potato crops. His model accounted very well for the relatively high level of tolerance of late maturing cultivars (Figure 1). The physiological explanation for the tolerance by late cultivars, brought out nicely by the model, was that late cultivars produce a greater leaf area that dilutes the hormone more than in early cultivars, and that late cultivars only start tuber bulking when most of the root invasion by nematodes has ended.

Schans' model is very complex but still lacks some of the damage mechanisms identified by Haverkort and Trudgill (1995). In the remainder of this chapter we will show how a simpler model can be developed, which can give a more comprehensive treatment of damage, despite its simplicity.

Physiological effects to be incorporated in a model for potato cyst nematodes

Crop growth rate can be analysed as the product of light interception by the crop and the efficiency with which intercepted light is used to produce biomass (Monteith 1977). This type of analysis has also proved useful to study the effects of potato cyst nematodes on potato crop growth (Trudgill et al. 1990; Haverkort and Trudgill 1995). Nematodes primarily affect leaf area dynamics, i.e. light interception, but light-use efficiency is reduced as well. Haverkort and Trudgill (1995), reviewing the literature, identified four mechanisms through which cyst nematodes affect leaf area dynamics and LUE of potato crops:

1. Nematode infection reduces the specific leaf area (SLA);
2. Allocation of assimilates to growth of leaves and stems is decreased in favour of allocation to the root system;
3. Leaf senescence is accelerated;
4. Photosynthesis is affected resulting in a decreased LUE.

The extent to which these four mechanisms contribute to loss of yield will differ, so a simulation model need not necessarily contain all four. In fact, we will show

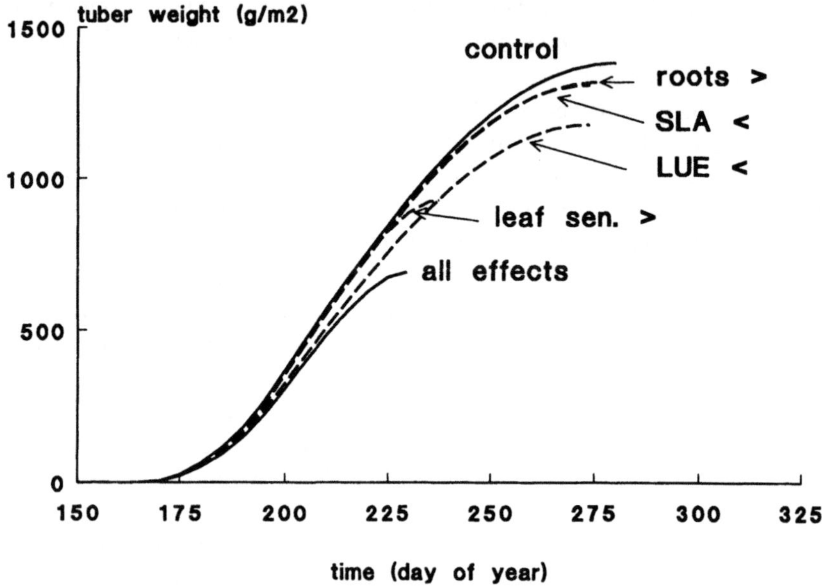

Figure 2. Simulated time course of tuber dry matter growth in a potato crop affected by various numbers of damage mechanisms: zero ("control"), one (increased allocation to roots, decreased SLA or LUE or accelerated leaf senescence: "roots>", "SLA<", "LUE<" and "leaf sen.>", respectively) or all four damage mechanisms simultaneously ("all effects").

how to use a simple simulation model to quantify the importance of the four mechanisms. For that purpose we applied the potato crop growth simulation model LINTUL (Spitters and Schapendonk 1990, where a program listing can be found). In LINTUL, crop growth rate is calculated as the product of light interception and LUE, both assimilate partitioning and leaf senescence depend on thermal time, and leaf area is calculated from leaf weight using a constant SLA. For the magnitude of the four damage effects we took a 10% reduction of SLA and LUE, and a doubling of leaf senescence rate and allocation to roots. These changes are typical for potato crops growing on heavily infested soils (Haverkort and Trudgill 1995; Van Oijen et al. 1995a). The model was not extended with a nematode submodel or with calculation of root length density, since the effects of nematodes were treated as forcing functions. A similar approach, also using standard crop growth models to identify the major damage mechanisms in a plant-nematode system, was used by Boote et al. (1993) for soybean root-knot nematodes.

The simulation results are shown in Figure 2. Obviously, acceleration of leaf senescence and, to lesser extent, reduction of LUE, account for most of the final yield loss. We conclude that the effect of nematode infection on SLA need not be incorporated in a potato cyst nematode model. Stimulated allocation of assimilates to roots also has little *direct* effect on crop growth potential (Figure 2). However, if we plan to couple the crop growth model to a simulation model of nematode population dynamics, root growth should be accurately simulated so the effect of nematodes on allocation of assimilates to roots must still be incorporated. For the same reason the effect of nematodes on the acceleration of root senescence (not discussed by Haverkort and Trudgill 1995) should also be included in the model.

A simple simulation model for potato cyst nematodes

We will now show how the damage mechanisms, listed in the previous paragraph, can be incorporated in a simple crop growth model. A preliminary version of such a model has been presented before (Van Oijen et al. 1993). For the crop growth model we choose LINTUL, already referred to above. A complete listing of this model is given by Spitters and Schapendonk (1990), but for convenience the six main equations of LINTUL are repeated here (see Table 1: Eqs. 1–6).

To calculate the rate of penetration of roots by second-stage juveniles we added three equations to the model (Table 1: Eqs. 7–9), which in fact constitute a simplified representation of the model of population dynamics given by Schans (1993). These equations are based on the following three assumptions: (1) the specific root length is constant; (2) root length is randomly distributed in the soil, and (3) rate of penetration of the root system by second-stage juveniles is proportional to both the nematode density in the soil and the rate of increase of soil volume colonized by roots.

Finally, we need four more equations to quantify the damage mechanisms selected in the previous paragraph (Table 1: Eqs. 10–13). We will consider these in more detail. Firstly, the acceleration of leaf senescence is assumed to be proportional to the ratio of crop growth rate and root length density (Eq. 10). This reflects the fact that nematode-induced reduction of root length may lead to an insufficient supply of nutrients to sustain growth and maintain leaf area (Haverkort and Trudgill 1995).

Secondly, we assume that root death is proportional to the density of recent penetrations by second-stage juveniles in the root system (Eq. 11). It is possible that later stages of the nematodes present in the roots may also contribute significantly to root damage. We did not quantify this possibility in the model since it would have involved increasing model complexity without major effects on model behaviour.

The third and fourth damage mechanisms incorporated in the model are increased allocation of assimilates to the roots (Eq. 12) and decreased LUE (Eq. 13). We assume that both mechanisms depend on the ratio of penetration of roots to the leaf area index. This represents, as in the model of Schans (1993), the hypothesis that both effects are caused by an hormonal messenger from roots to the shoot, the production of which, probably, is proportional to root penetration rate.

Obviously, the damage mechanisms could be modeled in a more comprehensive manner, by adding state variables for plant nutrient and

Table 1. Basic equations for a simple simulation model of growth of potato crops infested by cyst nematodes

(1)	LAI	$= W_l \times SLA$
(2)	$I_{intercepted}$	$= I_0 \times (1 - e^{-k \times LAI})$
(3)	CGR	$= I_{intercepted} \times LUE$
(4)	$d\,W_l/dt$	$= f_l \times CGR - Senescence_l$
(5)	$d\,W_r/dt$	$= f_r \times CGR - Senescence_r$
(6)	$d\,W_t/dt$	$= f_t \times CGR$
(7)	L_r	$= W_r \times SRL$
(8)	$V_{rhizosphere}$	$= V_{total} \times (1 - e^{-Lr \times c/Vtotal})$
(9)	Penetration	$= k_1 \times P_i \times d\,V_{rhizosphere}/dt$
(10)	$Senescence_l$	$= k_2 \times CGR/L_r$
(11)	$Senescence_r$	$= k_3 \times Penetration/W_r$
(12)	f_r	$= k_4 \times Penetration/LAI$
(13)	LUE	$= LUE_{max} - k_5 \times Penetration/LAI$
Where:	LAI	= Leaf Area Index ($m^2\,m^{-2}$)
	W_l, W_r, W_t	= Weight of leaves, roots, and tubers ($g\,m^{-2}$)
	SLA	= Specific Leaf Area ($m^2\,g^{-1}$)
	I_0, $I_{intercepted}$	= Incident and intercepted light ($MJ\,m^{-2}\,d^{-1}$)
	k	= Light extinction coefficient (−)
	CGR	= Crop Growth Rate ($g\,m^{-2}\,d^{-1}$)
	LUE, LUE_{max}	= Light-Use Efficiency under actual and optimum conditions ($g\,MJ^{-1}$)
	f_l, f_r, f_t	= Partitioning coefficients, dependent on thermal time (for f_r see also Eq. 12)
	L_r	= Root length ($m\,m^{-2}$)
	SRL	= Specific Root Length ($m\,g^{-1}$)
	$V_{rhizosphere}$, V_{total}	= Volume of rhizosphere and total rooting zone ($m^3\,m^{-2}$)
	c	= Cross-sectional area of the rhizosphere surrounding the roots (m^2)
	$Senescence_l$, $Senescence_r$	= Senescence rate of leaves and roots ($g\,m^{-2}\,d^{-1}$)
	Penetration	= Rate of penetration of roots by second-stage juveniles ($m^{-2}\,d^{-1}$)
	P_i	= Soil nematode population density (m^{-3})
	k_1, k_2, k_3, k_4, k_5	= proportionality constants (−, $m\,m^{-2}$, $g^2\,m^{-2}$, $m^2\,d$, $g\,MJ^{-1}\,m^2\,d$)

hormonal status and for nematodes in another than the second stage. However, we refrained from doing so because of the scarcity of data on these processes, the associated problems with model parameterization, and the experience that added complexity need not necessarily improve model performance.

Preliminary tests of the model

The first test of the model would be to check its ability to simulate plant-nematode relations observed in the field. For this purpose we collected

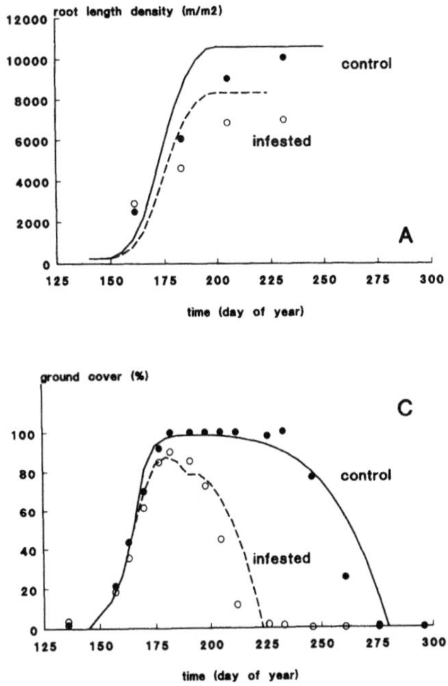

Figure 3. Simulated time courses for cultivar Darwina on infested soil and in a control treatment (points: experimental observations, lines: simulations).
A: root length; B: root penetration by second-stage juveniles; C: ground cover by green foliage; D: tuber dry matter growth.

information from a field experiment conducted in 1991 on a sandy soil infested with *G. pallida*, in the North of the Netherlands (Van Oijen et al. 1995a,b). In that experiment cultivar Darwina was subjected to the prevailing high nematode density (about 15 living juveniles cm^{-3}), and to a control treatment where most nematodes had been killed by nematicide. The model used local weather as input, and the crop growth parameters were maintained at standard values for Dutch growing conditions. The parameters that quantify the plant-nematode relationships (k_1-k_5 in Eqs. 9–13 of Table 1) were fitted to the measurements. Therefore, in this simulation only the control treatment represents an independent check of the model, or rather of its crop growth part, while the simulation of the interaction with the nematodes can only show the ability of the model to capture the complex response of the crop to the nematode in a few simple equations. The results of the simulations are given in Figure 3. We consider that the model simulates the time courses of root length, penetration of roots, ground cover and tuber growth quite well, for both the control treatment and the treatment with nematode infestation. These results do not suggest the need for a more complex model.

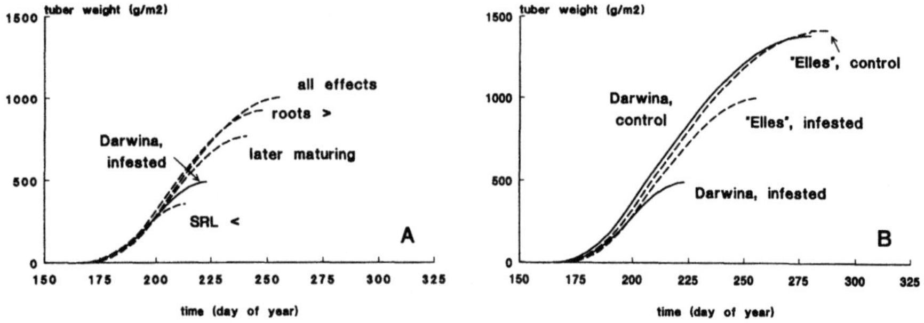

Figure 4. Simulated time courses of tuber growth for the nematode-sensitive cultivar Darwina, with or without hypothetical changes in plant characteristics. Each change corresponds to an observed characteristic of the nematode-tolerant cv. Elles.
A: the effect on growth in infested soil of changing one plant characteristic (increased allocation to roots, delayed maturation, decreased specific root length (SRL)), or all three combined; B: the effect of changing all three characteristics, simultaneously, on growth in infested soil and in a control treatment.

Application of the model to the evaluation of tolerance characteristics

As a further test of the model we analysed its ability to explain the high level of nematode-tolerance of cultivar Elles, from knowledge of plant characteristics. Elles differs from cultivar Darwina by virtue of its later maturity, its greater root-shoot ratio and its lower specific root length. In contrast to cultivar Darwina, Elles hardly suffered any loss of yield under the experimental conditions described earlier for cultivar Darwina (Van Oijen et al. 1995a). We used the simulation model to see whether the typical characteristics of cultivar Elles fully accounted for its maintenance of yield. The results of the simulation are shown in Figure 4. It seems that allocating a greater proportion of biomass to roots is indeed advantageous, as is the late maturation. Regarding the effect of late maturation, our simple model thus confirms the results of the complex model of Schans (Figure 1). The low specific root length of cv. Elles, on the other hand, only increases loss of yield. When all three characteristics are combined in one model run, we see that loss of yield is much reduced compared to cultivar Darwina. However, the simulated tolerance of nematodes is still only partial, whereas the measurements for cv. Elles showed complete tolerance. The difference between measurements and simulations can be explained in various ways. Cultivar Elles may show less stimulation of hormone production in response to nematode penetration, or simply be less sensitive to the signal. There are no data to prove or disprove this. Alternatively, root death and associated reduction of tissue nutrient concentration may be less pronounced in cultivar Elles than in Darwina, although this is not confirmed by experimental data (Van Oijen et al. 1995b). Whichever explanation is true, and irrespective of the validity of our model, the model does already indicate hypotheses for investigation.

Concluding remarks

In this chapter we have concentrated on the use of simulation models for examining the physiological processes underlying nematode damage, and the differences among cultivars therein. A simple model was shown to be useful in identifying the major damage mechanisms, and in assessing the importance of various host characteristics for tolerance of nematodes. The simulation models can, of course, also be used to investigate the effects of the environment on the damage relations. The relation between pre-planting density of cyst nematodes and final tuber yield is subject to many interactions with environmental factors and is, therefore, very variable (Trudgill 1986). Barker and Noe (1987) suggested the use of simulation models of plant-nematode systems to explain the variation.

We have shown that there is no need for explicitly incorporating nutrient or hormone relations into a model for potato cyst nematodes, even though the underlying damage mechanisms affect those relations. If one wants to use our model to explain observed interactions between, e.g. effects of nematode density and fertilization level (Trudgill 1980; 1986), it seems prudent to reparameterize the modeled relation between root length and rate of leaf senescence rather than to increase model complexity. It is our intention to apply the model in that manner.

Acknowledgements

We thank N. van Dijk for technical assistance in the preparation of this manuscript.

References

Barker K R, Noe J P (1987) Establishing and using threshold population levels. Pages 75–81 in Veech J A, Dickson D W (Eds.) Vistas on Nematology. Society of Nematologists, Hyattsville, USA.

Bird G W, Tummala R L, Gage S H (1985) The role of systems science and data management in nematology. Pages 160–173 in Barker K R, Carter C C, Sasser J N (Eds.) An Advanced Treatise on *Meloidogyne*, Vol. II: Methodology. North Carolina State University Graphics, Raleigh, NC, USA.

Boote K J, Batchelor W D, Jones J W, Pinnschmidt H, Bourgeois G (1993) Pest damage relations at the field level. Pages 277–296 in Penning de Vries F W T, Teng P, Metselaar K (Eds.) Systems Approaches for Agricultural Development. Kluwer Academic Publishers, Dordrecht, The Netherlands.

Boote K J, Jones J W, Mishoe J W, Berger R D (1983) Coupling pests to crop growth simulators to predict yield reductions. Phytopathology 73:1581–1587.

Caswell E P, MacGuidwin A E, Milne K, Nelsen C E, Thomason I J, Bird G W (1986) A simulation model of *Heterodera schachtii* infecting *Beta vulgaris*. Journal of Nematology 18:512–519.

Den Toom A L (1990) Simulation of the host/parasite system *Lolium perenne*/*Tylenchorynchus dubius*. 2. The effect of *T. dubius* on L. *perenne*. Netherlands Journal of Plant Pathology 96:211–225.

Elston D A, Phillips M S, Trudgill D L (1991) The relationship between initial population density of potato cyst nematode *Globodera pallida* and the yield of partially resistant potatoes. Revue de Nématologie 14:213–219.

Ferris H (1976) Development of a computer simulation model for a plant-nematode system. Journal of Nematology 8:255–263.

Ferris H (1978) Modification of a computer simulation model for a plant-nematode system. Journal of Nematology 10:198–201.

Freckman D W (Ed.) (1982) Nematodes in Soil Ecosystems. University of Texas Press, Austin, TX, The Netherlands.

Haverkort A J, Trudgill D L (1995) Crop physiological responses to infection by potato cyst nematodes (*Globodera* spp.). Pages 167–184 in Haverkort A J, MacKerron D K L (Eds.) Potato Ecology and Modelling of Crops under Conditions Limiting Growth. Kluwer Academic Publishers, Dordrecht, The Netherlands.

Jones F G W, Kempton R A, Perry J N (1978) Computer simulation and population models for cyst-nematodes (*Heteroderidae*: *Nematoda*). Nematropica 8:36–55.

Jones F G W, Perry J N (1978) Modelling populations of cyst-nematodes (*Heteroderidae*: *Nematoda*). Journal of Applied Ecology 15:349–371.

Jones F G W, Kempton R A (1980) Forecasting crop damage by nematodes: Nematode population dynamics. EPPO Bulletin 10:169–180.

McSorley R, Ferris J M, Ferris V R (1982) Model synthesis and validation: primary consumers. Pages 141–156 in Freckman D W (Ed.) Nematodes in Soil Ecosystems. University of Texas Press, Austin, TX, USA.

Monteith J L (1977) Climate and the efficiency of crop production in Britain. Philosophical Transactions of the Royal Society of London, Series B, 281:277–294.

Schans J (1991) Reduction of leaf photosynthesis and transpiration rates of potato plants by second-stage juveniles of *Globodera pallida*. Plant, Cell and Environment 14:707–712.

Schans J (1993) Population Dynamics of Potato Cyst Nematodes and Associated Damage to Potato. Ph.D. Thesis, Agricultural University Wageningen, The Netherlands.

Schmidt K (1992) Zeitdiskrete Modelle zur Vorhersage der Populationsdynamik des Rübenzystennematoden *Heterodera schachtii* (Schmidt) in Abhängigkeit von der Fruchtfolge und des Temperaturmusters. Ph.D. Thesis, University of Bonn, Germany.

Spitters C J T, Schapendonk A H C M (1990) Evaluation of breeding strategies for drought tolerance in potato by means of crop growth simulation. Plant and Soil 123:193–203.

Trudgill D L (1980) Effect of *Globodera rostochiensis* and fertilisers on the mineral nutrient content and yield of potato plants. Nematologica 26:243–254.

Trudgill D L (1986) Yield losses caused by potato cyst nematodes: a review of the current position in Britain and prospects for improvements. Annals of Applied Biology 108:181–198.

Trudgill D L, Marshall B, Phillips M (1990) A field study of the relationship between pre-planting density of *Globodera pallida* and the growth and yield of two potato cultivars of differing tolerance. Annals of Applied Biology 117:107–118.

Van Haren R J F, Hendrikx E M L, Atkinson H J (1994) Growth curve analysis of sedentary plant parasitic nematodes in relation to plant resistance and tolerance. Pages 172–184 in Grasman J, Van Straten G (Eds.) Predictability and Nonlinear Modelling in Natural Sciences and Economics. Kluwer Academic Publishers, Dordrecht, The Netherlands.

Van Keulen H, Penning de Vries F W T, Drees E M (1982) A summary model for crop growth. Pages 87–97 in Penning de Vries F W T, Van Laar H H (Eds.) Simulation of Plant Growth and Crop Production. Simulation Monograph. Pudoc, Wageningen, The Netherlands.

Van Oijen M, De Ruijter F J, Ammerlaan F H M (1993) Simulation of root growth reduction following compaction of nematode infested soil. Pages 135–136 in European Association for Potato Research, Abstracts 12th Triennial Conference, Paris, France.

Van Oijen M, De Ruijter F J, Van Haren R J F (1995a) Responses of field-grown potato cultivars to cyst nematodes at three levels of soil compaction. I. Leaf area dynamics, photosynthesis and crop growth (Submitted).

Van Oijen M, De Ruijter F J, Van Haren R J F (1995b) Responses of field-grown potato cultivars to cyst nematodes at three levels of soil compaction. II. Root length dynamics and nutrient uptake (Submitted).

Wallace H R (1987) Effects of nematode parasites on photosynthesis. Pages 253–259 in Veech J A, Dickson D W (Eds.) Vistas on Nematology. Society of Nematologists, Hyattsville, USA.

Ward S A, Rabbinge R, Den Ouden H (1985) Construction and preliminary evaluation of a simulation model of the population dynamics of the potato cyst-nematode *Globodera pallida*. Netherlands Journal of Plant Pathology 91:27–44.

13. A growth model for plants attacked by nematodes

C.H. SCHOMAKER[1], T.H. BEEN[1] and J.W. SEINHORST[2]

[1] *DLO Research Institute for Plant Protection (IPO-DLO), P.O. Box 9060, 6700 GW Wageningen, The Netherlands*

[2] *Middelberglaan 6, 6711 JH Ede, The Netherlands*

Abstract. The relation between small and medium initial population densities and the relative total plant weight was derived as cross sections at right angles to the time axis of a growth model with three dimensions: time after planting t, relative total plant weight Y and relative growth rate r_p/r_0. The relative growth rate is the (constant) ratio between the growth rate r_p of plants of a certain weight at a nematode density P and the growth rate r_0 of (younger) plants of the same weight without nematodes. Therefore, the ratio between the time after planting that plants need to reach a certain weight in the absence of nematodes and at nematode density P, $t_0/t_p = r_p/r_o$ (2).

The relative growth rate $r_p/r_o = k + (1 - k)0.95^{P/T-1}$ for $P > T$ and $= 1$ for $P \leq T$ (3). Formally, k is the minimum relative growth rate as $P \to \infty$. As a result the arbitrary equation $y = m + (1 - m)0.95^{P/T-1}$ for $P > T$ and $= 1$ for $P \leq T$ (6) also applies to the relation between small and medium initial population densities and relative total plant weight. T is the tolerance limit, below which growth and yield are not reduced by nematodes; m is the relative minimum yield.

The relations between small and medium initial population densities of potato cyst nematodes and relative tuber weight of potatoes can be derived from the growth model in an analogous way. However, there is one complication: tuber initiation does not start at the same haulm weight in plants with and without nematodes, but at the smaller haulm weight the larger the nematode density. As a consequence, tuber weights of plants with a certain total weight at nematode density P are not equal to those of plants with the same total weight without nematodes, but $r_p\Delta t$ units of weight larger, Δt being the difference between the actual time of tuber initiation and the time total plant weight becomes the same as that of plants without nematodes at the initiation of tuber formation.

Relative total and tuber weights of plants with 'early senescence' and at large nematode densities are smaller than estimated by the model and Eq. 2. This indicates that at large initial population densities growth reducing mechanism(s) become active that were not operating at smaller densities.

Introduction

I shall take the simpleminded view that a theory is just a model of the universe or a restricted part of it, and a set of rules that relate quantities in the model to observations that we make. It exists only in our minds and does not have any other reality (whatever that may mean). A theory is good theory if it satisfies two requirements: It must accurately describe a large class of observations on the basis of a model that contains only few arbitrary elements, and it must make definite predictions about the results of future observations. For example,

A.J. Haverkort and D.K.L. MacKerron (eds.), Potato Ecology and Modelling of Crops under Conditions Limiting Growth, 197–214.

> *Aristotle's theory that everything was made out of four elements ( ) was simple enough to qualify, but did not make any definite predictions. On the other hand, Newton's theory of gravity was based on an even simpler model ( ). Yet it predicts the motions of the sun, the moon and the planets to a high degree of accuracy.* (Stephen Hawking 1988, A brief History of Time).

In many biological and nematological studies one is not very particular about these requirements. Several nematologists or biologists have formulated mathematical relations between pre-plant nematode densities and crop yield that were indeed based on observations – although mostly limited ones – but that lacked a theoretical base and predictive value. Examples are the linear and log-linear functions (Lownsberry and Peters 1956; Hesling 1957; Seinhorst 1960; Hoestra and Oostenbrink 1962; Brown 1969), the quadratic curve (Peters 1961); the inverse linear function (Elston et al. 1991) and an exponential function (Van Haren et al. 1993). These equations do not represent a vision of the mechanisms involved in nematode-plant interactions. Calculations of limits of variables or parameters in some of them produce bizarre results.

Others (e.g. Evans and Haydock 1990; Trudgill 1992) tried to build theories on the effect of nematode attack on plants without using a model. This approach makes it difficult to study a (biological) problem methodically or to detect flaws in one's chain of reasoning. Moreover, there is no efficient transfer of these theories to fellow-nematologists for further development or criticism.

An approach which at least partly satisfies one of Hawking's requirements is the 'comprehensive simulation model'. This type of model tries to describe an often-limited research area with detailed information in a large number of sub-models and parameters. They were first applied in nematology by Ward et al. (1985) in a preliminary model on the population dynamics of potato cyst nematodes and were later extended by Schans (1993), who tried to describe the population dynamics of potato cyst nematodes and the associated damage to potatoes with information about some plant physiological processes and their changes with environmental conditions. Only because of the arbitrary nature of the sub-models and the multitude of parameters, which, although treated as constants, are more often variable than not, these models did not contribute much to theorem building and have a limited predictive value. In the model of Schans this was the more so as description of plant physiology was confined to short-term measurements of photosynthesis, respiration and transpiration on one or a few leaves, which did not explain differences in yield, at a limited range of nematode densities which, moreover, were not always constant during plant growth. Besides, as external conditions during crop growth are indispensable data in these models, they are unsuitable as a base for recommendations about control measures for nematode pests, which, unlike those for fungi or insects, must be taken, at the latest, at the time of sowing or planting of the crop to be protected (nematostatics), but generally much earlier (nematicides, crop rotation, resistant cultivars). Therefore, these simulation models would not qualify as good models.

In nematology Seinhorst's models, especially those for potato cyst nematodes (Seinhorst 1986a,b; 1993) come closest to the classical approach recommended by Hawking. The models are based on a theory which describes the principal mechanisms of nematode/nematode and nematode/plant interactions, they don't conflict with a large number of observations in different countries on different nematode species over a large period of time. The integrated, simplified, stochastic versions of the models (Been et al. 1994) deduced from more complicated ones that suit mainly scientific purposes, contain only a few parameters with known distribution functions and allow predictions to be made within sufficiently narrow limits, albeit with some constraints. The predictive value of the models, which makes them suitable for an advisory system for the control of potato cyst nematodes on farmer's fields, will be demonstrated by Been et al. (1994), elsewhere in this book; the theorem of one of Seinhorst's models: the effect of nematodes on growth and yield of plants, is explained in this article.

Causes of yield reduction by nematodes

Crop returns are reduced by nematode attack as a result of reduction of crop weight per unit area (which is mostly equivalent to average weight of marketable product per plant) and reduction of the value of the product per unit weight. For example, carrots attacked by rootknot nematodes (*Meloidogyne*) may be worthless because of branching and deformation of the tap root, although they have the same weight per unit area as carrots without nematodes. Onions of normal weight but infested with few stem nematodes (*Ditylenchus dipsaci*) at harvest will, nevertheless, be lost in storage. Attack of crop plants by potato cyst nematodes (*Globodera rostochiensis*, *G. pallida*) does not only reduce potato tuber weight, but may also reduce tuber size. However, potato cyst nematodes attacking potatoes and almost all root infesting nematodes attacking crop plants of which the above ground parts are harvested, hardly ever affect the value per unit weight of harvested product, unless crop weight is reduced so much by large initial population densities that growing it was uneconomical from the start. Therefore, prediction of crop reduction by these nematodes can, in general, be based on models of the relation between nematode density at sowing or planting and average weight of single plants at harvest. In the following the term 'yield' will be avoided. The yield in the agronomic sense must be derived from individual plant weights, taking the restrictions mentioned above into account.

To make a model of the relation between population density before planting (P) and the proportion (y) of uninfested plants (onions, flower bulbs) or relative plant weight (y) after a certain period of growth of the plants (the ratio between the yield of plants at density P and the yield of plants without nematodes, grown under the same conditions) a theory was developed about the mechanisms involved. The theory was tested by comparing it with the patterns of observed relations between nematode density and plant growth and plant weight at a certain time after sowing or planting. This also allows the estimation of values of system parameters under various experimental conditions.

Seinhorst's (1965) model for the relation between the population densities of stem nematodes (*Ditylenchus dipsaci*) at sowing and the proportion of infested onion plants is identical to that of Nicholson (1933) for the infection of caterpillar pupae by parasite wasps. The chain of reasoning is as follows: Only infested and non-infested onions are distinguished. The degree of infestation of single plants is irrelevant as only nematode-free onions are marketable. Only three assumptions (which in a slightly different formulation also apply to the root infesting species) were needed to formulate the model:

1. The average nematode is the same at all densities. This means that population density does not affect the average size or activity of the nematodes.
2. Nematodes do not affect each other's behaviour. They do not attract or repel each other directly or indirectly.
3. Nematodes are distributed randomly over the plants in a certain area.

If now a density of one nematode per unit area or weight of soil attacks a proportion d of the onion plants in a certain area (and leaves a proportion $1 - d$ uninfested), then a density of two nematodes per unit area or weight of soil will attack, on the average, a proportion d of already damaged plants (which has no additional effect on yield reduction), plus a proportion d of the still uninfested proportion $(1 - d)$, leaving a proportion $1 - d - d(1 - d) = (1 - d)^2$ of the plants uninfested. Generalized: a nematode density P leaves a proportion

$$y = (1 - d)^P = z^P \qquad (1)$$

of the onions uninfested. In Figure 1 values of z^P are plotted, not against P, but against log P. This has not only the advantage that the shape of the curves is the same for all z ($d(\log P)/dy$ is only determined by y), but also that, if P is estimated by counting nematodes in a soil sample, the variance of log P depends less on true P than the variances of P. The values of z are determined by conditions that affect the efficiency of the nematodes in finding and penetrating plants. In patchy infestations of stem nematodes these conditions for attack appeared to be more favourable in the centre of the patch than towards the borders, resulting in an increase of z with increase of the distance from this centre. This results in persistency of the patchiness. The model also applies when nematodes spread from randomly distributed infested plants to neighbouring ones, which results in overlapping circular patches of infested plants.

Patchy infestations of nematodes do not always develop this way. Infestation foci with potato cyst nematodes, for instance, occur by introduction of cysts in fields: small numbers of cysts are mostly transmitted by seed potatoes and larger numbers of cysts by agricultural machinery. After multiplication on potatoes they are spread by tillage and harvesting machines. As a result the patches have the same shape in all potato growing areas in the Netherlands (Seinhorst 1982; 1988; Schomaker and Been 1992). If good host plants are grown in short rotations, patchy infestations can become larger until the whole field is infested more or less uniformly. Examples of such infestations are found in south-east Groningen and east Drenthe, where starch potatoes are grown every other year.

Reduction of plant growth

Tylenchid root infesting nematodes (for instance cyst nematodes, root knot nematodes and *Pratylenchus* species) generally reduce crop yield in a less direct way than stem nematodes. The rate of growth and development of attacked plants is reduced, resulting in smaller weights than of plants without nematodes at given times after sowing or planting or, in exceptional cases, in reaching the same final weight as of plants without nematodes later. In general such a delay of the ripening of the crop is prevented by the external conditions at the end of the growing season of the plants.

Seinhorst (1979b; 1986b) based a growth model on two simple concepts: one of the nature of the plant; an element that increases in weight in the course of time, and one of the nature of plant parasitic nematodes: elements that reduce the rate of increase of this weight and, in principle, more so the larger the population density. Further the following principles are applied in addition to those mentioned above for the model on the relation between stem nematode density and proportion of onions attacked:

4. Root infesting nematodes are distributed randomly in the soil.
5. Nematodes enter the roots of plants randomly in space and time. Therefore the average number of nematodes entering per quantity of root and time is constant. This number is proportional to the nematode density P (number of nematodes per unit weight or volume of soil).
6. If the growth rate of plants at a given time t after sowing or planting is the increase in total weight (Y) per unit time (dY/dt), represented by r_0 for plants without nematodes and r_P for plants at nematode density P, then the ratio r_p/r_0 for plants of the same total weight (and, therefore, of different age) without nematodes and at nematode density P, is constant throughout the growing period. Further, according to Figure 2, $r_0 = \tan\alpha = \Delta Y/\Delta t_0$ and $r_P = \tan\beta = \Delta Y/\Delta t_p$. Therefore,

$$r_P/r_0 = t_0/t_P \tag{2}$$

The relation between population density of the nematodes and its total effect on the growth rate of the plants is also considered to be according to Nicholson's competition model (Eq. 1, Figure 1). Eq. 1 is a continuous function for $0 \leq P \leq \infty$. However, all sufficiently accurate observations on the relation between the population density P of various nematode species, including potato cyst nematodes, on the weight of various plant species demonstrate the existence of a minimum density T, below which the nematodes do not reduce plant weight. It may be concluded that at densities $< T$ they also do not reduce the growth rate of plants. This is corroborated by the results of the few experiments

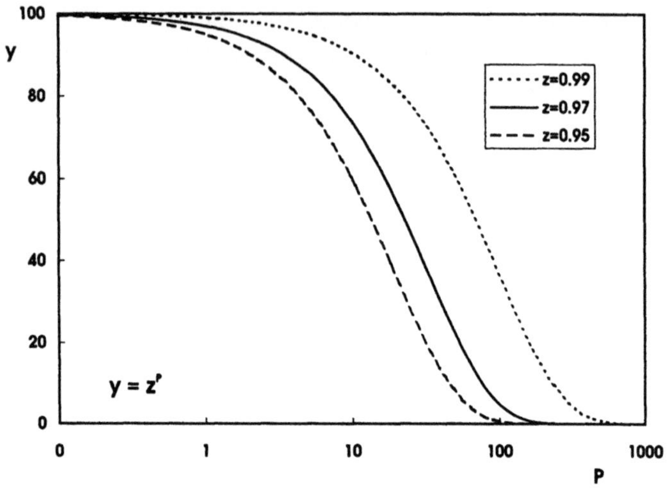

Figure 1. The relation $y = z^P$ (Eq. 1) between the initial population density (P) of stem nematodes and the proportion y of the onions that is not attacked.

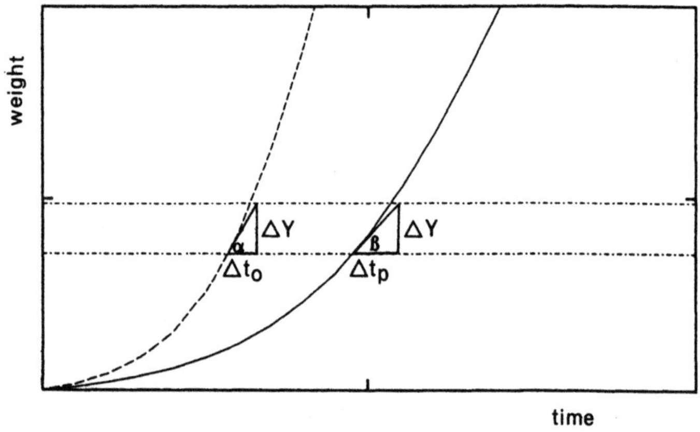

Figure 2. Growth curves of plants without nematodes and at nematode density P. Y = total plant weight; t = time after sowing or planting; t_0 and t_p are the times plants without and with nematodes need to reach the same total weight Y; r_0 (= $\tan\alpha = \Delta Y/\Delta t_0$) and r_p (= $\tan\beta = \Delta Y/\Delta t_p$) are the growth rates of plants without nematodes and at nematode density P respectively.

in which growth rates were actually determined. Moreover, at large nematode densities plant weight approached zero in only few of these experiments and growth rates never did. Therefore, Eq. 1 is adapted by replacing P by $P - T$ and introducing the minimum relative growth rate $r_P/r_0 = k$ for $P \rightarrow \infty$. The second equation constituting the model then becomes:

$$r_P/r_0 = k + (1 - k)z^{P-T} \text{ for } P > T$$
$$\text{and} \quad r_P/r_0 = 1 \text{ for } P \leq T \qquad (3)$$

in which z is a constant smaller than 1. The value of k is independent of nematode density and time after sowing or planting, but may vary between experiments. Growth curves of plants for different nematode densities can be derived from a growth curve of plants without nematodes with the help of the Eqs. 2 and 3. These curves may vary in shape, as long as they are continuous and the growth rate decreases monotonously from shortly after sowing or planting. Figure 3 gives an impression of the three dimensional model with axes for total plant weight Y, relative nematode density P/T and time after planting t for a given value of k.

A simple model for the relation between nematode density and relative plant weight

The primary results of experiments are almost always weights of plants attacked by known nematode densities at a given time after sowing or planting. To investigate whether these relations are in accordance with the model, they must be compared with cross sections orthogonal to the time axis of the model, through growth curves of plants for ranges of densities P/T and different values of k. These cross sections were in close accordance with the now arbitrary equation:

$$y = m + (1 - m)z^{P-T} \text{ for } P > T \tag{4}$$

in which m is the minimum relative plant weight, z is a constant < 1 with the same or a slightly smaller value than in Eq. 3 and T is the tolerance limit with the same value as in Eq. 3.

Seinhorst (1986b) gives the relations in pot experiments between P/T of different tylenchid nematode species, including *Heterodera avenae* (Seinhorst 1981), *G. rostochiensis* and *G. pallida* (Greco et al. 1988) and the relative dry plant weight (y) several months after sowing or planting. Values of $y' = (y - m)/(1 - m) = z^{P-T}$ (5) in thirteen pot experiments are plotted against P/T (Figure 4). The relation between average y' and P/T is in close accordance with $y' = z^{P-T}$ which suggests that the deviations of individual observations in the different experiments from values according to this relation were due to experimental error. Average relative plant and tuber weights of different resistant and susceptible potato cultivars at different times after planting were in close agreement with those according to Eq. 5 with $T = 1.8$ eggs/g soil. As Eqs. 4 and 5 are identical, we may conclude that the results of Figure 4 and the other experiments are well described by the model, which apparently applies to tylenchid nematodes that feed and multiply in as different ways as cyst nematodes, *Pratylenchus penetrans* and *Tylenchorhynchus dubius*. It also indicates that the general relation between nematode density of tylenchids and relative plant weight of attacked plants some time after sowing or planting is independent of the shape of the growth curve of plants without nematodes and of external conditions. Although values of z and T differed between nematode species and plant species and m also between experiments with the same

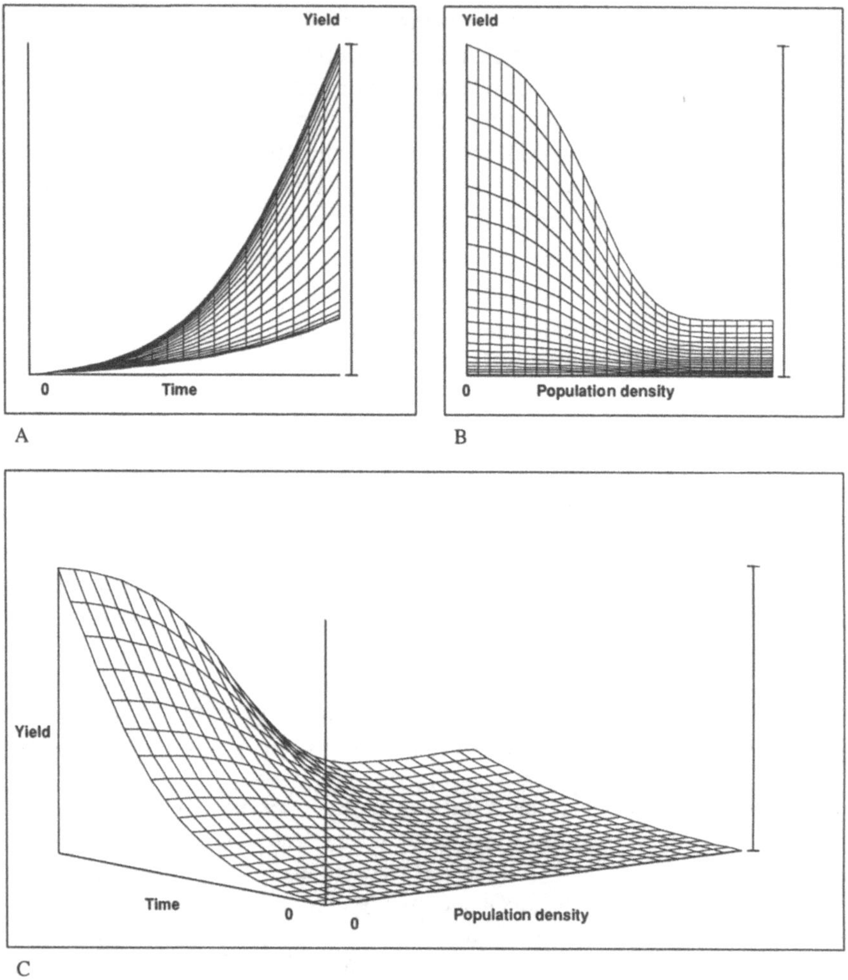

Figure 3. Surface plots of the three-dimensional model which represents the relation between total weight (Y) and nematode density P as cross sections at right angles with the time axis (t) of growth functions of plants at different nematode densities, $Y(r_p,t)$, and that without nematodes, $Y(r_0,t)$:
a at 0° rotation, which shows the relation between Y and t at different nematode densities.
b at 270° rotation, which exposes the relation between Y and population density P.
c at 230° rotation, which illustrates the relation between (Y,P) and (Y,t).
All rotations are clockwise. The growth rates of plants of the same weight without nematodes (r_0) and at nematode densities P (r_p) are related by Eqs. 1 and 2.

combination of nematode and plant species, z^T differed too little from 0.95 between experiments to distinguish the variation from that caused by experimental error. Therefore, for fitting curves according to Eq. 3 to experimental data, z^T is generally assumed to be 0.95, which transforms Eq. 4 to:

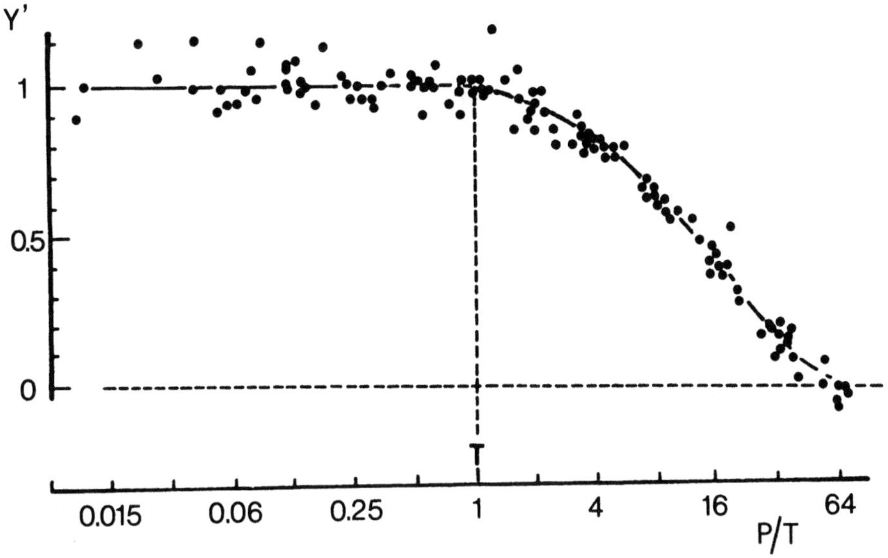

Figure 4. The relation between nematode density P/T at sowing or planting and the relative weight y' of plants after a certain growing time in thirteen experiments with several tylenchid nematode species and several plant species. $y' = (y - m_x)/(1 - m_x) = 0.95^{P/T-1}$, in which $y = m_x + (1 - m)0.95^{P/Tx-1}$ for the x-th combination of nematode species and plant species and all values T_x are coinciding.

$$y = m + (1 - m)0.95^{P/T-1} \tag{5}$$

The value of T depends on nematode species and pathotype and on the plant species, but seems in general not to be affected by external conditions. For potatoes with potato cyst nematodes, planted in spring, T was about 1.8 eggs/g soil. An exception is the sensitivity to daylength of the value of T for potato cyst nematodes on potatoes (Been and Schomaker 1986; Greco and Moreno 1992). In contrast, m for a given nematode species varies between cultivars or plant species within experiments and for the same cultivar between experiments under the influence of so far unknown external conditions. It is not established whether there are consistent differences between cultivars or even plant species. Therefore, T is the main measure of the degree of tolerance of a plant species or cultivar to a certain nematode species or pathotype, whereas the importance of m remains to be investigated. The shortcoming of most discussions on differences in tolerance between cultivars or crop plants is, that they do not refer to these two parameters.

Implications of the model

The similarity between very different nematode species in their reaction upon nematode attack is not restricted to its effect on the size of the plant. Nematode densities up to at least 16T, but generally more than 32T, affect neither water consumption per unit weight of plant and per unit duration of time, nor dry matter content (Seinhorst 1981). There are indications that this also applies to shoot to root weight ratios (Seinhorst 1979a; Been and Schomaker 1986). The only difference between plants without nematodes and those at small and medium nematode densities of the same weight stated so far is that the latter may be slightly taller.

The model implies comparison of growth rates, as affected by nematode attack, of plants of the same weight and not of the same age, because the latter also differ in developmental stage. The importance of this consequence of the model is evident from the investigation on the effect of *H. avenae* on the shoot to root ratio of young oat plants (Seinhorst 1979a) and on the relation between weight of potato plants with and without *G. pallida* and nitrate and potassium content (Been and Schomaker 1986).

Seinhorst (1986b) rated the mechanisms by which nematodes should reduce growth rates of plants (extraction of nutrients, mechanical damage to root tissue resulting in a hampered uptake of water and minerals, obstruction of plant vessels causing wilting, extraction of food), suggested in the literature, as myths and fairy tales. A more probable mechanism is the production of a growth reducing substance only during the penetration of the nematodes in the roots but not any more when they have settled. Because of the constant number of nematodes penetrating per unit quantity of root and per unit duration of time, the growth reducing stimulus will then remain constant per unit weight of plant. It remains to be investigated how the reduction of the efficiency per nematode with the increase of population density comes about. Further major problems to be investigated are the way growth reduction is prevented up to a certain nematode density and what determines the ratio k, keeping it independent of T.

Strictly taken, the model only applies to nematode species with a single generation per growing season, as of potato cyst nematodes, oat cyst nematodes and *Meloidogyne naasi*. However, Figure 3 demonstrates that it also applies at least to nematodes with small rates of reproduction (e.g. ten to twenty fold in a growing season). Seinhorst (in press) shows that the reason why may not be simple.

The effect of potato cyst nematode attack on tuber growth

After the start of tuber initiation the rate of weight increase of the haulms decreases strongly to change finally to a decrease. In the model increase of haulm weight is assumed to stop shortly after the initiation of tuber growth, after which increase of total weight is due entirely to that of tuber weight.

If the effect of nematode attack on potato plants would only be retardation of development, tuber initiation of plants without and with nematodes would start at the same haulm (= total) weight of plants, irrespective of the retardation of the development. If t_{s0} is the time tuber initiation of plants without nematodes starts, then tuber initiation at nematode density P would start at a time

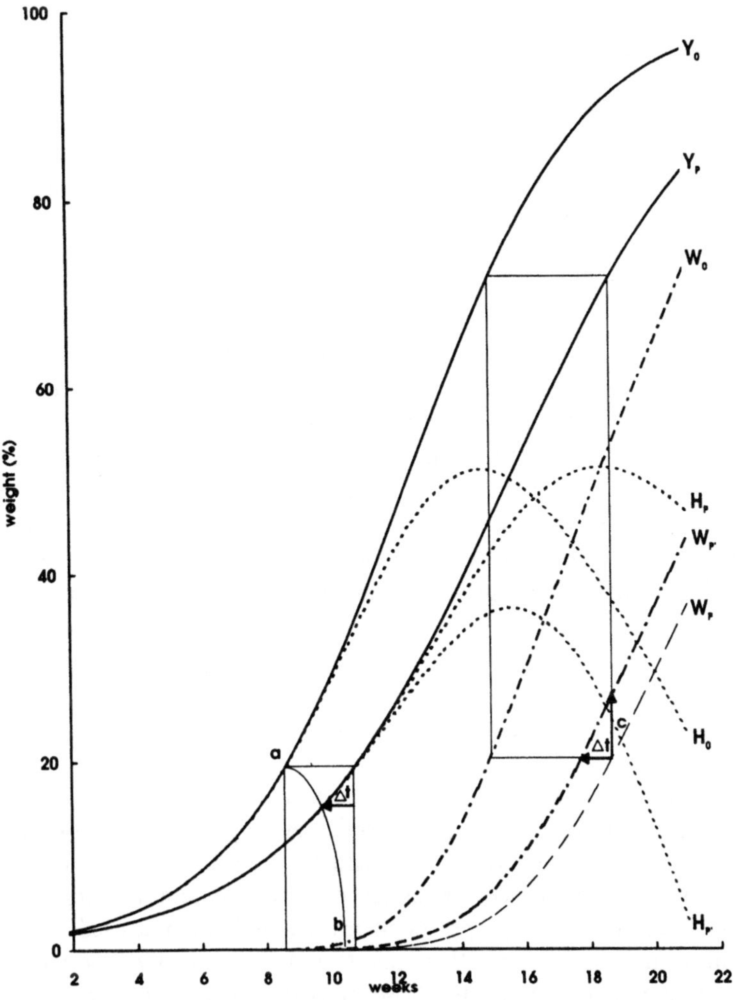

Figure 5. Growth curves for total, haulm and tuber weight of potatoes with and without nematodes. The growth rates for plants of the same total weight or the same tuber weight (measured some time after tuber initiation) with and without nematodes are related by Eq. 2: $r_p/r_0 = t_0/t_p$. The curves for tuber weights of plants of the same total weight with and without nematodes are related by $W'_P = W_0 + r_p\Delta t$. In the figure, the curve ab represents the relation between haulm weight and the time tuber growth starts; $c = r_p\Delta t$; Y_0 and Y_p represent total weight of plants without nematodes and at density P respectively; W_0 and W_p are tuber weights of plants without nematodes and at nematode density P if tuber initiation had started at the same haulm weight as in plants without nematodes; H_0 *and* H_P are haulm weights of plants without nematodes and at nematode density P, if tuber intiation had started at the same haulm weight as in plants without nematodes. H_P represents the actual haulm weight at nematode density P.

$$t_{sP} = r_0\, t_{s0}/r_P \tag{6}$$

However, the ratio between tuber weights of plants of different age but of the same total weight at nematode density P and without nematodes (W_P/W_0) increases most probably from a density T, and there are indications that the initiation of tuber growth is delayed less than according to Eq. 7 (Seinhorst et al. submitted) and therefore starts at a time

$$t'_{sP} = (r_0\, t_{s0}/r_P) - \Delta t \qquad (7)$$

Δt increases with r_P/r_0 in such a way that t'_{sP} approaches a maximum of t_{s0} + about 2 weeks, possibly because the external conditions inducing the initiation of tuber growth become stronger than the delaying effect of nematode attack (relation between r_P/r_0 and t'_{sP} according to line *ab* in Figure 5). As a result tuber weights of plants at nematode density P (W_P') are $r_p\, \Delta t$ units larger than tuber weights of plants without nematodes of the same total weight.

The relation at given times t between nematode density and relative tuber weights in a model, constructed as described above, are again according to Eq. 6 up to $P/T = 50$ eggs/g soil and with $k > 0.4$ ($m > 0.05$) from the time tuber to total dry weight ratios larger than 0.5 are reached and with the same value of T but a smaller one of m than for total plant weight. Within these ranges of tuber to total dry weight ratio's and values of P/T and k actual relations between nematode density and relative tuber weights were according to those in the model, if not affected by growth reduction by other mechanisms than that described by the model.

More than one mechanism of growth reduction

According to Seinhorst (1981) nematodes cause two kinds of growth reduction: the 'first mechanism of growth reduction' operating at all population densities, and a 'second mechanism of growth reduction' additional to that of the 'first mechanism', with a noticeable effect only at medium to large nematode densities. The model only applies to growth reduction caused by the 'first mechanism' that retards growth of plants and, occasionally, also increases the length of haulms compared to those of plants of the same weight without nematodes. As long as only the first mechanism is active, water consumption during short periods is proportional to plant weight and, therefore, relative water consumption at different nematode densities and times after sowing or planting is a measure of relative plant weight. Actual plant weights are these relative weights times the actual weights of plants of the same age without nematodes, determined at the same time (Seinhorst 1981).

The 'second mechanism' reduces water consumption per unit plant weight and the (active) uptake or excretion of K^+ and Na^+ and increases the (passive) uptake of Ca^{2+} and dry matter content of plants (Seinhorst 1981; Been and Schomaker 1986). There probably is a negative correlation between age of the plant and nematode density at which the effect of the 'second mechanism' becomes noticeable. For potato cyst nematodes this density rarely is as small as $16T$, but more commonly, also for other nematode species on other plant species $> 32T$. Contrary to the 'first mechanism' it tends to advance the initiation of tuber growth.

A 'third mechanism of growth reduction' is 'early senescence' of potato plants attacked by *G. pallida*: a sudden ending of the increase of haulm length and weight. The time after planting at which it occurs may be negatively correlated with nematode density. The earliest occurrence was nine weeks after planting in the early cultivar Ehud and the smallest nematode density $25T$ in cultivar Darwina. Not all cultivars are equally sensitive (Seinhorst et al. submitted). The cause of 'early senescence' is unknown.

The estimation of *t* and *m*

There are strong indications that the value of T for potatoes planted in spring is not affected by differences in external conditions and can, therefore, be determined in pot experiments in both greenhouse and (much more laborious) field experiments. The only requirement of greenhouse tests is that large enough pots are used to guarantee about the same root density in the soil as in the field and to prevent the plants from becoming pot bound, which affects the relation between nematode density and plant weight and obscures the true value of T (Seinhorst and Kozlowska 1977). According to the model and the results of pot experiments, T for total weight, a short time after planting, suffices as an estimate of T for final tuber weight, which allows the use of smaller pots. Experiments should be done with ranges of nematode (egg) densities with ratios not larger than 1 to 2 between successive densities and a sufficient number of densities $< T$ and pots to provide an accurate estimate of plant weight at $P < T$. The largest density in the range should be about $30T$. The accuracy of the estimates is mainly a matter of uniformity of plant material and growth conditions (light, water content) and carefully filling and handling of the pots.

At best m is characteristic of a combination of nematode and plant species with a very large variability between experiments. Therefore, a small number of tests in greenhouses, as will suffice to obtain a reliable estimate of T, will not produce such an estimate for a combination of a potato cultivar and a potato cyst nematode pathotype. However, it may be possible to derive differences, if they exist, between host varieties from the results of greenhouse experiments. In addition, of at least some host varieties per nematode pathotype, a sufficient number of values of m must be estimated in field experiments to establish a distribution function.

As growing potatoes at large potato cyst nematode densities is uneconomical, the relation between nematode density and growth and yield reduction by the 'second mechanism' can be ignored in a model for advisory purposes.

Estimation of T and m in field experiments is much more laborious than estimation in pot experiments. The same range of densities is needed and nematode density must be the only variable. Ranges of nematode densities in patches of potato cyst nematode infestation (but not necessarily of other species) come closest to this requirement, if free of other causes of variation of tuber yield with intractable spatial distributions. Ranges of nematode densities cannot be created by applying different dosages of nematicide or other biocide, as has often been done, as this has unpredictable effects on crop yield, other than by killing nematodes. The ranges of nematode densities, that are of interest for the estimation of the parameters in question, must be determined in sufficiently large samples from each field plot to guarantee a coefficient of variation of egg counts not larger than 15% (density differences 1:2 just distinguishable). According to Seinhorst (1988) soil samples of 4 kg per plot are then needed to estimate population densities of 1 egg per g of soil, given a coefficient of variation of the number of eggs per cyst of 16% and a negative binomial distribution of egg densities in samples from small plots with a coefficient of aggregation of 50 for a kg soil. As the coefficient of variation per unit weight of soil is negatively correlated with nematode density, required sample size also is. For instance, to estimate densities of 0.5 or 0.25 eggs/g soil with the same accuracy, eggs from soil samples of 10 and 20 kg must be counted respectively. Another requirement is that plots must on the one hand be small (e.g. 1 m^2) to reduce the effect of medium scale density variation, whereas on the other hand a large enough area per small density interval must be available to guarantee a small variability of tuber weight per unit area. Again, there must be a sufficient number of plots at densities smaller than T to estimate the maximum yield accurately.

Therefore, to avoid unnecessary handling of large soil samples, T and m should, as much as possible, be estimated from greenhouse experiments. Field trials are most efficiently used to confirm of falsify these estimates or ratios of estimates for different combinations of pathotypes and cultivars under more natural external conditions.

Final considerations

As a result of the striking conformity between the effects of all root infesting nematodes investigated upon the growth of attacked plants, whatever the host status of the plants, the ways the nematodes attack, and the reactions of the affected tissues, the growth model discussed above applies to small and medium population densities of all tylenchids. It also results in the same formal relation during the year of sowing or planting between nematode density at sowing or planting and relative total plant weight and, in potatoes, also relative tuber weight, from some time after planting, as long as later generations of species with large rates of reproduction (e.g. *Meloidogyne* species) do not cause additional growth reduction. This is corroborted by a large number of pot experiments and the few field experiments, that were suitable for the purpose. It allows a characterization of the sensitivity to growth and yield reduction (the tolerance) of a crop plant (annuals and perennials during the first year after sowing or planting) by small and medium population densities at sowing or planting of a particular nematode species by the values of two parameters, which is preferable to the (customary) single characterization without reference to nematode density. The most important of the two parameters, the tolerance limit *T*, most probably is insenitive to variation in external conditions that normally occur during the period the crop is grown. That of potatoes to potato cyst nematode attack depends on change of day length during the plant growth. The values of these parameters, especially that of the tolerance limit, are key factors in the calculation of combinations of cost of reduction by control measures of a given nematode population density, expected at the time of sowing or planting of a crop to be protected, and the cost of crop loss, caused by the surviving nematodes, resulting in maximum net returns. These calculations are again formally the same for all tylenchid nematodes, annual crops and control measures.

On the other hand the model gives strong indications of the nature of the mechanisms resulting in the growth reduction by nematode attack and is, therefore, a base for investigations on its biochemistry, the counteraction of this reduction, resulting in the existence of tolerance limits, and the adding up of the effects of very large numbers of nematodes to a maximum reduction of the growth rate of the attacked plant by an amount which, in general, is smaller than the actual growth rate of plants without nematodes. So far, the physiological mechanism(s) resulting in growth reduction by nematode attack have hardly been investigated seriously. Guided by this growth model, this type of research, which may open new ways for the management of these nematodes, could make a better progress.

References

Been T H, Schomaker C H (1986) Quantitative analyses of growth, mineral composition and ion balance of the potato cultivar Irene infested with *Globodera pallida* (Stone). Nematologica 32:339–355.

Been T H, Schomaker C H, Seinhorst J W (1994) An advisory system for the management of potato cyst nematodes (*Globodera* spp.). Pages xxx in Haverkort A J, McKerron D K L (Eds.) Ecology and Modelling of Potato Crops under Conditions Limiting Growth. Kluwer Academic Publishers, Dordrecht, The Netherlands.

Brown E B (1969) Assessment of the damage caused to potatoes by the potato cyst eelworm, *Heterodera rostochiensis* Woll. Annals of Applied Biology 103:471–476.

Elston D A, Phillips M S, Trudgill D L (1991) The relationship between initial population density of potato cyst nematode *Globodera pallida* and the yield of partially resistant potatoes. *Revue de Nématologie* 14:213–219.

Evans K, Haydock P P J (1990) A review of tolerance by potato plants of cyst nematode attack, with consideration of what factors may confer tolerance and methods of assaying and improving it in crops. Annals of Applied Biology 117:703–740.

Greco N, Di Vito M, Brandonisio A, Giordano I, De Marinis G (1988) The effect of *Globodera rostochiensis* and *G. pallida* on potato yield. Nematologica 28:379–386.

Greco N, Moreno L J (1992) Influence of *Globodera rostochiensis* on yield of summer, winter and spring sown potato in Chile. Nematropica 22:165–173.

Hawking S (1988) A Brief History of Time. Space Time Publications. Bantam Press, Toronto, Canada.

Hesling J J (1957) *Heterodera major* O. Schmidt 1930 on cereals – a population study. Nematologica 2:285–299.

Hoestra H, Oostenbrink M (1962) Nematodes in relation to plant growth. IV *Pratylenchus penetrans* (Cobb) on orchard trees. Netherlands Journal of Agricultural Science 10:286–296.

Lownsberry B F, Peters B G (1956) The relation of the tobacco cyst nematode to tobacco growth. Phytopathology 45:163–167.

Nicholson A J (1933) The balance of animal populations. Journal of Animal Ecology 2:132–178.

Peters B G (1961) *Heterodera rostochiensis* population density in relation to potato growth. Journal of Helminthology. R. T. Leiper Supplement:141–150.

Schans J (1993) Population Dynamics of Potato Cyst Nematodes and Associated Damage to Potato. Ph.D. Thesis, Wageningen Agricultural University, Wageningen, The Netherlands.

Schomaker C H, Been T H (1992) Sampling strategies for the detection of potato cyst nematodes: Developing and evaluating a model. Pages 182–194 in Gommers F J, Maas P W Th (Eds.) Nematology from Molecule to Ecosystem. Dekker and Huisman, Wildervank, The Netherlands.

Seinhorst J W (1960) Over het bepalen van door aaltjes veroorzaakte opbrengst-vermindering bij cultuurgewassen. Mededelingen Landbouw Hoogeschool Gent 25:1025–1040.

Seinhorst J W (1965) The relation between nematode density and damage to plants. Nematologica 11:137–154.

Seinhorst J W (1979a) Effect of *Heterodera avenae* on the shoot:root ratio of oats. Nematologica 25:495–497.

Seinhorst J W (1979b) Nematodes and growth of plants: Formalization of the nematode-plant system. Pages 231–256 in Lamberti F, Taylor C E (Eds.) Root-Knot Nematodes (*Meloidogyne* Species) Systematics, Biology and Control. Academic Press, London, UK/New York/San Francisco, USA.

Seinhorst J W (1981) Water consumption of plants attacked by nematodes and mechanisms of growth reduction. Nematologica 27:34–51.

Seinhorst J W (1982) The relationship in field experiments between population density of *Globodera rostochiensis* before planting potatoes and the yield of potato tubers. Nematologica 28:277–284.

Seinhorst J W (1986a) The development of individuals and populations of cyst nematodes on plants. Pages 101–118 in Lamberti F, Taylor C E (Eds.) Cyst Nematodes. Plenum Press, New York, USA/London, UK.

Seinhorst J W (1986b) Effects of nematode attack on the growth and yield of crop plants. Pages 191–210 in Lamberti F, Taylor C E (Eds.) Cyst Nematodes. Plenum Press, New York, USA/London, UK.

Seinhorst J W (1988) The Estimation of Densities of Nematode Populations in Soil and Plants. Växtskyddrapporter. Jordbruk 51. Research Information Centre of the Swedish University of Agricultural Sciences (Ed.), Uppsala, Sweden.

Seinhorst J W (1993) The regulation of numbers of cysts and eggs per cyst produced by *Globodera rostochiensis* and *G. pallida* on potato roots at different initial egg densities. Nematologica 39:104–114.

Seinhorst J W (1995) The reduction of the growth and weight of plants added by a second and later generation to that caused by a first generation. Nematologica (in press).

Seinhorst J W, Kozlowska J (1977) Damage to carrots by *Rotylenchus uniformis*, with a discussion on the cause of increase of tolerance during the development of the plant. Nematologica 23:1–23.

Seinhorst J W, Schomaker C H, Been T H (1995) Effect of attack by *Globodera rostochiensis* and *G. pallida* on growth, weight and water consumption of various susceptible and resistant potato cultivars. II Effects on tuber and total plant weight. Journal of Fundamental and Applied Nematology (submitted).

Trudgill D (1992) Mechanisms of damage and tolerance in nematode infested plants. Pages 182–194 in Gommers F J, Maas P W Th (Eds.) Nematology from Molecule to Ecosystem. Dekker and Huisman, Wildervank, The Netherlands.

Van Haren R J F, Hendrikx E M L, Atkinson H J (1993) Growth curve analysis of sedentary plant parasitic nematodes in relation to plant resistance and tolerance. Pages 1–11 in Grasman J, Van Straten G (Eds.) Predictability and Nonlinear Modelling in Natural Sciences and Economics. Proceedings in Nonlinear Science. Wiley, New York, USA.

Ward S A, Rabbinge R, Den Ouden H (1985) Construction and preliminary evaluation of a simulation model of the population dynamics of the potato cyst nematode *Globodera pallida*. Netherlands Journal of Plant Pathology 91:27–44.

14. Factors involved in the development of potato late blight disease (*Phytophthora infestans*)

J.G. HARRISON
Scottish Crop Research Institute, Invergowrie, Dundee DD2 5DA, Scotland, UK

Abstract. Late blight, caused by infection by the fungus *Phytophthora infestans*, is the most serious disease of potatoes and can completely destroy a crop. Blight is controlled by frequent applications of fungicides to the foliage, but there is increasing concern over the liberal use of agrochemicals. Loss of tuber yield depends on the amount of green foliage destroyed by the disease which, in turn, depends on the timing of the onset of blight and the rate of increase in blighted foliage. Modelling potato blight attempts to describe disease onset and progress using as input the many factors that affect blight.

Sexual reproduction of *P. infestans*, resulting in the formation of oospores, is probably important in the survival of the fungus in the field and in increasing the genotypic diversity of *P. infestans*. However, most of the build-up of disease results from the asexual reproductive cycle that consists of spore germination, infection of foliage, colonisation of foliage, sporulation, and spore dispersal and survival. The different stages in the asexual cycle of reproduction are affected by environmental factors that are largely determined by weather conditions and genotypic and physiological features of the pathogen (*P. infestans*) and the host. Most factors affecting blight interact with each other in complex ways, and their effects are difficult to quantify. Furthermore, many factors change continually and also vary horizontally and vertically within a crop, so the development of a realistic model for potato late blight is not easy.

Introduction

Late blight (blight) is the most serious disease of potatoes and can completely destroy a crop. Blight first appeared in Europe in Belgium in June 1845 and spread rapidly. Within a few months it had affected the potato crop throughout Ireland with devastating effects (Bourke 1965). Most people depended on potatoes as their main food and millions died from starvation or emigrated to escape the famine. Blight is caused by a phytopathogenic fungus, *Phytophthora infestans*, that infects and kills the leaves, stems and tubers of the potato plant. Fungicides have been used to control blight since the nineteenth century and today control depends on their frequent application, although genetic resistance of the potato plant to blight is increasingly important. The liberal application of agrochemicals has been questioned in recent years and there is concern over their potentially harmful effects on both man and the environment. Effective control of blight with a minimum of fungicides is therefore desirable. Fungicides

A.J. Haverkort and D.K.L. MacKerron (eds.), Potato Ecology and Modelling of Crops under Conditions Limiting Growth, 215–236.

are usually sprayed on to the haulm of a potato crop every 7–14 days to control blight, so there may be up to ten applications during the growing season (Krause et al. 1975). However, it has long been known that the date of blight first appearing in a crop and its subsequent progress depend, *inter alia*, on the weather. Blight is associated with cool, moist conditions and leaf infection occurs only if the foliage is wet (Rotem et al. 1971). The destructive nature of blight and the realisation that its progress depends on the weather provided the stimulus for the development of blight prediction systems that could be used to schedule the application of fungicides. Van Everdingen (1926) produced the first system for forecasting the onset of the disease, based on temperature, rain, cloudiness and the duration of dew. Subsequently there have been many attempts to formulate new systems and to improve existing ones (Harrison 1992) with the consequence that the forecasting of blight is now more advanced than that of any other plant disease.

Forecasting systems led to the development of computer programmes e.g. BLITECAST (Krause et al. 1975) from which computer-based simulations describing disease development were later developed (Bruhn and Fry 1981). With ever increasing sophistication, some systems now consider residual fungicide activity and tolerance of *P. infestans* to fungicides (Levy et al. 1991), and others use weather forecasts (Raposo et al. 1993). Computer-based models describing the behaviour of blight have recently been linked with those describing crop growth to simulate host-pathogen interactions and provide estimates of both disease progress and crop production (Van Oijen 1992). Computer simulations can be used to assess the consequences of altering one or more factors on blight severity and its effects on crop growth and yield. Although computer simulation models produce results much more quickly and with a lower labour input than do field experiments the simulations do depend for their accuracy on a realistic assessment of the factors they include, and on the inclusion of all factors that affect the parameter(s) under investigation. Because the factors affecting blight are not fully understood the simulation models all contain approximations, assumptions and guesses, so their potential for error is great. This paper considers some of the many interacting factors that should be taken into account in modelling blight development, in a synthesis of existing knowledge.

Life cycle of *P. infestans*

Knowledge of both the life cycle of *P. infestans* and the related cycle of the disease it causes are essential in order to understand how blight is affected by the various factors that influence its development. The life cycle of *P. infestans* is shown in Figure 1 which is largely self-explanatory. The asexual part of the cycle, i.e. the production of sporangia and the release and then germination of zoospores to produce hyphae, predominates in the build-up of disease. The sexual stage with the production of oospores that may survive for long periods

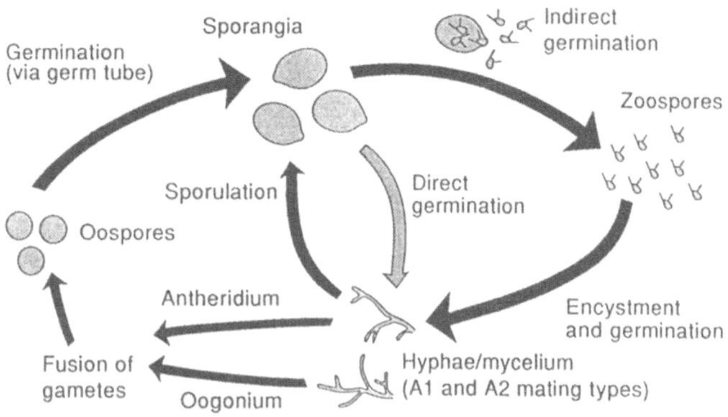

Figure 1. Life cycle of *Phytophthora infestans.*

in field soil normally requires the presence of both the A1 and A2 mating types. Its importance outside Central America has been uncertain until recently, but there is now compelling evidence that sexual reproduction frequently occurs, in the field, in Europe (Drenth 1994). Genetic recombination occurring during sexual reproduction provides a mechanism to increase the genotypic diversity of *P. infestans* so that it can adapt more readily to adversity, for example, by becoming tolerant of fungicides or overcoming host resistance. Genetic recombination may occur by parasexual mechanisms (Leach and Rich 1969), but these probably play a minor role in increasing the diversity of *P. infestans.*

Disease cycle and epidemic development

The disease cycle (Figure 2) reflects the life cycle of the pathogen as it occurs on the host, with leaf infection, lesion expansion, sporulation, and spore dispersal and survival largely determining disease development.

Sporangia arising on shoots growing from infected tubers, often in cull piles (Van der Zaag 1956), are blown to a new crop and form the primary inoculum. Blight is often initiated by a single sporangium or several sporangia each of which produces a leaf lesion. The lesions expand and sporangia are formed from a few days after infection (Figure 3) with eventually up to about 10^6 sporangia present on a single lesion (Harrison and Lowe 1989). Sporangia are dispersed by rain and wind (Fitt and Shaw 1989) and infect more leaves, so the unrestricted progress of blight follows a compound interest pattern (Van der Plank 1963) and the rate of increase in the amount of blighted foliage is explosive (Figure 4). Cycles of infection and lesion expansion are shown in Figure 4 as irregularities in the approximately sigmoidal disease progress curve. Mathematical models and computer simulations of blight progress attempt to describe this curve but its form is determined by a multitude of factors, some favouring blight and others not (Figure 5). Spores washed into the soil can infect tubers that act as an inoculum source for subsequent crops.

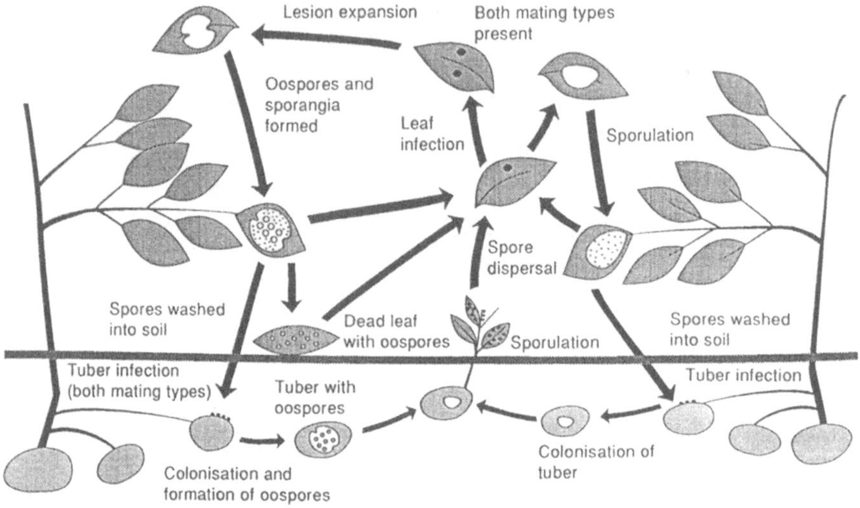

Figure 2. Disease cycle of potato late blight.

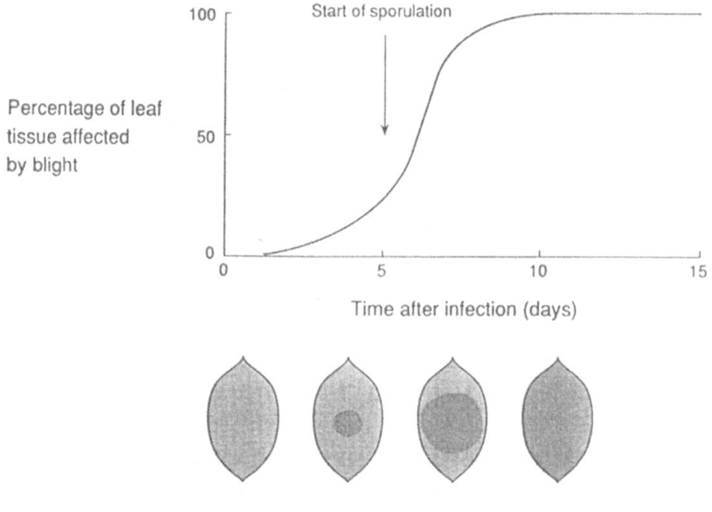

Figure 3. Increase in area of single blight lesion on a potato leaf under constant conditions.

Effects of blight on crop growth and yield

Effects of blight on crop growth and tuber yield depend on the disease progress curve (Figure 6). In the absence of senescence, the potential for photosynthesis of uninfected tissue is not significantly affected by the pathogen in other parts of the plant (Van Oijen 1990) and the loss of harvested yield is almost entirely due to a reduction in leaf area duration caused by the disease (Haverkort and

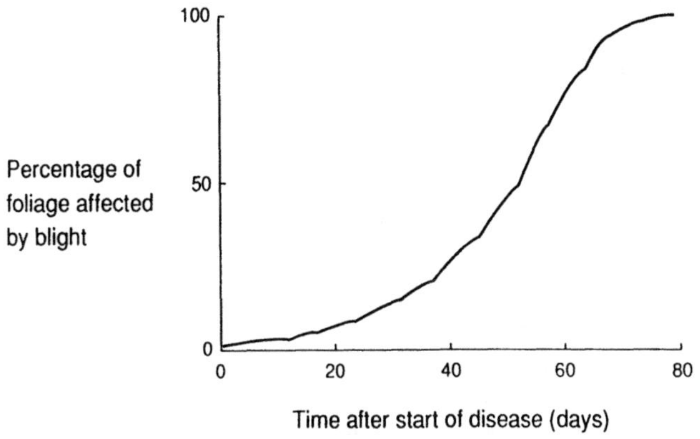

Figure 4. Hypothetical increase in blight lesion area on potato foliage of whole crop.

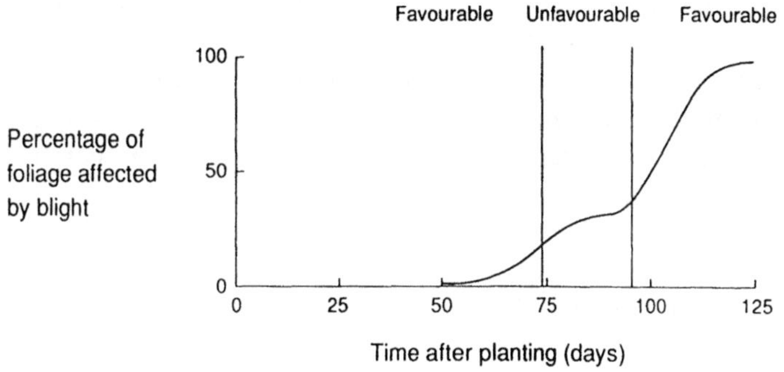

Figure 5. Disease progress curve during favourable and unfavourable periods for development of potato blight.

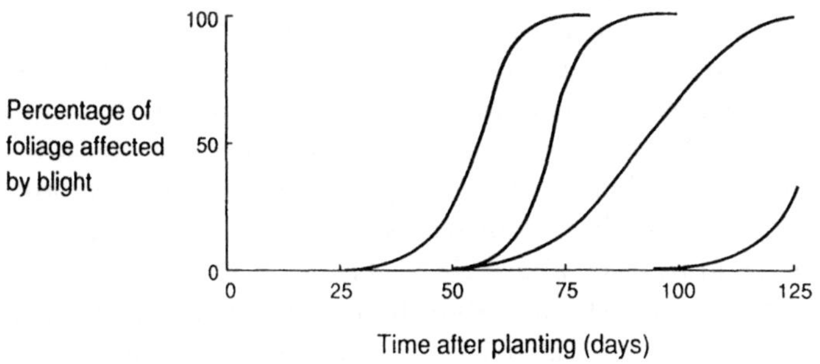

Figure 6. Disease progress curves showing effects of a) time of initiation of potato blight. b) rate of blight progress.

Bicamumpaka 1986; Van Oijen 1990; 1991a,b). Van Oijen (1991b) considered that the area under the disease progress curve was closely and inversely related to yield. Clearly, if blight develops only slowly, or if it starts late in the season after the bulk of the potential tuber yield has been formed, then losses at harvest will be relatively small (Figure 6). Infected tubers, however, frequently rot during storage (Dowson and Rudd Jones 1951) so blight can also cause losses after harvest.

Factors affecting blight and the asexual cycle of *P. infestans*

The factors that affect the timing of the onset of blight and its subsequent progress can be divided into three broad categories: environmental factors, determined mainly by weather conditions; biotic factors; and fungicides. Biotic factors can be sub-divided conveniently into those of the host, i.e. the potato plant, and those of the pathogen, i.e. *P. infestans*. However, it cannot be over-emphasised that effects on blight are the result of complex *interactions* between factors. In order to understand how each factor affects blight, it is necessary to explore the interactions, although the resulting study may sometimes be related only tenuously to field conditions. The dynamic nature of the field environment further complicates matters. Weather changes continuously, as do the physiology of the host and pathogen, and often also the genotype of the pathogen. It may be difficult to relate results from experiments conducted under controlled conditions to a crop in a field.

Because the asexual life cycle dominates disease development, and is usually its sole determinant, I will discuss the effects of various factors on infection and colonisation of foliage by *P. infestans*, sporulation, and spore dispersal and survival.

Environmental factors

Temperature. Sporangia can germinate either directly by the formation of a germ tube, or indirectly by releasing zoospores but leaves are not infected from direct germination (Schöber and Ullrich 1985). The optimum temperature for indirect germination appears to be between 12 and 16 °C (Crosier 1934; Yamamoto and Tanino 1961) although zoospores are released slowly at temperatures approaching 0 °C. At higher temperatures the proportion of sporangia germinating directly increases, although indirect germination does occur at temperatures well above 20 °C. The length of time zoospores remain motile before they encyst and start to germinate is highly dependent on temperature. They swim for up to 24 h at temperatures close to 0 °C, but for only 15–20 min at 24–25 °C. Zoospores are extremely sensitive to desiccation and die almost immediately unless they are in water. They depend for their survival on the plant surface remaining wet until they have infected so, temperature and duration of surface water together determine whether or not infection occurs. Crosier (1934) and Rotem et al. (1971) observed germination of zoospores between about 3 and 28 °C. It was fastest at 21–24 °C. Germ tube elongation was most rapid at about 21 °C and infection could occur within 2–3 h of inoculation.

It is unrealistic to separate germ tube growth from colonisation of plant tissues by hyphal growth and, indeed, they have similar cardinal temperatures, with hyphae growing slowly at 2–4 °C and not at all above about 28–30 °C (Zan 1962). Lesions readily visible to the naked eye do not normally appear until at least 2–3 days after inoculation, even at temperatures that favour hyphal growth. The production of sporangia (sporulation), a process that can be considered to be modified hyphal growth, has a similar minimum temperature to germination and colonisation of tissues, but the maximum temperature of 25--26 °C for sporulation is a few degrees below that for the other processes. The optimum temperature for both the speed of formation of sporangia and their final abundance has been variously reported to be 16–22.5 °C. Fresh sporangia can be formed under optimum conditions a few hours after the removal of old spores from a lesion.

Temperature probably has little direct effect on spore dispersal but it is a major factor in the survival of *P. infestans*. The low temperatures required to kill the fungus would completely destroy a potato crop, so they are irrelevant to disease development during the growing season. Effects of temperature on survival are therefore of concern only at high temperatures. Survival of sporangia in air depends on an interaction of factors, chiefly temperature, humidity, irradiation and the physiological state of spores, determined by factors such as their age and water content. For example, a change in temperature, caused by a change in irradiation, affects the relative humidity and the rate of water loss from sporangia and so their sensitivity to the potentially damaging effects of high temperatures. Sporangia die rapidly in air above about 25 °C (Smith 1915), and Wallin (1953) reported that temperatures as low as 20 °C reduced their viability. However, survival at different temperatures is highly dependent on humidity. Rotem and Cohen (1974) found that sporangia deposited on leaf surfaces, where the humidity is usually high due to transpiration, survived adverse conditions well. There is little information on direct effects of temperature on zoospore survival but the drying of plant surfaces, resulting in desiccation of zoospores, is itself affected by temperature. Once established in plant tissues, *P. infestans* is remarkably tolerant of high temperatures. While it may not grow at temperatures above 30 °C, it can survive much higher temperatures, even those above 40 °C for short periods and recommence growth and development when the weather becomes more favourable (Rotem and Cohen 1974), although its capacity to produce sporangia may be impaired by exposure to high temperatures (Kable and MacKenzie 1980). It is important to appreciate that leaf and stem temperatures can be considerably higher than air temperatures, especially when foliage is exposed to direct sunlight (Ansari and Loomis 1959).

Infection of tubers is encouraged by moderate soil temperatures and is markedly decreased above 18 °C (Sato 1979).

In the interaction between temperatre and other factors such as duration of surface moisture and humidity, temperature limits are broad when other factors are favourable and narrow when they are unfavourable (Rotem et al. 1971).

Temperature also has an indirect effect on asexual development of *P. infestans* through its effects on the potato plant, considered later in this chapter, and loss of fungicide activity.

Humidity. Humidity is a major determinant of blight development, yet it is impossible to control humidity precisely, or to measure it in the zone of air where infection and sporulation occur as no instrument is small enough to measure humidity adjacent to plant surfaces. Transpiration humidifies the air adjacent to leaves so, although the air above a crop may be relatively dry, it is close to saturation next to an unwilted leaf. The gradient of decreasing humidity away from a leaf surface depends on the rate at which transpired water vapour is swept away which in turn depends on air speed. Further, leaves heated by sunlight warm the adjacent air and lower its relative humidity (RH), further complicating the estimation of humidity next to a leaf. The net result of these effects is that we do not know precisely how humidity affects microorganisms on plant surfaces. However, sporangia and zoospores germinate only in the presence of liquid water so humidity influences infection indirectly through its effects on the rate of drying of surface moisture.

Once *P. infestans* has become established within the potato plant, RH between 80 and 100% has no effect on colonisation (Harrison and Lowe 1989). Lower humidities may cause desiccation of severely infected tissues, limiting the availability of water to the pathogen and reducing its growth rate.

There are numerous reports that sporangia are formed only at high humidities (e.g. De Weille 1963) but the precise requirements close to a plant surface are unknown. Harrison and Lowe (1989) demonstrated an interaction between ambient humidity and air speed in determining numbers of sporangia formed on leaflets. Abundant sporangia were formed at low air speeds when the ambient humidity was 90–100% RH, but not at 85 or 80% RH. At higher air speeds numerous sporangia were formed at 100% RH, but not at 95–80% RH. These results suggest that sporangia are formed only if the humidity immediately surrounding the pathogen is at or very close to saturation. Harrison and Lowe also investigated sporulation when the humidity alternated between 80 and 100% RH. Numbers of sporangia formed were intermediate between those at each humidity continuously and higher than the number formed at a constant 90% RH.

Hirst (1958) thought that variations in humidity aid dispersal of sporangia, although the changes in RH that he observed were largely the result of temperature changes. Rotem and Cohen (1974) considered that low humidities enhance dispersal.

Humidity has a large effect on survival of spores. Sporangia of *P. infestans* can survive for several days in saturated air (Rogoshin and Filippov 1983) and survive well on leaf surfaces, but viability is lost rapidly in drier air (Rotem and Cohen 1974) and more so at high than at low temperatures. Warren and Colhoun (1975) considered that survival of sporangia depended on a surrounding film of water, without which sporangia lost their viability after only 5 min in air at 95% RH. Minogue and Fry (1981) presented evidence that the germination ability of partially desiccated sporangia depended on the rate of subsequent rehydration. A low rate favoured survival, while only a few spores germinated after rapid rehydration. Survival of sporangia in relation to environmental conditions almost certainly depends on spore age (considered later in this chapter).

Zoospores die rapidly in the absence of liquid water (Lapwood 1968) and there are no quantitative data on their survival at different humidities.

Precipitation and dew. Rain, drizzle, mist, overhead irrigation and dew formation result in wet foliage, a prerequisite for infection. Dew usually forms at night as temperatures fall. The rate at which it forms on leaves is related to temperature, ambient humidity and wind speed (Collins and Taylor 1961) and, conversely, the rate at which surface water, however deposited, evaporates is also related to temperature (determined partly by irradiation intensity), ambient humidity and wind speed. These meteorological variables therefore interact to determine the duration of surface moisture and the likelihood of sporangia releasing zoospores and of zoospores germinating and infecting the host. Duration of surface moisture itself interacts with inoculum density and temperature to determine whether or not germination and infection occur. Longer periods of wetness are required for germination as the temperature deviates from the optimum, and high spore loads inhibit infection when the wet period is short (Rotem et al. 1971). A short period during which the leaves are dry, initiated within the first 3 h after inoculation, substantially reduces the number of lesions that develop, but temporary surface dryness occurring later generally has less effect, presumably because infection is better established (Hartill et al. 1990). Subsequently, surface water probably has little or no effect on colonisation of plant tissues.

Although formation of sporangia occurs after rain or dew, liquid water is not required (Harrison and Lowe 1989). Sporulation is undoubtedly encouraged by the high humidities close to foliage associated with surface moisture, rather than by the liquid water itself. Indeed, abundant water on foliage may suppress sporulation of *P. infestans* (Rotem et al. 1978).

Precipitation and wind are the main factors controlling the dispersal of sporangia of *P. infestans*. Rain, drizzle and overhead irrigation wash spores from leaves and stems into the soil where they can remain infective for several weeks (Lacey 1965) and may infect tubers (Lapwood 1977) or may be splashed on to leaves to initiate lesions. Intensity and duration of precipitation determine the extent to which these occur. Sporangia are readily incorporated into water falling on them. Splash droplets containing spores, produced when water droplets hit sporangia on a leaf or stem are responsible for transporting the pathogen over short distances within a crop (MacKenzie et al. 1983). Fitt et al. (1986) found that fungal spores were dispersed in splash droplets in still air as far as 1 m from infected leaves. Numbers of spores declined exponentially with distance from the source and decreased rapidly with increasing height. Fewer than 10% of spores were splashed to a height of more than 300 mm. Large raindrops dispersed more spores than did smaller drops. Duration and intensity of precipitation and the size distribution of water droplets are therefore important in determining the extent of dispersal in this manner. Hirst (1958) thought that about half the foliage blight lesions in a crop originated from rainwater dripping from one leaf to another. A small proportion of droplets splashed from foliage and containing spores may be launched into the prevailing air flow. The role of wind in the dispersal of *P. infestans* is dealt with in the following section. Rain is important in removing sporangia from the air (Hirst 1953).

Wind. Wind is not known to affect colonisation of tissues by *P. infestans* but modifies infection of aerial plant parts indirectly through drying surface moisture. Wind affects sporulation through its effect on humidity close to foliage depending *inter alia* on the amount of haulm and the geometry and orientation of individual leaves. In the more sheltered, lower parts of the canopy wind speed is less than at the top of the haulm so there is usually a gradient of decreasing humidity with increasing height from the soil surface, reflected in a tendency for fewer spores to be produced in the upper parts of the canopy. However, turbulence increases with wind speed, reducing gradients of both humidity and temperature, so that the environment within the canopy becomes more uniform at higher wind speeds. Wind speed just above a crop also varies with time – wind is usually gusty, and in space depending on the topography of the land and shelter from trees and buildings. This dynamic interaction of factors that determines air movement within a crop and thus the duration of surface moisture and conditions suitable for sporulation and germination, makes it difficult to quantify from weather data the sporulation and infection that are likely to have occurred. Conditions may favour sporulation and infection in some parts of a field and not in others.

The principal effect of wind is on dispersal of sporangia but few sporangia are detached by wind alone, even at speeds of up to 15 m/sec (Stepanov 1935; Hirst 1958). Wind can spread rain-splash droplets containing spores within the crop or to adjacent fields (Hirst and Stedman 1960) often resulting in a fan-shaped pattern of disease from spreading a locus of infection. Being blown by the wind is the sole means by which spores can travel long distances. Once they have entered the air above a crop, droplets containing sporangia can be carried many kilometres (Robinson 1976) and to heights up to 1 km, particularly in a strong wind. The importance of wind in long distance spread of blight, however, depends on sporangia remaining infective, favoured by cool, moist, dull conditions.

Irradiation. The spectrum of solar radiation has both direct and indirect effects on *P. infestans* and is an important factor reducing the survival of sporangia (Rotem et al. 1970). The indirect effects are caused by the radiation heating plants, soil and the air adjacent to them, reducing RH, and increasing the rate of evaporation (see the earlier discussion). Bashi et al. (1982) thought that the death rate was aggravated by high temperatures and fully turgid sporangia may be less damaged by exposure to sunlight than partially desiccated spores.

Information on direct effects of solar radiation on the asexual development of *P. infestans* is sparse. At low intensities, the UV radiation may increase the proportion of sporangia that are able to germinate and infect a potato plant, but it can kill spores at high intensities (De Weille 1963; 1964). There are conflicting reports of effects of visible radiation on production of sporangia and its effects on colonisation may be due solely to changes in the physiology of the host which are considered later.

Biotic factors – the potato plant

Genetic resistance. There are two forms of genetic resistance of the potato plant to blight, controlled by major and minor genes respectively, although the distinction between them is sometimes unclear. Major and minor genes are often present together in a potato plant. Major gene resistance, also known as race-specific resistance, usually acts at the infection stage and can effectively result in total resistance to blight. So far, at least eleven major resistance genes (R1 to R11) have been identified each of which can lead to immunity to blight. However, for every host R-gene, *P. infestans* can possess a corresponding virulence gene (virulence genes or factors 1 to 11) that can completely overcome the complementary R-gene and lead to disease susceptibility. A strain, or isolate, of *P. infestans* with virulence gene 1 (i.e. that can overcome resistance gene R1 in the potato plant) is referred to as race 1, etc. A plant may not possess any R-genes (RO) so that it is susceptible to all races of *P. infestans* or it may possess one or more R-genes. Similarly *P. infestans* may have no virulence genes (race 0), or one or more may be present. No blight will develop in the field unless there is a race of *P. infestans* compatible with the crop genotype. However, *P. infestans* readily adapts to overcome R-gene resistance (Malcolmson 1969), which is therefore short-lived, so the incorporation of R-genes into potato genotypes has largely been discounted for blight control, although many potato cultivars do possess R-genes. There is evidence that not all R-genes are expressed to the same extent. In particular R10 is sometimes incompletely expressed (Wastie et al. 1993) so that *P. infestans* without virulence factor 10 can infect and colonise the tissues albeit at a reduced rate. The expression of major gene resistance may be partially dependent on the physiology of the potato plant which itself may depend on the environment and plant age (Stewart 1990). Although the resistance of both foliage and tubers is controlled by R-genes, there is often only a poor correlation between foliage resistance and tuber resistance (Stewart et al. 1992). For example, genes R2 and R3 provide leaf immunity to races of *P. infestans* without virulence factors 2 and 3, but the same isolates may grow in the tubers. In contrast, R1 appears to provide comparable immunity in leaves and tubers (Roer and Toxopeus 1961). These differences may be due to physiological/biochemical factors, e.g. only tubers can produc sesquiterpenoid phytoalexins (Rohwer et al. 1987), or to the involvement of other genetic factors.

Minor-gene resistance is thought to be controlled by many genes whose effect is additive. It can reduce the proportion of zoospores that infect the foliage, but its main effect is in reducing the rate of colonisation (Colon et al. 1992) so it does not confer immunity. However, the relative importance of the different resistance components depends on cultivar. Being under multi-genic control, minor gene resistance is much more durable than R-gene resistance and offers the best potential for long-term blight control using a minimum of fungicides (Shtienberg et al. 1994). However, it is sometimes difficult to separate the effects of minor and major gene resistance (Gees and Hohl 1988). There is also evidence for synergism between the two forms of resistance (Darsow et al. 1987). It has become increasingly apparent in recent years that expression of minor gene resistance of some cultivars is strongly influenced by the environment (Simmonds and Wastie 1987). This genotype-x-environment interaction was investigated by Harrison et al. (1994) who identified three environmental components (photoperiod, light intensity and temperature) each of which could effectively switch on or off minor gene resistance in leaves. Depending on the cultivar, resistance could be switched by one or more environmental factors, or be unaffected by any of them. The possibility that R-gene resistance may be similarly influenced by the environment cannot be excluded. Minor genes can also determine the resistance of tubers to *P. infestans* (Bjor and Mulelid 1991). Again, foliage and tuber resistance may not be closely correlated.

Neither major- nor minor-gene resistance appears to have a direct effect on sporulation of *P. infestans* once infection has occurred.

The structure of the leaf canopy, which is largely genetically determined, can also affect blight susceptibility by, for example, impeding air circulation resulting in slow drying of foliage and high humidities within the canopy (Bonde et al. 1940).

Physiology. The physiological state of a potato plant has a large influence on blight susceptibility. Although the mechanisms involved are poorly understood, certain factors have been identified that affect infection and colonisation. Effects on sporulation are usually indirect and simply reflect the extent of colonisation.

The physiological age of a plant, itself determined by growing conditions, particularly temperature and the length of the growing period, affects the susceptibility of leaves to infection. Leaves of young plants are highly susceptible to infection. As plants age the leaves first become more resistant to infection and later become more susceptible again (Stewart 1990). Carnegie and Colhoun (1982) reported the self-evident effect that the most blight resistant cultivars showed the greatest changes in susceptibility. Most reports relating differences in blight susceptibility to plant age do not distinguish between effects on infection, colonisation and sporulation (e.g. Darsow et al. 1988).

Leaf position, which is generally confounded with leaf age, also affects susceptibility to infection. Fry and Apple (1986) thought that the oldest leaves were most susceptible but Warren et al. (1971) found that, in mature plants, leaves in the middle zone of a stem were the most resistant, and that leaf position had no effect in young plants. Populer (1978) reported that effects of leaf position on susceptibility to infection depended on cultivar.

Rate of colonisation of leaf tissues similarly depends on leaf position and age. Carnegie and Colhoun (1982) reported that the rate of lesion expansion generally increased linearly from the apex to the base of a plant, but that on young plants these differences were absent.

As with foliage, the age of tubers affects their susceptibility to blight (Pathak and Clarke 1987).

The maturity class of a potato cultivar is another factor that modifies the incidence of blight with differences probably related to foliage growth (Van Oijen 1991b). Early varieties are generally less attacked by blight than later ones (Kolbe 1982), possibly because growth is often completed and the crop harvested before inoculum levels become high enough to cause extensive disease. Resistance to infection, colonisation of tissues or sporulation are probably unaffected by maturity class.

Mineral nutrition of the host affects its susceptibility to blight and Rotem and Sari (1983) suggested that infection of leaves, colonisation and sporulation are all affected by host nutrition. Plants grown with high levels of nitrogen are particularly susceptible to both foliage and tuber blight and have higher rates of lesion expansion (Carnegie and Colhoun 1983; Kurzawinska 1989) with genotypic differences (Main and Gallegly 1964). Awan and Struchtemeyer (1957) reported that large applications of phosphorus or potassium decreased the size of leaf lesions, with P having the greater effect. Nutrient levels in leaves can change lesion diameter by a factor of 2–3 which, presumably, reflects on the potential number of sporangia that can be formed. Borys (1964) found that mineral nutrition, especially chloride and sulphate ions, strongly affected the susceptibility of leaves to infection, the effects depending on leaf age and cultivar. Some of the effects of mineral nutrition, especially N, on blight susceptibility may be indirect, by encouraging luxurious haulm growth or by delaying leaf senescence.

The carbohydrate status of potato plants has been implicated as a factor affecting blight susceptibility (Grainger 1979) but its importance remains unclear.

Carnegie and Colhoun (1980) reported that potato plants under water stress had increased resistance to infection by *P. infestans*, but they offered no explanation of this observation.

Virus infection. Infection of potato plants with any of several viruses tested increased the resistance to blight of leaves (Pietkiewicz 1974) and tubers (Fernandez de Cubillos and Thurston 1975). The presence of virus increases resistance to both infection and colonisation by *P. infestans*, delays the onset of sporulation and reduces the number of sporangia formed. However,

Richardson and Doling (1957) reported that the rolled leaf surfaces of plants infected with potato leaf roll virus remained wet for longer than the flat surfaces of uninfected leaves, encouraging infection, so that more blight lesions developed in the field, although in laboratory tests, small leaf pieces from virus-infected plants were less susceptible to infection than pieces of virus-free leaves.

Surface microflora.
Naturally-occurring bacteria that colonise the surface of tubers have recently been shown to be antagonistic to *P. infestans* (S.A. Clulow, personal communication) but there is no known effect, either as promotors or inhibitors, of microorganisms that occur naturally on the phylloplane.

Biotic factors – P. infestans

Genotype. The importance of virulence factors in overcoming R-gene resistance to blight in potato foliage and tubers has been explained. *P. infestans* can acquire new virulence factors either through mutation (Denward 1970) or through somatic variation (Caten and Jinks 1968). Even in the absence of sexual reproduction, new races can be produced when colonies of two different races are grown together (Leach and Rich 1969; Malcolmson 1970) or when a single isolate is passaged through hosts with different R-genes (Graham et al. 1961). There is a strong selection pressure in the field favouring races of *P. infestans* that are pathogenic on the crop, so the population structure of the pathogen may change during the growing season (Caten 1970; Shattock et al. 1977).

The aggressiveness of *P. infestans*, a characteristic that is independent of virulence, depends on the isolate of the pathogen and has a major effect in determining rates of infection and colonisation of leaves and tubers, and generation time (period between inoculation and sporulation) on leaves (Latin et al. 1981). Aggressiveness, like most physiological parameters, is difficult to quantify. It is also rather unstable and can increase or decrease over many generations of an isolate (Jinks and Grindle 1963). Caten (1970) suggested that directional selection for rapid growth and profuse sporulation operates in the field. He provided evidence for adaptation to specific cultivars, supporting the report of Latin et al. (1981) that minor gene resistance and isolate (aggressiveness) interact to determine blight severity. The results of Bjor and Mulelid (1991) using tubers without R-genes also suggested that adaptation of *P. infestans* to specific cultivars may occur and that minor gene resistance to blight may eventually become eroded.

Spielman et al. (1992) related individual components of pathogen fitness (aggressiveness) to disease progress. They concluded that sporulation capacity was most closely related and infection frequency was least related.

P. infestans can become adapted to the climate of a region. For example, the optimum temperature for sporulation of isolates from Egypt is several degrees higher than that of British isolates (Shaw 1987). Such differences, presumably genetically-determined, may account for apparent discrepancies in reports on the sensitivity of *P. infestans* to temperature.

There are numerous reports of fungicide tolerance in *P. infestans* (e.g. Holmes and Channon 1984). Populations of *P. infestans* in the field are normally heterogeneous, particularly late in the growing season, and consist of both fungicide-susceptible and fungicide-tolerant types in a dynamic equilibrium. The use of a fungicide shifts the balance between the two in favour of strains that tolerate the fungicide (Levy et al. 1983). In the absence of fungicide, metalaxyl-resistant isolates of *P. infestans* showed strong competitive ability and their proportion increased from 10 to 100% after 8–10 sporulation cycles (Kadish and Cohen 1988). This strong competitive ability was attributed to a rapid rate of colonisation of leaves, rather than to differences in infection or sporulation capacity. However, metalaxyl-resistant isolates may not survive in tubers as well as sensitive isolates (Kadish and Cohen 1992).

Physiology. There is a paucity of information on possible effects of the physiological state of *P. infestans* on blight development, despite the fact that effects of the environment clearly have a physiological basis. Moreover, the genotype of *P. infestans* has a strong influence on its physiology, so environmental effects, genetics and physiology are all inter-related.

Physiological and biochemical changes associated with ageing may render all stages of the pathogen more vulnerable to adverse conditions, so that, for example, old sporangia may lose their viability and die more rapidly than freshly-produced spores. The author has observed that zoospore release is slower from old than from young sporangia. It therefore follows that old sporangia would require a prolonged period of surface wetness for infection. Bashi et al. (1982) reported that sporangia dispersed late in the day were more infectious than those that had been dispersed earlier. De Weille (1963) considered that newly-formed sporangia needed a period of maturation before they would germinate. As mature spores age their environmental requirements for germination may become more rigorous so that zoospores can be released only within a narrow temperature range. Effects of ageing or nutritional status on colonisation of the host and on sporulation are speculative.

Inoculum density. The amount of inoculum, effectively number of sporangia, obviously has a major bearing on disease development. In the absence of inoculum, no disease develops at all. Under a given set of conditions the amount of blight that develops on foliage is directly proportional to inoculum quantity, but it also depends on the amount of healthy foliage available for infection and colonisation, which at any particular time, itself depends on preceding rates of disease progress and haulm growth. In the initial stages of blight, the inoculum depends on number and proximity of infected shoots, often in cull piles or refuse heaps, but sometimes in the crop itself or on volunteer plants in other fields (Van der Zaag 1956). The inoculum also depends on the suitability for sporulation and spore dispersal of weather conditions and on the direction of the prevailing wind. As blight progresses, sporangia formed within the crop dominate, although spores from sources outside the crop remain important in the introduction of new races, strains or a second mating type of *P. infestans*.

There is a marked diurnal periodicity in the concentration of airborne sporangia, with a maximum usually occurring during late morning (Hirst 1953), which, together with the suitability of conditions for infection, will influence the number of new lesions initiated.

The extent of tuber infection clearly depends on numbers of sporangia formed on the foliage.

Fungicides

Although the scope of this paper is restricted to the 'natural' ecology of blight, the almost universal application of fungicides for its control means that factors affecting their efficacy in slowing down blight development should be considered briefly. Again, many factors interact, this time to determine the effect of fungicides. The most important are fungitoxicity, itself dependent on fungicide tolerance by *P. infestans*, rate and frequency of application of fungicide and its distribution within the crop canopy, type of fungicide (e.g. systemic or contact), its redistribution and the rate of decline of fungicidal activity (determined mainly by precipitation and temperature, and the properties of the fungicide), and the rate of growth of new foliage. Again, some of these factors, such as the distribution and dynamics of fungicides, are difficult to quantify.

Factors affecting the sexual cycle of *P. infestans*

The potential importance of sexual reproduction of *P. infestans* in the epidemiology of late blight has already been stressed, but knowledge of factors that affect the production and germination of oospores is scant. While oospores are produced freely when hyphae of both mating types grow together, they can also be formed in the laboratory as a result of self fertilization of isolates of either mating type (Skidmore et al. 1984; Campbell et al. 1985). The importance of self-fertilization in the field, if it occurs at all, is not known. Chang and Ko (1990) claimed that *P. infestans* isolates of mating type A1 could change into type A2 when grown in the presence of metalaxyl. As the author knows of no other reports of a change of mating type, that report may be anomalous.

Harrison (1992) and Drenth (1994) found that oospores could be formed at temperatures from 5 to 25 °C. Numbers of oospores formed depended on host cultivar, with fewer in leaf tissues that were highly susceptible to blight and that rotted quickly, than in tissues that were moderately resistant (Drenth 1994). Harrison considered that the rate of oospore formation was closely related to the rate of hyphal growth. Smoot et al. (1958) observed germination of oospores from 12 to 25 °C. There is evidence that continuous illumination inhibits oospore formation (Romero and Gallegly 1963), and of a light requirement for germination (Shattock et al. 1986). Drenth (1994) hypothesized that oospores at or near the soil surface released zoospores during rainy conditions and that these could infect plant parts in contact with the soil, or be splashed to initiate infections higher in the canopy. Oospores readily survive temperatures as low as − 80 °C and up to 35 but not 40 °C (Drenth 1994), but the conditions favouring survival and length of time they can survive in the field are not known.

Conclusions

This paper has attempted to explain how development of potato late blight is controlled by many inter-dependent and constantly-changing factors. There may be others that remain to be discovered, particularly concerning the physiology and biochemistry of the host-pathogen interaction.

Modelling blight progress requires quantification of factors that affect it. While quantification of some environmental factors may not be straightforward, there are much greater problems in attempting to quantify biotic factors, particularly those that determine the degree of susceptibility to blight. A further major difficulty in attempting to develop a realistic model to describe blight progress is the heterogeneity within a field, with disease developing at different rates depending on location. Addressing these two problems offers the greatest potential for improving the modelling of blight. Systems already exist for the precise quantification of infection, colonisation, sporulation and, perhaps less precisely, spore dispersal and survival. A systematic study is required of effects of the biotic factors identified earlier, including interactions between biotic factors and between them and environmental factors, on these components of blight development. Similarly, effects of fluctuating environmental conditions on these same parameters need to be investigated, paying particular attention to possible lag phases resulting from carry-over effects of previous environments. It should be stressed, however, that each of these would be a major undertaking best tackled by large co-ordinated groups of workers. Although no mathematical model will ever be totally accurate, there is still considerable scope for improving existing systems.

References

Ansari A Q, Loomis W E (1959) Leaf temperatures. American Journal of Botany 46:713–717.

Awan A B, Struchtemeyer R A (1957) The effect of fertilization on the susceptibility of potatoes to late blight. American Potato Journal 34:315–319.

Bashi E, Ben-Joseph Y, Rotem J (1982) Inoculum potential of *Phytophthora infestans* and the development of potato late blight epidemics. Phytopathology 72:1043–1047.

Bjor T, Mulelid K (1991) Differential resistance to tuber late blight in potato cultivars without R-genes. Potato Research 34:3–8.

Bonde R, Stevenson F J, Clark C F (1940). Resistance of certain potato varieties and seedling progenies to late blight in the tubers. Phytopathology 30:733–748.

Borys M W (1964) Influence of calcium, magnesium, chloride and sulphate nutrition on the resistance of potato leaves to *Phytophthora infestans* de Bary. Acta Microbiologica Polonica 13:221–226.

Bourke P M A (1965) The Potato, Blight, Weather, and the Irish Famine. Ph.D. Thesis, University of Dublin, Dublin, Ireland.

Bruhn J A, Fry W E (1981) Analysis of potato late blight epidemiology by simulation modelling. Phytopathology 71:612–616.

Campbell A M, Duncan J M, Malcolmson J F (1985) Production of oospores in vitro by self fertilization in single isolates of A1 mating types of *Phytophthora infestans*. Transactions of the British Mycological Society 84:533–535.

Carnegie S F, Colhoun J (1980) Differential leaf susceptibility to *Phytophthora infestans* on potato plants of cv. King Edward. Phytopathologische Zeitschrift 98:108–117.

Carnegie S F, Colhoun J (1982) Susceptibility of potato leaves to *Phytophthora infestans* in relation to plant age and leaf position. Phytopathologische Zeitschrift 104:157–167.

Carnegie S F, Colhoun J (1983) Effects of plant nutrition on susceptibility of potato leaves to *Phytophthora infestans*. Phytopathologische Zeitschrift 108:242–250.

Caten C E (1970) Spontaneous variability of single isolates of *Phytophthora infestans*. II. Pathogenic variation. Canadian Journal of Botany 48:897–905.

Caten C E, Jinks J L (1968) Spontaneous variability of single isolates of *Phytophthora infestans*. I. Cultural variation. Canadian Journal of Botany 46:329–348.

Chang T T, Ko W H (1990) Effect of metalaxyl on mating type of *Phytophthora infestans* and *P. parasitica*. Annals of the Phytopathological Society of Japan 56:194–198.

Collins B G, Taylor R J (1961) Conditions governing the onset of dew on large leaves. Australian Journal of Applied Science 12:23–29.

Colon L T, Pieters M M J, Budding D J (1992) Field experiments on components of resistance to Phytophthora infestans (Mont.) de Bary in wild Solanum species. Pages 25–29 in Rousselle-Bourgeois F, Rousselle P J (Eds.) Proceedings of the Joint Conference of the EAPR Breeding and Varietal Assessment Section and the EUCARPIA Potato Section, Landerneau, France, 12–17 Jan. 1992. INRA, Ploudaniel, France.

Crosier W (1934) Studies in the Biology of *Phytophthora infestans* (Mont.) de Bary. Cornell University Agricultural Experiment Station Memoir No. 155.

Darsow U, Goebel S, Goetz E, Oertel H, Schueler K (1987) R-Gene und relative Resistenz der Kartoffelknolle gegen *Phytophthora infestans* (Mont.) de Bary. Archiv für Züchtungsforschung 17:387–397.

Darsow U, Junges W, Oertel H (1988) Die Bedeutung der Prädisposition für die Laborprüfung von Kartoffelblättern auf relative Resistenz gegenüber *Phytophthora infestans* (Mont.) de Bary. Archiv für Phytopathologie und Pflanzenschutz 24:109–119.

Denward T (1970) Differentiation in *Phytophthora infestans*. II. Somatic recombination in vegetative mycelium. Hereditas 66:35–47.

De Weille G A (1963) Laboratory results regarding potato blight and their significance in the epidemiology of blight. European Potato Journal 6:121–130.

De Weille G A (1964) Forecasting Crop Infection by the Potato Blight Fungus. Koninklijk Nederlandsch Meteorologisch Instituut, Mededelingen en Verhandelingen No. 82.

Dowson W J, Rudd Jones D (1951) Bacterial wet rot of potato tubers following *Phytophthora infestans*. Annals of Applied Biology 38:231–236.

Drenth A (1994) Molecular Genetic Evidence for a New Sexually Reproducing Population of *Phytophthora infestans* in Europe. Doctoral Thesis, Landbouwuniversiteit Wageningen, Wageningen, The Netherlands.

Fernandez de Cubillos C, Thurston H D (1975) The effect of viruses on infection by *Phytophthora infestans* (Mont.) de Bary in potatoes. American Potato Journal 52:221–226.

Fitt B D L, Creighton N F, Lacey M E, McCartney H A (1986) Effects of rainfall intensity and duration on dispersal of *Rhynchosporium secalis* conidia from infected barley leaves. Transactions of the British Mycological Society 86:611–618.

Fitt B D L, Shaw M W (1989) Transports of blight. New Scientist 123 (1677):41–43.

Fry W E, Apple A E (1986) Disease management implications of age-related changes in susceptibility of potato foliage to *Phytophthora infestans*. American Potato Journal 63:47–56.

Gees R, Hohl H R (1988) Cytological comparison of specific (R_3) and general resistance to late blight in potato leaf tissue. Phytopathology 78:350–357.

Graham K M, Dionne L A, Hodgson W A (1961) Mutability of *Phytophthora infestans* on blight-resistant selections of potato and tomato. Phytopathology 51:264–265.

Grainger J (1979) Scientific proportion and economic decisions for farmers. Annual Review of Phytopathology 17:223–252.

Harrison J G (1992) Effects of the aerial environment on late blight of potato foliage – a review. Plant Pathology 41:384–416.

Harrison J G, Lowe R (1989) Effects of humidity and windspeed on sporulation of *Phytophthora infestans* on potato leaves. Plant Pathology 38:585–591.

Harrison J G, Lowe R, Williams N A (1994) Effects of temperature and light on non-race-specific resistance of potato leaflets to late blight. Plant Pathology 43: 733–739.

Hartill W F T, Young K, Allan D J, Henshall W R (1990) Effects of temperature and leaf wetness on the potato late blight. New Zealand Journal of Crop and Horticultural Science 18:181–184.

Haverkort A J, Bicamumpaka M (1986) Correlation between intercepted radiation and yield of potato crops infested by Phytophthora infestans in central Africa. Netherlands Journal of Plant Pathology 92:239–247.

Hirst J M (1953) Changes in atmospheric spore content: diurnal periodicity and the effects of weather. Transactions of the British Mycological Society 36:375–393.

Hirst J M (1958) New methods for studying plant disease epidemics. Outlook on Agriculture 2:16–26.

Hirst J M, Stedman O J (1960) The epidemiology of *Phytophthora infestans*. II. The source of inoculum. Annals of Applied Biology 48:489–517.

Holmes S J I, Channon A G (1984) Studies on metalaxyl-resistant *Phytophthora infestans* in potato crops in south-west Scotland. Plant Pathology 33:347–354.

Jinks J L, Grindle M (1963) Changes induced by training in *Phytophthora infestans*. Heredity 18:245–264.

Kable P F, MacKenzie D R (1980) Survival of *Phytophthora infestans* in potato stem lesions at high temperatures and implications for disease forecasting. Plant Disease 64:165–167.

Kadish D, Cohen Y (1988) Competition between metalaxyl-sensitive and metalaxyl-resistant isolates of *Phytophthora infestans* in the absence of metalaxyl. Plant Pathology 37:558–564.

Kadish D, Cohen Y (1992) Overseasoning of metalaxyl-sensitive and metalaxyl-resistant isolates of *Phytophthora infestans* in potato tubers. Phytopathology 82:887–889.

Kolbe W (1982) Importance of potato blight control exemplified by Höfchen long-term trial (1943–1982), and historical development. Pflanzenschutz-Nachrichten Bayer 35:247–290.

Krause R A, Massie L B, Hyre R A (1975) Blitecast: A computerized forecast of potato late blight. Plant Disease Reporter 59:95–98.

Kurzawinska H (1989) Effect of agrotechnical factors on late blight [Phytophthora infestans (Mont.) de Bary] in selected cultivars of potato. Folia-Horticulturae 1:63–73.

Lacey J (1965) The infectivity of soils containing *Phytophthora infestans*. Annals of Applied Biology 56:363–380.

Lapwood D H (1968) Observations on the infection of potato leaves by *Phytophthora infestans*. Transactions of the British Mycological Society 51:233–240.

Lapwood D H (1977) Factors affecting the field infection of potato tubers of different cultivars by blight (*Phytophthora infestans*). Annals of Applied Biology 85:23–42.

Latin R X, MacKenzie D R, Cole Jr H (1981) The influence of host and pathogen genotypes on the apparent infection rates of potato late blight epidemics. Phytopathology 71:82–85.

Leach S S, Rich A E (1969) The possible role of parasexuality and cytoplasmic variation in race differentiation in *Phytophthora infestans*. Phytopathology 59:1360–1365.

Levy Y, Cohen Y, Benderly M (1991) Disease development and buildup of resistance to oxadixyl in potato crops inoculated with *Phytophthora infestans* as affected by oxadixyl and oxadixyl mixtures: experimental and simulation studies. Journal of Phytopathology 132:219–229.

Levy Y, Levi R, Cohen Y (1983) Buildup of a pathogen subpopulation resistant to a systemic fungicide under various control strategies: a flexible simulation model. Phytopathology 73:1475–1480.

MacKenzie D R, Elliott V J, Kidney B A, King E D, Royer M H, Theberge R L (1983) Application of modern approaches to the study of the epidemiology of diseases caused by *Phytophthora*. Pages 303–313 in Erwin D C, Bartnicki-Garcia S, Tsao P H (Eds.) Phytophthora. Its Biology, Taxonomy, Ecology and Pathology. The American Phytopathological Society, St Paul, Minnesota, USA.

Main C E, Gallegly M E (1964) The disease cycle in relation to multigenic resistance of potato to late blight. American Potato Journal 41:387–400.

Malcolmson J F (1969) Factors involved in resistance to blight (*Phytophthora infestans* (Mont.) de Bary) in potatoes and assessment of resistance using detached leaves. Annals of Applied Biology 64:461–468.

Malcolmson J F (1970) Vegetative hybridity in *Phytophthora infestans*. Nature 225:971–972.

Minogue K P, Fry W E (1981) Effect of temperature, relative humidity, and rehydration rate on germination of dried sporangia of *Phytophthora infestans*. Phytopathology 71:1181–1184.

Pathak N, Clarke D D (1987) Studies on the resistance of the outer cortical tissues of the tubers of some potato cultivars to *Phytophthora infestans*. Physiological and Molecular Plant Pathology 31:123–132.

Pietkiewicz J (1974) Effect of viruses on the reaction of potato to *Phytophthora infestans*. I. Characteristic of the reaction to *Ph. infestans* of plants infected with potato viruses X, Y, S, M and leafroll. Phytopathologische Zeitschrift 81:364–372.

Populer C (1978) Changes in host susceptibility with time. Pages 239–262 in Horsfall J G, Cowling E B (Eds.) Plant Disease. An advanced Treatise. Volume 2. Academic Press, New York, USA.

Raposo R, Wilks D S, Fry W E (1993) Evaluation of potato late blight forecasts modified to include weather forecasts: A simulation analysis. Phytopathology 83:103–108.

Richardson D E, Doling D A (1957) Potato blight and leaf-roll virus. Nature 180:866–867.

Robinson R A (1976) The components of parasite population flexibility. Pages 51–54 in Yaron B (Co-ordinating Ed.) Plant Pathosystems (Advanced Series in Agricultural Sciences 3). Springer-Verlag, Berlin, Germany.

Roer L, Toxopeus H J (1961) The effect of R-genes for hypersensitivity in potato-leaves on tuber resistance to Phytophthora infestans. Euphytica 10:35–42.

Rogoshin A N, Filippov A V (1983) [Distribution and conidial viability of *Phytophthora infestans* (Mont.) d.By. in the air above infected potato fields]. Mikologiya i Fitopatologiya 17:225–227. (Summary in: Review of Plant Pathology 63:75.)

Rohwer F, Fritzmeier K H, Scheel D, Hahlbrock K (1987) Biochemical reactions of different tissues of potato (*Solanum tuberosum*) to zoospores or elicitors from *Phytophthora infestans*. Accumulation of sesquiterpenoid phytoalexins. Planta 170:556–561.

Romero S, Gallegly M E (1963) Oogonium germination in *Phytophthora infestans*. Phytopathology 53:899–903.

Rotem J, Cohen Y (1974) Epidemiological patterns of *Phytophthora infestans* under semi-arid conditions. Phytopathology 64:711–714.

Rotem J, Cohen Y, Bashi E (1978) Host and environmental influences on sporulation in vivo. Annual Review of Phytopathology 16:83–101.

Rotem J, Cohen Y, Putter J (1971) Relativity of limiting and optimum inoculum loads, wetting durations, and temperatures for infection by *Phytophthora infestans*. Phytopathology 61:275–278.

Rotem J, Palti J, Lomas J (1970) Effects of sprinkler irrigation at various times of the day on development of potato late blight. Phytopathology 60:839–843.

Rotem J, Sari A (1983) Fertilization and age-conditioned predisposition of potatoes to sporulation of and infection by *Phytophthora infestans*. Zeitschrift für Pflanzenkrankheiten und Pflanzenschutz 90:83–88.

Sato N (1979) Effect of soil temperature on the field infection of potato tubers by *Phytophthora infestans*. Phytopathology 69:989–993.

Schöber B, Ullrich J (1985) Keimung der Sporangien von *Phytophthora infestans* (Mont.) de Bary auf Kartoffelblatt- und Knollengewebe. Potato Research 28:527–530.

Shattock R C, Janssen B D, Whitbread R, Shaw D S (1977) An interpretation of the frequencies of host specific phenotypes of *Phytophthora infestans* in North Wales. Annals of Applied Biology 86:249–260.

Shattock R C, Tooley P W, Fry W E (1986) Genetics of *Phytophthora infestans*: Characterization of single-oospore cultures from A1 isolates induced to self by intraspecific stimulation. Phytopathology 76:407–410.

Shaw D S (1987) The breeding system of *Phytophthora infestans*: the role of the A_2 mating type. Pages 161–174 in Day P R, Jellis G J (Eds.) Genetics and Plant Pathogenesis. Blackwell Scientific Publications, Oxford, UK.

Shtienberg D, Raposo R, Bergeron S N, Legard D E, Dyer A T, Fry W E (1994) Incorporation of cultivar resistance in a reduced-sprays strategy to suppress early and late blights on potato. Plant Disease 78:23–26.

Simmonds N W, Wastie R L (1987) Assessment of horizontal resistance to late blight of potatoes. Annals of Applied Biology 111:213–221.

Skidmore D I, Shattock R C, Shaw D S (1984) Oospores in cultures of *Phytophthora infestans* resulting from selfing induced by the presence of *P. drechsleri* isolated from blighted potato foliage. Plant Pathology 33:173–183.

Smith J W (1915) The effect of the weather upon the yield of potatoes. Monthly Weather Review. United States Department of Agriculture 43:222–236.

Smoot J J, Gough F J, Lamey H A, Eichenmuller J J, Gallegly M E (1958) Production and germination of oospores of *Phytophthora infestans*. Phytopathology 48:165–171.

Spielman L J, McMaster B J, Fry W E (1992) Relationships among measurements of fitness and disease severity in *Phytophthora infestans*. Plant Pathology 41:317–324.

Stepanov K M (1935) Dissemination of infectious diseases of plants by air currents. Bulletin of Plant Protection (Leningrad), Series 2, 8:1–68.

Stewart H E (1990) Effect of plant age and inoculum concentration on expression of major gene resistance to *Phytophthora infestans* in detached potato leaflets. Mycological Research 94:823–826.

Stewart H E, Wastie R L, Bradshaw J E, Brown J (1992) Inheritance of resistance to late blight in foliage and tubers of progenies from parents differing in resistance. Potato Research 35:313–319.

Van der Plank J E (1963) Plant Diseases: Epidemics and Control. Academic Press, New York, USA.

Van der Zaag D E (1956) Overwintering en epidemiologie van *Phytophthora infestans*, tevens enige nieuwe bestrijdingsmogelijkheden. Tijdschrift over Plantenziekten 62:89–156.

Van Everdingen E (1926) Het verband tusschen de weersgesteldheid en de aardappelziekte (*Phytophthora infestans*). Tijdschrift over Plantenziekten 32:129–140.

Van Oijen M (1990) Photosynthesis is not impaired in healthy tissue of blighted potato plants. Netherlands Journal of Plant Pathology 96:55–63.

Van Oijen M (1991a) Light use efficiencies of potato cultivars with late blight (*Phytophthora infestans*). Potato Research 34:123–132.

Van Oijen M (1991b) Identification of the Major Characteristics of Potato Cultivars which Affect Yield Loss Caused by Late Blight. Doctoral Thesis, Landbouwuniversiteit Wageningen, Wageningen, The Netherlands.

Van Oijen M (1992) Evaluation of breeding strategies for resistance and tolerance to late blight in potato by means of simulation. Netherlands Journal of Plant Pathology 98:3–11.

Wallin J R (1953) The production and survival of *Phytophthora infestans* on tomato and potato plants. Phytopathology 43:505–508.

Warren R C, Colhoun J (1975) Viability of sporangia of *Phytophthora infestans* in relation to drying. Transactions of the British Mycological Society 64:73–78.

Warren R C, King J E, Colhoun J (1971) Reaction of potato leaves to infection by *Phytophthora infestans* in relation to position on the plant. Transactions of the British Mycological Society 57:501–514.

Wastie R L, Bradshaw J E, Stewart H E (1993) Assessing general combining ability for late blight resistance and tuber characteristics by means of glasshouse seedling tests. Potato Research 36:353–357.

Yamamoto M, Tanino J (1961) Physiological studies on the formation and germination of sporangia of *Phytophthora infestans* (Mont.) De Bary. Forschungen auf dem Gebiet der Pflanzenkrankheiten (Kyoto) 7 (2):7–22.

Zan K (1962) Activity of *Phytophthora infestans* in soil in relation to tuber infection. Transactions of the British Mycological Society 45:205–221.

15. Simulation models of potato late blight

M. VAN OIJEN
Research Institute for Agrobiology and Soil Fertility (AB-DLO), P.O. Box 14, 6700 AA Wageningen, The Netherlands
Present address: Wageningen Agricultural University, Dept. of Theoretical Production Ecology, P.O. Box 430, 6700 AK Wageningen, The Netherlands

Abstract. Plant disease epidemiology has its roots in the study of infectious diseases in man. The oldest mathematical model of a human disease stems from the 18th century (Bernoulli 1760), but it was not till two centuries later that mathematical analysis of plant diseases was initiated (Van der Plank 1963). Potato late blight featured prominently in Van der Plank's analyses, and by now there is a rich literature of blight models. Before the mid 1980s blight models focused on the life cycle of the fungus, and on effects of the environment on the various stages in that life cycle. In recent years more emphasis has been given to the role of host growth in epidemic development and loss of yield. Sophisticated pathogen-crop combination models are now available.

Potato late blight modelling has served two purposes mainly. Firstly, blight models were used to evaluate strategies for disease control, especially the scheduling of fungicide application. Secondly, blight models have been used to explain differences in loss of yield among cultivars, and among various temporal and spatial patterns of disease development.

Many models have been made and their behaviour differs strongly. The sensitivity of the models to changes in parameters or inputs depends largely on the structure of the model. The models may be put to better use if more attention is paid to correct initialization and parameterization, and if comprehensive sensitivity analyses are carried out.

Introduction

Potato late blight, caused by the fungus *Phytophthora infestans*, is a major infectious disease of potatoes and other *Solanaceae spp*. The life cycle of the fungus, in which leaf infection is followed by a latent period before lesion growth and sporulation start and spore dispersal leads to new leaf infections, has been described in some detail by Harrison (1995). The life cycle strongly resembles that of many infectious diseases of humans. Present-day models of potato late blight, therefore, still show many similarities to much older models developed for medical purposes. In this chapter we will briefly describe the medical origins of plant disease epidemiological models before discussing recent developments and prospects of blight modelling. We will not strictly follow a chronological order, nor restrict ourselves completely to potato late blight, but the key blight modelling studies that have appeared since 1963 are listed chronologically in Table 1.

A.J. Haverkort and D.K.L. MacKerron (eds.), Potato Ecology and Modelling of Crops under Conditions Limiting Growth, 237–250.

Table 1. Key simulation model studies of *P. infestans*. References are listed chronologically, with indication of new features of potato late blight modeled in the particular study, or areas of improvement on older models. The list is not claimed to be exhaustive

Reference	Keywords
Van der Plank (1963)	Mathematical analysis of epidemics
Waggoner (1968)	Computer simulation of fungal life cycle, Meteorology, EPDEM
Sparks (1980)	Stem lesions
Stephan and Gutsche (1980)	Forecasting, SIMPHYT
Bruhn and Fry (1981)	Tuber yield, Contact fungicides, Cultivar resistance
Minogue and Fry (1983)	Spatial distribution of disease
Paysour and Fry (1983)	Interplot interference in experiments
Fohner et al. (1984)	Fungicide scheduling
Michaelides (1985)	Sensitivity to weather
Milgroom and Fry (1988)	Buildup of fungal resistance to metalaxyl
Ferrandino (1989)	Host growth, Spore distribution, Spatial aggregation of disease
Van Oijen (1989)	Lesion growth, Resistance components
Shtienberg et al. (1989)	Fungicide scheduling against early and late blight simultaneously
Kluge and Gutsche (1990)	Validation in field studies
Levy et al. (1991)	Fungicide mixtures
Michaelides (1991)	Plant density, Overhead sprinkling irrigation
Van Oijen (1991b)	Upward spread of disease through leaf layers, Induced resistance
Van Oijen (1992a)	Host maturity class, Tolerance components
Van Oijen (1992b)	Initialization, Parameterization, Multi-parameter changes
Raposo et al. (1993)	Incorporation of weather forecasts
Shtienberg et al. (1994)	Fungicide scheduling

Human disease epidemiology: the development of SEIR-models

The oldest mathematical model of an infectious disease was devised by Bernoulli (1760). The purpose of his study still sounds very modern although the name of his subject disease happily does not: Bernoulli used his model to evaluate vaccination strategies against smallpox.

Almost two centuries later the first model appeared that, in its structure, resembled the life cycle of infectious disease agents, and therefore may be called the first disease *simulation* model (Kermack and McKendrick 1927). This model, consisting of two linked differential equations, was applied to infectious diseases in human populations of fixed size. In the model *susceptible*, *infectious* and *removed* individuals were distinguished. Susceptible individuals became infectious at a rate directly proportional to the product of the densities of susceptible and infectious individuals. The proportionality constant, later termed *daily multiplication factor* (*DMFR*) in the botanical literature, thus was a measure of rate of production, dispersal and infectivity of infectious propagules. Infectious individuals were removed at a constant relative rate accounting for loss of infectiousness, isolation or death. The reciprocal of the removal rate was the average *infectious period* of diseased individuals. The infectious period thus was implicitly assumed to be exponentially distributed.

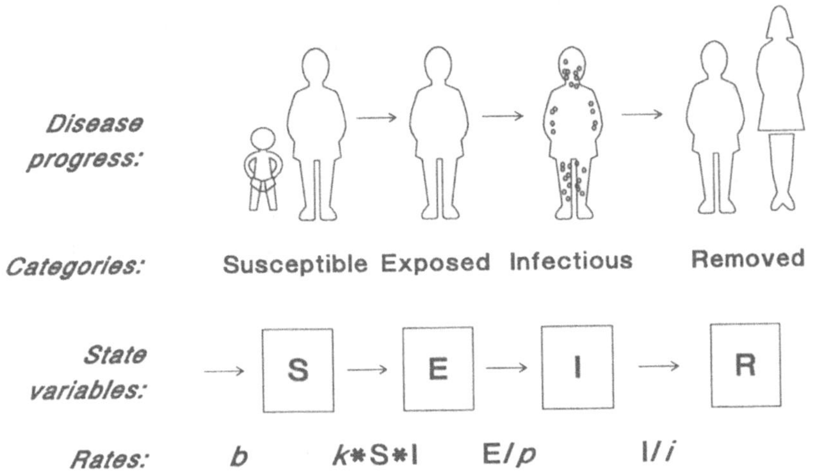

Figure 1. General structure of SEIR-models. Disease progresses by the transition of susceptible individuals (S) to subsequent categories of exposed (E), infectious (I) and recovered or removed (R) individuals. The picture shows terms often used for the transition rates, and their dependence on birth rate (b), contact frequency (k), latent period (p) and infectious period (i).

The model of Kermack and McKendrick has been termed the first 'SIR-model', after the initial letters of the three categories of individuals distinguished in the population (see Hethcote 1976). A fourth category was later distinguished: individuals that had already been *exposed* to the disease, but were not yet infectious (SEIR-models, Anderson and May 1982). The period before the start of infectiousness was called the *latent period.* A schematic representation of the structure of SEIR-models, including the most common transition rate terms, is given in Figure 1.

SEIR-models in plant disease epidemiology

Van der Plank (1963) introduced mathematical analysis into plant disease epidemiology. He applied SIR- and SEIR-models to plant diseases, including potato late blight. The latent and infectious periods in his SEIR-model, however, were not exponentially distributed, but constant for all disease units, i.e. lesions. His model is therefore formulated as a time-delayed differential-difference equation, later termed the 'paralogistic' equation (Zadoks and Schein 1979).

Preliminary analysis of the paralogistic equation was carried out by Van der Plank (1963) himself. The analysis was hampered by the fact that the paralogistic equation is analytically less tractable than the original SEIR-model from human epidemiology, even though its dynamics are not dissimilar (Jeger 1986). Some analytical results can be derived if the removal term is left out, which reduces the paralogistic to a 'para-exponential' SEI-model, describing

Table 2. Definitions of the five components of resistance to infectious diseases of plants, as used in this chapter. Note: the daily multiplication factor (*DMFR*), not listed here, is a higher-level parameter that represents the product of sporulation intensity, efficiency of spore dispersal and subsequent infection

Resistance component		Epidemiological meaning
Infection Efficiency	(e: %)	Fraction of sporangia, landed on plant tissue, that successfully initiate a new lesion
Latent Period	(p: d)	Interval between infection and start of lesion expansion and sporulation
Lesion Growth rate	(g: m d^{-1})	Radial rate of expansion of lesions
Infectious Period	(i: d)	Duration of sporulation of lesion area
Sporulation Intensity	(s: m^{-2} d^{-1})	Rate of production of sporangia per unit area of sporulating lesion

only the initial phase of epidemics. Oort (1968) presented a sensitivity analysis of the exponential rate in the initial phase by varying the latent and infectious periods and *DMFR*. He found the strongest sensitivity for changes in the latent period. The parameters tested, and some others, are generally referred to as 'resistance components', for together they determine the level of host resistance to the pathogen (Table 2). Oort's analysis, in fact, constituted the first use in plant disease epidemiology of mathematical models for testing the sensitivity of disease progress to the levels of resistance components. Unfortunately, the initial settings of the model parameters did not correspond to values appropriate for real epidemics, but more realistic and extensive numerical analyses were given by Zadoks (1971; Rabbinge et al. 1989), pointing to a major importance of the latent period in the analyses of 1971, and of *DMFR* in 1989.

Waggoner (1968) presented the first computer simulation model for potato late blight, in which the effect of environmental conditions on the resistance components was extensively modeled. He showed that a wealth of information on late blight could be integrated in one simulation model, but he did not apply his model to any practical or scientific problems. Nevertheless, Waggoner's intricate model served as a paradigm for later plant disease modelers, so that crop disease simulation models, generally, are still rich in parameters and can only be analyzed numerically.

Alternative distributions for the latent and infectious periods

As indicated earlier, latent and infectious periods were assumed to be exponentially distributed in the human disease SEIR-models, while in the first plant disease SEIR-models both periods were supposed to have zero variance among lesions. Alternative distributions were later introduced. Berger and Jones (1985) used distributed delays for the latent period in their model, but kept the infectious period fixed. The method of distributed delays offers a range of distributions, from a step function to almost normal distributions. Berger and Jones (1985) used four delay intervals and thus approached a normal distribution quite closely. This seems to be a realistic distribution for the latent period of fungal leaf diseases (Shaner 1980). Knudsen et al. (1987) used distributed delays for both latent and infectious period.

It is still unclear how important the distributions of latent and infectious period are in epidemic models. Van der Plank (1963) thought that the distribution had little effect on calculated disease progress, while Berger and Jones (1985) had the opposite view.

Stem lesions of potato late blight represent a special category of lesions with extremely long infectious periods, albeit often accompanied by low sporulation intensity. Their special epidemiological role as survival mechanisms under adverse weather conditions was incorporated in the forecasting model of Sparks (1980).

Modelling lesion growth

Zadoks (1977) mentioned five components that should be included in an analysis of the epidemiological importance of the various stages in the fungal life cycle, viz. the resistance components: infection efficiency, latent and infectious period, sporulation intensity and *lesion growth rate* (Table 2). However, a common feature of the models discussed so far is the absence of lesion growth rate as a separate resistance component. This may have historical reasons: no analogon for lesion growth was present in the earlier models for human diseases. Lesions are assumed to occupy a fixed lesion area from the beginning of their latent period. For example, a fixed lesion area of 0.3 cm^2 was used by Michaelides (1985) in his simulation model of *P. infestans*. Michaelides treated sporangium dispersal in great detail, thus complicating the experimental determination of parameter values for his model, while losing realism by oversimplifying lesion growth.

Incorporating lesion growth rate in epidemiological models might be expected to be very important in diseases with indeterminate lesion growth, i.e. without fixed final lesion sizes, as in potato late blight. Berger and Jones (1985) incorporated a constant relative growth rate of the total diseased leaf area in their model. This neglects both the increasing limitation of susceptible host area when the disease progresses, and the fact that even individual lesions generally have decreasing relative area growth rates. This also applies to potato late blight: blight lesions have a constant radial growth rate (Gees and Hohl 1988; Van Oijen 1989) so that lesion areas increase as a quadratic function of time instead of exponentially. A submodel of lesion expansion, with a dynamically changing lesion size distribution, was incorporated in a model of potato late blight by Van Oijen (1989), in which circular lesions were Poisson-distributed over leaflets, while their diameters changed by a radial growth rate that was proportional to the fraction of free leaf area on infected leaflets. Lesions in most pathosystems are not distributed randomly, so the Poisson-distribution might better be replaced by a negative binomial distribution (Waggoner and Rich 1981). Lapwood (1961) reported that even within potato leaves and leaflets lesions of late blight are not randomly distributed. He found relatively many lesions on the distal leaflet and on tips and edges of leaflets. Waggoner and Rich (1981) further suggested abandoning the direct proportionality between lesion formation rate and the product of densities of susceptible and infectious sites. Such nonlinear incidence rates are investigated intensively in human epidemiology (Liu et al. 1987).

Spatial heterogeneity

Epidemiological models usually simulate 'general epidemics', defined by Zadoks and Schein (1979) as epidemics that develop homogeneously in space from homogeneously distributed initial disease. Gradually, however, more models now incorporate the spatial aspect of development of epidemics. Minogue and Fry (1983) modeled the spatial dynamics of potato late blight, and found that blight foci expand with a constant velocity, determined by host resistance and fungicide use. Paysour and Fry (1983) used a model to calculate the level of interplot interference in experiments with potato late blight. They examined the effect of plot shapes, sizes and distances, and showed that a square plot shape minimizes interplot interference. Ferrandino (1989) also made a spatial model of potato late blight, which he used to examine the relation between spatial distribution of disease and yield loss. From his model study he concluded that tuber yield loss is maximal if the disease shows a highly aggregated pattern.

The vertical distribution of disease in the crop may be of epidemiological importance. Potato late blight usually starts in the lower leaf layers and gradually spreads to the top of the canopy (Lapwood 1961; Van Oijen 1991c). It is not yet clear whether this is caused by higher susceptibility of older leaves, better microclimatic conditions for infection in the lower canopy or greater deposition of sporangia low in the canopy. Björling and Sellgren (1955) found 2–4 times as many sporangia deposited on middle and bottom leaves as on top leaves, both in incipient and well-developed epidemics. A preliminary model of the gradual upward spread of blight infection was discussed by Van Oijen (1991b). He showed that resistant cultivars are characterized more by retarded upward spread of the pathogen than by reduced rates of spread within leaf layers.

Host growth

Until a few years ago, only few blight models simulated host growth in a realistic manner. Many models simply considered host leaf area as a forcing function of time (e.g. Bruhn and Fry 1981; Michaelides 1985). Slightly more sophisticated is the approach in which leaf area grows towards an asymptote according to a logistic function of the uninfected leaf area (Berger and Jones 1985; Van Oijen 1989).

Recent years have witnessed a trend to include host growth by means of a more realistic submodel. Several authors have based their host growth submodel on the principle, outlined by Monteith (1977), of a linear relation between crop growth rate and the amount of light intercepted by green leaf area. Ferrandino (1989), Michaelides (1991) and Van Oijen (1992a) all built models of potato late blight according to this principle. The slope of the relation, i.e. the light-use efficiency, is unaffected by late blight (Haverkort and Bicamumpaka 1986; Van Oijen 1991a). So, the fungus does not impair host photosynthesis (Van Oijen 1990), and a host crop growth model need not include sophisticated subroutines for calculating effects of infection on leaf photosynthesis parameters (Rossing et al. 1992).

Waggoner (1990) gave equations for interception of light by green leaf area, when the disease is heterogeneously distributed horizontally or vertically. These equations may conveniently be incorporated in the models based on light interception and utilization.

An important object of study for potato late blight models is the relation between host size and disease escape. Most models show that whenever the susceptible leaf area decreases below a certain threshold, the area of infectious tissue can no longer increase, and that some susceptible tissue will remain uninfected when the epidemic has ended (Kermack and McKendrick 1927; Van der Plank 1963). The size of this fraction of the host area escaping the disease depends on a nonlinear function of all resistance components that influence the epidemic rate, thus pointing to a close link between host characteristics that determine escape and those that determine resistance (Van Oijen 1989).

Stochasticity

The early models of human infectious disease, e.g. the model of Kermack and McKendrick (1927), were deterministic (Bailey 1975). Gradually, however, stochasticity has been incorporated and is now featured in most human epidemiological models (Becker 1979). In contrast, most models of plant disease epidemics are still deterministic (Gilligan 1985). The results of these models may differ from those of stochastic models, in spite of the large number of disease units involved, because of the non-linearity of pathosystems (Rouse 1991). Preliminary stochastic models have been presented by Teng et al. (1977) and Sall (1980), for barley leaf rust and grape powdery mildew, respectively. In these models, parameter values for a subset of the resistance components are drawn from uniform or normal probability distributions. In some recent models, e.g. the blight model of Ferrandino (1989), spore dispersal is treated stochastically to account for variable wind patterns not incorporated in the model.

The stochastic models provide estimates of the variation in rates of disease progress. Such estimates of variation are useful if the model is used in disease forecasting, because they set boundaries to the reliability of the forecasts. However, if the model is used for explaining system behaviour in terms of its components or to assess the importance of resistance components for breeding purposes, this variation is unwanted because it obscures the relation between the individual components and rate of disease progress, and thus complicates the identification of the major components.

Blight models applied in fungicide scheduling

Much effort, costs and environmental hazards are involved in the regular spraying of potato crops with fungicide against *P. infestans* (Harrison 1995). To assist in this activity, many mathematical models have been made to forecast the development of blight epidemics depending on past and expected weather conditions. With few exceptions (Kluge and Gutsche 1990), most of these models are of empirical nature, directly linking spray advice to weather variables, without explicitly modelling fungal population dynamics, let alone the relation with the host crop. Therefore, these models fall beyond the scope of this chapter, but recent reviews on this topic are available (Bajic 1988). However, simulation models have been applied in their own way to the question of fungicide scheduling. Rather than function as forecasters themselves, simulation models have repeatedly been used to evaluate the effectiveness of empirical strategies for fungicide scheduling (Fohner et al. 1984; Shtienberg et al. 1994). Unfortunately, the general conclusion from these simulation studies was that little economic gain may be expected from refraining from spraying regularly, weekly or biweekly. Apparently, late blight is too explosive a disease to warrant taking the risks of skipping sprays while weather forecasts are still unreliable.

A special role of simulation models in guiding use of fungicides has been to assess the risk for buildup of resistance against systemic fungicides like metalaxyl (Milgroom and Fry 1988; Levy et al. 1991). Levy and coworkers (1991) showed in a modelling study that the use of fungicide mixtures may delay the buildup of resistance in the pathogen population.

Blight models applied in resistance breeding

Straightforward multiplication of the resistance components can be considered as the simplest model for assessing their combined effect on the initial exponential development of epidemics (Zadoks 1977). Van der Zaag (1959) indeed found that the ranking of potato cultivars for partial resistance to *P. infestans* closely followed the ranking of their multiplied resistance components. However, the analyses of Van der Plank (1963) and Oort (1968) have shown that

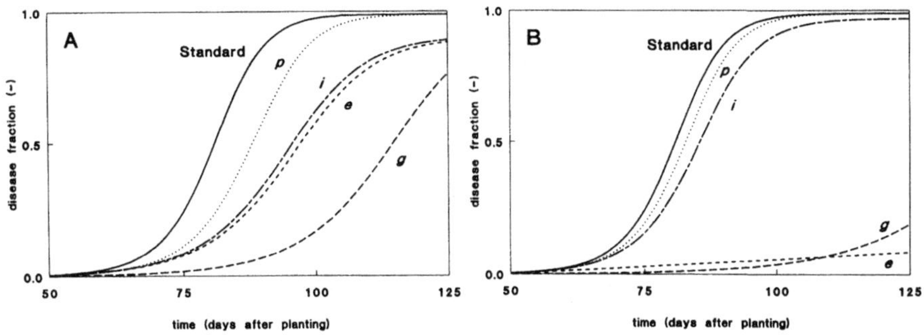

Figure 2. Simulated disease progress curves of potato late blight in various host cultivars. (A) A standard susceptible cultivar (cv. Bintje) and hypothetical partially resistant genotypes with half the standard value for infectious period (*i*), infection efficiency (*e*), or lesion growth rate (*g*), or a doubled value for latent period (*p*); (B) the same standard susceptible cultivar, and genotypes in which *i*, *e*, *g* or *p* is set to the most resistant value observed among current cultivars. (Source: Van Oijen 1992b.)

this method unjustifiably considers all components to have the same effect on total resistance, and the method has not been applied frequently.

Van Oijen (1992a,b) used his model to identify the resistance components that should be improved to breed more resistant cultivars. A sensitivity analysis of the model showed that disease progress rate was most sensitive to changes in lesion growth rate, followed by infection efficiency and infectious period, and, finally, the latent period (Figure 2A). However, this result in itself did not warrant advising breeders to concentrate solely on decreasing lesion growth rate. Breeding perspectives depend not only on the intrinsic importance of plant characters, but also on the scope for changing the characters, i.e. the available genetic variation. Therefore a second sensitivity analysis was done, in which the genetic variation in the different resistance components was taken into account. This more realistic analysis pointed to both lesion growth rate and infection efficiency as prime targets for improving the level of partial resistance to potato late blight in breeding programmes (Figure 2B).

Problems with use of blight models: initialization and parameterization

When the performance of epidemiological models is tested, the status of the initial inoculum is rarely considered. However, the way a model is initialized affects its behaviour in various ways. For example, model epidemics starting from equally-aged *latent* lesions differ strongly from epidemics started from *infectious* lesions, if the latent period is not treated as a constant (Jeger 1986). When the model is initialized with *too many* lesions, sensitivity of disease progress rate to changes in infection efficiency can be underestimated (Van Oijen 1992b). Other problems with initialization are found in models with time delays, as the paralogistic equation of Van der Plank (1963), where the level of disease must be defined for some period (equal to the maximum delay in the model) preceding the simulated period.

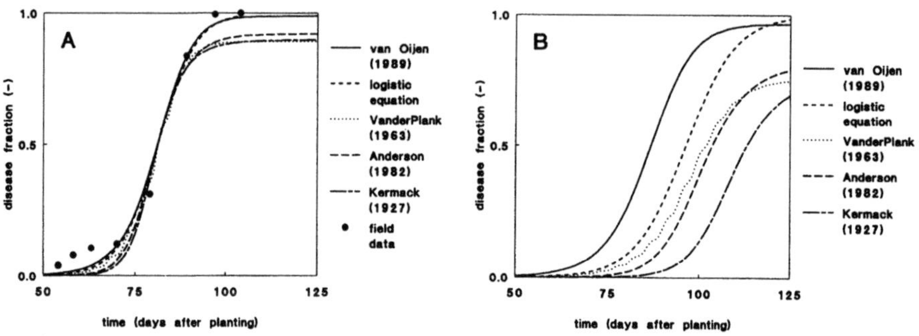

Figure 3. Simulated disease progress curves of potato late blight, generated by five different models. (A) Best fit of the models on disease progress data for the susceptible cultivar Bintje (field measurements 1988); (B) the effect of reducing the infection efficiency parameter by 25% in each of the five models. (Source: Van Oijen 1992b.)

Parameterizing epidemiological models can be difficult if the model parameters correspond poorly to quantities that are actually measured. Methods of measuring resistance components often aim at maximum discrimination between cultivars or treatments, or ease of measurement. For modelling purposes, however, components should be measured in a way that corresponds best to their function in the life cycle of the pathogen. Therefore, measurements of lesion *growth rate* are more useful than measurements of lesion *size* at an arbitrary time after inoculation. Determining the total sporangium production *per unit of area* of diseased leaf tissue is better than the more common practice of quantifying sporulation as number of sporangia washed off *per leaflet or lesion*, again at arbitrary times after inoculation. Only measurement of sporulation intensity on an area basis allows a clear separation of sporulation intensity from lesion size (i.e. integrated lesion growth rate) and infectious period in fungal diseases with expanding lesions.

Rate of disease progress is generally quantified differently in measurements and component models, thus complicating model validation. Total visibly diseased tissue is measured, while latent, infectious and removed tissue are modeled. Models with a fixed size for all lesions, including latent ones, give special problems here, since in actual epidemics latent lesions occupy much less leaf area than visible lesions. The paralogistic equation of Van der Plank (1963), for example, calculates the total infected (latent + infectious + removed) leaf area, whereas only infectious and removed leaf area can be observed. This is a further argument in favour of models that explicitly include lesion growth, where latent lesions can realistically have zero or negligible area.

The sensitivity of the models to changes in parameters is often used as a measure of the importance of those parameters for disease progress rate. However, one should use this approach with caution, because sensitivity to parameter changes strongly depends on model structure. The sensitivity, therefore, is a feature of the model and may not reflect true sensitivity of the epidemic. This was shown in a comparison of five different models of potato late blight (Van Oijen 1992b). In each model, the parameter that quantifies infection efficiency was reduced by 25%. When infection efficiency was thus altered, disease progress rate was reduced in all models, but to widely differing extent (Figure 3).

Concluding remarks

As described in the preceding paragraph, many modelling studies have suffered from inadequate parameterization and insufficient validation. Even well-validated models, however, may be put to better use. Too often a sensitivity analysis for resistance components is missing (Jeger and Groth 1985), although the usefulness of such analyses for guiding breeding efforts has been emphasized repeatedly (Zadoks 1971; 1977). Whenever such a sensitivity analysis has been performed, the analysis has usually been restricted to single-parameter changes. Simultaneous changes of more than one resistance component should be evaluated too (Van Oijen 1992b). These multi-parameter changes should take into account the fact that some parameters may not vary independently in the real pathosystem, if they are genetically or physiologically linked.

Lesion growth rate should be explicitly included in models of fungal leaf diseases with indefinitely expanding lesions. The importance of this component has been shown by the few models that do include it: rate of disease progress calculated by these models shows great sensitivity to changes in the lesion growth rate.

If a disease spans a long period of the host growing season, which applies to most fungal leaf diseases, especially when partially resistant cultivars are evaluated, the disease is incorrectly simulated by models that consider the host leaf area to be constant. For such diseases the effect of the pathogen on host growth should be explicitly modeled, primarily if the model is to be used for analysing effects of host characteristics on loss of yield due to the disease. In the case of potato late blight, simulation of tuber growth and infection may require more attention than so far.

Despite the cautionary remarks given above, it is still clear that simulation models of potato late blight have been used successfully in many respects. Blight models have been useful in epidemiological analysis, for the evaluation of fungicide application strategies, and for the identification of host characteristics that improve resistance. Probably these areas of modelling, aiming at explanation and evaluation rather than forecasting, will remain the proper territories for blight simulation models in the near future.

References

Anderson R M, May R M (1982) Directly transmitted infectious diseases: control by vaccination. Science 215:1053–1060.

Bailey N T J (1975) The Mathematical Theory of Infectious Diseases and its Applications. Charles Griffin & Company Ltd., London, UK.

Bajic S (1988) A Survey on Potato Blight *Phytophthora Infestans* (Mont) de Bary Forecasting. The Meteorological Office, Bracknell, UK.

Becker N (1979) The uses of epidemic models. Biometrics 35:295–305.

Berger R D, Jones J W (1985) A general model for disease progress with functions for variable latency and lesion expansion on growing host plants. Phytopathology 75:792–797.

Bernoulli D (1760) Essai d'une nouvelle analyse de la mortalité causée par la petite vérole, et des avantages de l'inoculation pour la prévenir. Histoire de l'Academie Royale des Sciences (Paris) avec des Mémoires de Mathematique et Physique:1–45.

Björling K, Sellgren K A (1955) Deposits of sporangia and incidence of infection by *Phytophthora infestans* on upper and lower surfaces of potato leaves. Acta Agriculturae Scandinavica V:375–386.

Bruhn J A, Fry W E (1981) Analysis of potato late blight epidemiology by simulation modelling. Phytopathology 71:612–616.

Ferrandino F J (1989) Spatial and temporal variation of a defoliating plant disease and reduction in yield. Agricultural and Forest Meteorology 47:273–289.

Fohner G R, Fry W E, White G B (1984) Computer simulation raises question about timing protectant fungicide application frequency according to a potato late blight forecast. Phytopathology 74:1145–1147.

Gees R, Hohl H R (1988) Cytological comparison of specific (R3) and general resistance to late blight in potato leaf tissue. Phytopathology 78:350–357.

Gilligan C A (1985) Introduction. In Gilligan C A (Ed.) Mathematical Modelling of Crop Disease. Advances in Plant Pathology 3:1–10.

Harrison J G (1995) Factors involved in the development of potato late blight disease (*Phytophthora infestans*). Pages 215–236 in Haverkort A J, MacKerron D K L (Eds.) Potato Ecology and Modelling of Crops under Conditions Limiting Growth. Kluwer Academic Publishers, Dordrecht, The Netherlands.

Haverkort A J, Bicamumpaka M (1986) Correlation between intercepted radiation and yield of potato crops infested by *Phytophthora infestans* in central Africa. Netherlands Journal of Plant Pathology 92:239–247.

Hethcote H W (1976) Qualitative analyses of communicable disease models. Mathematical Biosciences 28:335–356.

Jeger M J (1986) Asymptotic behaviour and threshold criteria in model plant disease epidemics. Plant Pathology 35:355–361.

Jeger M J, Groth J (1985) Resistance and pathogenicity: Epidemiological and ecological mechanisms. Pages 310–372 in Fraser R S S (Ed.) Mechanisms of Resistances to Plant Diseases. Martinus Nijhoff Publishers, The Hague, The Netherlands.

Kermack W O, McKendrick A G (1927) A contribution to the mathematical theory of epidemics. Proceedings of the Royal Society of London, Series A 115:700–721.

Kluge E, Gutsche V (1990) Prediction of late blight of potato with the help of simulation models – Results 1982–1988. Archiv für Phytopathologie und Pflanzenschutz 26:265–281.

Knudsen G R, Spurr H W, Johnson C S (1987) A computer simulation model for *Cercospora* leaf spot of peanut. Phytopathology 77:1118–1121.

Lapwood D H (1961) Potato haulm resistance to *Phytophthora infestans*. III. Lesion distribution and leaf destruction. Annals of Applied Biology 49:704–716.

Levy Y, Cohen Y, Benderly M (1991) Disease development and buildup of resistance to oxadixyl in potato crops inoculated with *Phytophthora infestans* as affected by oxadixyl and oxadixyl mixtures: Experimental and simulation studies. Journal of Phytopathology 132:219–229.

Liu W M, Hethcote H W, Levin S A (1987) Dynamical behavior of epidemiological models with nonlinear incidence rates. Journal of Mathematical Biology 25:359–380.

Michaelides S C (1985) A simulation model of the fungus *Phytophthora infestans* (Mont) De Bary. Ecological Modelling 28:121–137.

Michaelides S C (1991) A dynamic model of the interactions between the potato crop and *Phytophthora infestans*. EPPO Bulletin 21:515–525.

Milgroom M G, Fry W E (1988) A simulation analysis of the epidemiological principles for fungicide resistance management in pathogen populations. Phytopathology 78:565–570.

Minogue K P, Fry W E (1983) Models for the spread of disease: some experimental results. Phytopathology 73:1173–1176.

Monteith J L (1977) Climate and the efficiency of crop production in Britain. Philosophical Transactions of the Royal Society of London, Series B 281:277–294.

Oort A J P (1968) A model of the early stage of epidemics. Netherlands Journal of Plant Pathology 74:177–180.

Paysour R E, Fry W E (1983) Interplot interference: A model for planning field experiments with aerially disseminated pathogens. Phytopathology 73:1014–1020.

Rabbinge R, Zadoks J C, Bastiaans L (1989) Population models. Pages 83–97 in: Rabbinge R, Ward S A, Van Laar H H (Eds.) Simulation and Systems Management in Crop Protection. Simulation Monographs 32. Pudoc, Wageningen, The Netherlands.

Raposo R, Wilks D S, Fry W E (1993) Evaluation of potato late blight forecasts modified to include weather forecasts: A simulation analysis. Phytopathology 83:103–108.

Rossing W A H, Van Oijen M, Van der Werf W, Bastiaans L, Rabbinge R (1992) Modelling the effects of foliar pests and pathogens on light interception, photosynthesis, growth rate and yield of field crops. Pages 161–180 in Ayres P G (Ed.) Pests and Pathogens: Plant Responses to Foliar Attack. Bios Scientific Publishers, Oxford, UK.

Rouse D I (1991) Stochastic modelling of plant disease epidemic processes. Pages 647–665 in Arora D K, Rai B, Mukerji K G, Knudsen G R (Eds.) Handbook of Applied Mycology. Volume 1: Soil and Plants. Marcel Dekker, New York, USA.

Sall M (1980) Uses of stochastic simulation: grape powdery mildew example. Zeitschrift für Pflanzenkrankheiten und Pflanzenschutz 87:397–403.

Shaner G (1980) Probits for analyzing latent period data in studies of slow rusting resistance. Phytopathology 70:1179–1182.

Shtienberg D, Doster M A, Pelletier J R, Fry W E (1989) Use of simulation models to develop a low-risk strategy to suppress early and late blight in potato foliage. Phytopathology 79:590–595.

Shtienberg D, Raposo R, Bergeron S N, Legard D E, Dyer A T, Fry W E (1994) Incorporation of cultivar resistance in a reduced-sprays strategy to suppress early and late blights on potato. Plant Disease 78:23–26.

Sparks W R (1980) Blight: A computer model relating the progress of potato blight to weather. Unpublished Agricultural Memorandum No. 899. National Meteorological Library, Bracknell, UK.

Stephan S, Gutsche V (1980) Ein algorithmisches Modell zur Simulation der *Phytophthora*-Epidemie (SIMPHYT). Archiv für Phytopathologie und Pflanzenschutz 16:183–191.

Teng P S, Blackie M J, Close R C (1977) A simulation analysis of crop yield loss due to rust disease. Agricultural Systems 2:189–198.

Van der Plank J E (1963) Plant Diseases: Epidemics and Control. Academic Press, New York, USA.

Van der Zaag D E (1959) Some observations on breeding for resistance to *Phytophthora infestans*. European Potato Journal 2:278–286.

Van Oijen M (1989) On the use of mathematical models from human epidemiology in breeding for resistance to polycyclic fungal leaf diseases of crops. Pages 26–37 in Louwes K M, Toussaint H A J M, Dellaert L M W (Eds.) Parental Line Breeding and Selection in Potato Breeding. Pudoc, Wageningen, The Netherlands.

Van Oijen M (1990) Photosynthesis is not impaired in healthy tissue of blighted potato plants. Netherlands Journal of Plant Pathology 96:55–63.

Van Oijen M (1991a) Light use efficiencies of potato cultivars with late blight (*Phytophthora infestans*). Potato Research 34:123–132.

Van Oijen M (1991b) Identification of the Major Characteristics of Potato Cultivars which Affect Yield Loss Caused by Late Blight. Ph.D. Thesis, Wageningen Agricultural University, Wageningen. The Netherlands.

Van Oijen M (1991c) Leaf area dynamics of potato cultivars infected to various extent by *Phytophthora infestans*. Netherlands Journal of Plant Pathology 97:345–354.

Van Oijen M (1992a) Evaluation of breeding strategies for resistance and tolerance to late blight in potato by means of simulation. Netherlands Journal of Plant Pathology 98:3–11.

Van Oijen M (1992b) Selection and use of a mathematical model to evaluate components of resistance to *Phytophthora infestans* in potato. Netherlands Journal of Plant Pathology 98:192–202.

Waggoner P E (1968) Weather and the rise and fall of fungi. Pages 45–66 in Lowry W P (Ed.) Biometeorology. Oregon State University Press, Corvallis, USA.

Waggoner P E (1990) Defoliation, disease and growth. Pages 149–180 in Goudriaan J, Penning de Vries F W T, Rabbinge R, Van Keulen H, Van Laar H H (Eds.) Theoretical Production Ecology: Hindsights and Perspectives. Simulation Monographs. Pudoc, Wageningen, The Netherlands.

Waggoner P E, Rich S (1981) Lesion distribution, multiple infection, and the logistic increase of plant disease. Proceedings of the National Academy of Sciences USA 78:3292–3295.

Zadoks J C (1971) Systems analysis and the dynamics of epidemics. Phytopathology 61:600–610.

Zadoks J C (1977) Simulation models of epidemics and their possible use in the study of disease resistance. Pages 109–118 in International Atomic Energy Agency (Ed.) Induced Mutations Against Plant Diseases: Proceedings of a Symposium. Vienna, Austria.

Zadoks J C, Schein R D (1979) Epidemiology and plant disease management. Oxford University Press, New York, USA.

16. Life cycle and ecology of *Verticillium dahliae* in potato

L. MOL[1] and A.J. TERMORSHUIZEN[2]

[1] *Department of Agronomy, Wageningen Agricultural University, Haarweg 333, 6709 RZ Wageningen, The Netherlands*

[2] *Department of Phytopathology, Wageningen Agricultural University, P.O. Box 8025, 6700 EE Wageningen, The Netherlands*

Abstract. *Verticillium dahliae* is a serious pathogen in most countries where potato is grown. The density of microsclerotia of *V. dahliae* in soil mainly depends on the cropping history. Plant roots can be colonised if microsclerotia germinate in the vicinity of the root tip. Colonisation is followed by systemic infection of the vascular system of the plant. During colonisation of the root cortex and systemic infection of the plant interactions with many soil organisms. Cultural practices can lessen the colonisation of the roots and the severity of the disease.

In the vascular system of the plant, *V. dahliae* is dispersed by conidia and mycelial growth. Wilting symptoms appear after the reactions of the plant to the presence of the pathogen. Yield reduction is mainly caused by closure of the stomata and early senescence of the canopy after blockage of the vascular system of the haulm.

The fungus forms microsclerotia on dead plant tissue. External climatological factors and haulm killing practices have a large influence on the number of microsclerotia formed per unit haulm material. Since *V. dahliae* has a very broad host range, attention should be paid to the control of this pathogen in all crops in a rotation. In some crops host specificity has been found, but this is a gradual property.

Introduction

Diseases caused by *Verticillium dahliae* Kleb. are wide-spread throughout the world, wherever susceptible crops are grown, and are of economic importance in most countries (Pegg 1984). *Verticillium dahliae* is supposed to be the major component of the potato early dying complex that damages crops by causing early maturity of the crop by wilting. Hosts include all dicotylodonous plants, with the most important crops affected being potato, cotton, egg-plant, tomato, mint, and olive. *Verticillium dahliae* survives by means of microsclerotia (MS). The high survival potential of MS and the wide host range make *V. dahliae* endemic to many agricultural soils (Powelson 1970).

Verticillium dahliae is classified as a soil invading or root inhabiting fungus (Powelson 1970). These fungi are characterised by a parasitic phase on the living host plant (Figure 1: I), and by a saprophytic phase after the death of the host (Powelson 1970) (Figure 1: II). The two phases will be discussed separately.

A.J. Haverkort and D.K.L. MacKerron (eds.), Potato Ecology and Modelling of Crops under Conditions Limiting Growth, 251–264.

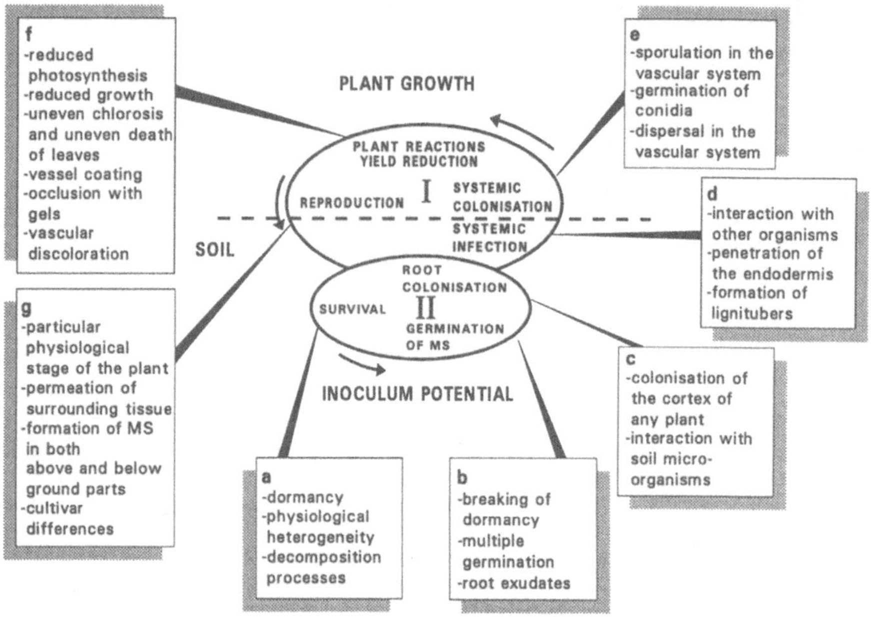

Figure 1. Schematic life cycle of *V. dahliae* showing a parasitic phase on a host (I), and a saprophytic phase after plant death and in soil (II). The scheme is explained in the text. MS = microsclerotia.

Dynamics of the soil population of *V. dahliae*

A schematic representation of processes concerning the MS population in the soil is shown in Figure 1: II. The population of MS in soil varies in composition and density depending on the cropping history of the field. After germination of a microsclerotium, the hypha may infect a plant root. Due to their low competitive saprophytic ability, the majority of the hyphae are not successful in reaching a plant root and they die. The next paragraphs will be focused on: survival of MS (Figure 1: a), germination of MS (Figure 1: b), and colonisation of plant roots (Figure 1: c).

Survival of MS (Figure 1: a)

Ben-Yephet and Szmulewich (1985) reported that MS of *V. dahliae* survive longer in the field than in the laboratory. After 5 years of storage of air-dried soil samples at 20–25 °C no *V. dahliae* could be detected, whereas 4% of the original population density remained viable after 7 years of crop rotation. Wilhelm (1955) found that *V. dahliae* persisted for 14 years in field soil with no hosts present. The long persistence of the fungus in the field is probably due to its ability to colonise and produce new MS on the root systems of nearly all plant species including monocots (Martinson and Horner 1962). The density of MS in soil was monitored by Itoh et al. (1989) following Chinese cabbage by using infested soil. A first order rate equation was fitted to the observed decrease in the number of MS which was more rapid at higher temperatures. A linear relation was observed between the logarithmic value of the half life and the temperature in the range of 5–31 °C.

Survival of MS of *V. dahliae* appears to be best under air-dry conditions (Coley-Smith and Cooke 1971). The mechanism for survival has not been elucidated. There is a clear relation between the presence of melanin in organisms and their persistence (Bloomfield and Alexander 1967). Temperature may exert an indirect influence on survival through direct effects on germination of MS (Coley-Smith and Cooke 1971) and decomposition of the plant material they are embedded in to achieve their release into the soil (Hancock and Benham 1980). The influences of the temperature and decomposition are entangled. Decomposition is the result of more than one process, and an increase in temperature, within a certain range, will accelerate most biological processes.

The release of MS from colonised plant debris lasts at least one year following its incorporation into the soil and consequently affects the inoculum density in soil. In two cultivated cotton-field soils studied by Evans et al. (1967), the density of MS in soil declined throughout the growing season, but increased again at harvest time when mechanical damage to colonised cotton plants caused the release of fresh MS. There was a further increase in the soil inoculum density when the dead cotton stalks were returned to the soil in preparation for the next crop. Huisman and Ashworth (1976) observed the soil inoculum density of *V. dahliae* in eight commercial fields with different cropping histories at about monthly intervals for 3.5–4.0 years. High inoculum densities in soil persisted under continuous cultivation of cotton. The inoculum density usually increased rapidly following one year of cotton, with a higher inoculum density occurring in the second year regardless of the susceptibility of the subsequent crop. The slow release of MS from decomposing plant debris probably causes a delay in the build-up of the inoculum density in soil. It is not known how the survival of MS embedded in plant material compares to the survival of released 'free' MS. Presumably, the death rate is higher in bound MS because of the higher microbial activity in and around the decomposing plant debris.

Germination of microsclerotia (Figure 1: b)

Microsclerotia of *V. dahliae* have no constitutive dormancy where development is delayed because of an endogenous property of the propagule (Pegg 1974), but they do have a requirement during germination for exogenous nutrients (Emmatty and Green 1969) from root exudates of either hosts or non-hosts (Schreiber and Green 1963; Mol 1994).

Concentrations of exogenous nutrients can be influenced by organic amendments to the soil, which influence general soil microbial activity and concentration of soluble compounds in the soil solution. Increased soil microbial activity has been reported to reduce the proportion of germination of microsclerotia and growth of germ-tubes of *V. dahliae* in the rhizosphere (Jordan et al. 1972). In greenhouse and field experiments, amendments with chopped barley or oat straw at several rates reduced the inoculum density of *V. dahliae* in soil and disease incidence in potato plants (Tolmsoff and Young 1959; Harrison 1976). The effect of organic amendments has not been elucidated, but it has been suggested to be an increase in production of antifungal substances by the enriched microflora of the rhizosphere (Curl and Truelove 1986). However, the interactions between the pathogen and the microflora in the rhizosphere are probably more important, since there is still doubt that antibiotic substances produced in soil can endure microbial degradation for long enough to act as effective inhibitors (Curl and Truelove 1986).

Microsclerotia have the ability to germinate more than once. Farley et al. (1971) showed that MS germinated and sporulated every time after re-moistening soil with a sucrose solution or water up to nine times. Germination percentage and the number of germ-tubes decreased with succeeding germinations. Where a susceptible crop is planted in a field in soil naturally infested with *V. dahliae*, the behaviour of the MS is likely to be different from that in experiments where artificial inoculum is applied (Menzies and Griebel 1967). Unless several years have elapsed since the last susceptible crop was grown, the inoculum may be mostly in the form of MS embedded in partly decomposed plant residues. Some of these MS may already have undergone a cycle of germination and be depleted, while others may be freshly exposed during cultivation operations and may go through a peak of germination as the host seedling roots are growing in their vicinity (Menzies and Griebel 1967).

Colonisation of plant roots by V. dahliae (Figure 1: c)

Germination hyphae of MS may infect susceptible hosts by penetration of the root cortex followed by systemic invasion of the xylem vessels (Powelson 1970).

Infectious hyphae of *V. dahliae* emerging from MS penetrate roots primarily near the root tip and in the root-hair zone (Fitzell et al. 1980; Gerik and Huisman 1988). The density of colonisation at a distance of more than 1 cm from the root apex appeared to be constant (Gerik and Huisman 1988). Most of the colonies are removed when the roots are surface-sterilised, indicating that *V. dahliae* is predominantly restricted to superficial sites in the root cortex (Evans and Gleeson 1973).

Host-pathogen interactions

A schematic representation of the host-pathogen interactions is shown in Figure 1: I. After colonisation of the plant root, the fungus crosses physical barriers in the plant to reach the vascular system. During systemic infection, colonisation of the plant may result in resistance reactions, yield reduction, and, finally, production of MS on the plant debris. The next paragraphs will be focused on systemic infection of plant roots (Figure 1: d), systemic colonisation of the plant (Figure 1: e), plant reactions and yield reductions (Figure 1: f), and reproduction of *V. dahliae* (Figure 1: g).

Systemic infection of plant roots (Figure 1: d)

In response to the initial invasion by the fungus, the inner tangential wall of the epidermis becomes swollen (Bell 1973). Occasionally a gum-like encapsulation zone may be present in the outer cortical cells of the plant roots around the hyphae of *V. dahliae*; these formations are referred to as lignitubers (Griffiths 1973). Lignitubers result from the extrusion of vesicles from the protoplast into the area between the cell wall and plasmalemma. When the fungus starts to penetrate the cell wall, the vesicles aggregate and lose their individual identity as they form the lignituber. Subsequently, the fungal cell wall becomes surrounded by the lignituber, while the host plasmalemma apparently remains intact. In most cases the invading hyphae lyse, but in a few cases successful penetration occurs and the fungus proceeds through the layer of cortical cells (Bell 1973).

Fungi that successfully penetrate the cortex encounter a second barrier at the endodermis (Bell 1973). Most of the penetrating hyphae that were not stopped by lignitubers are so at the endodermis; the few hyphae that penetrate this barrier progress into the vascular tissue and invade the vessels by penetrating through the pits. Alternatively, hyphae may reach the vascular system by colonising the young root tip, where the endodermis still has to be formed. The systemic infection of the plant root is the property determining the suitability of the plant as a host. For example, susceptible mint (*Mentha* spp.) roots showed more extensive systemic invasion, while the number of root cortex infections was similar to that of resistant mint species (Lacy and Horner 1966).

Infection of plant roots by *V. dahliae* may be influenced by other organisms. In most cases, potato early dying can be attributed to a complex of soil microorganisms including *V. dahliae*, *Globodera* spp., *Pratylenchus* spp., *Meloidogyne* spp., *Colletotrichum coccodes*, *Fusarium* spp. and *Rhizoctonia solani* (Scholte 1989; Scholte and s'Jacob 1989). Other pathogenic agents that have been associated with potato early dying include *Erwinia carotovora* and Potato Virus X (Rouse 1985). It has been established that potato early dying is frequently caused by *V. dahliae* in combination with endoparasitic nematodes. In microplot studies on potato cv. Russet Burbank, Kotcon et al. (1985) did not find a significant contribution by either *C. coccodes* nor *R. solani* to the disease syndrome either alone or in combination with *V. dahliae* or *P. penetrans*. However, in potato cv. Amethyst yield loss by *V. dahliae* was almost doubled in the presence of *C. coccodes* (Scholte et al. 1985), and Scholte and s'Jacob (1989) found a three-factor interaction on yield loss between *Meloidogyne* spp. or *P. neglectus* and *R. solani* and *V. dahliae* with several potato cultivars.

The mechanism of the interaction between *V. dahliae* and *P. penetrans* that results in the synergistic expression of symptoms and yield loss has been subject of speculation. Green (1981) considered that the mechanism could be wounding from nematode feeding, which gives the fungus easier access to the vascular system. Alternatively he proposed a physiological mechanism where the plant becomes more susceptible to the fungus because of a translocated substance. Wheeler et al. (1992) compared a model in which yield loss of potato is proportional to both *V. dahliae* and *P. penetrans* with a model in which yield loss is proportional only to the population density of *V. dahliae* and the presence of *P. penetrans* leads to a more severe yield loss function than in the absence of the nematode. However, their comparison was regarded not conclusive considering the variability in the expression of potato early dying and confounding effects of environmental conditions.

In evaluating his mechanistic model, Johnson (1992) described potato crop losses caused by multiple biotic stress factors, excluding nematodes. He concluded that crop loss by multiple stress factors was less than the sum of losses from each stress factor acting alone. This was illustrated by a competitive defoliation between *V. dahliae* and *Phytophthora infestans*.

Systemic colonisation of the plant (Figure 1: e)

The process of production of conidia in the vascular system is still unclear. A hypothesis put forward by several authors is that directly after *V. dahliae* has penetrated the vascular system, it starts to produce conidia (Howell 1973; Schnathorst 1981). Conidia are produced by simple conidiophores or by budding (Tolmsoff 1973) and are passively distributed through the vascular system throughout the plant. Conidia may germinate, and penetrate vessel walls (Garber 1973; Tolmsoff 1973). More research concerning the process of sporulation and distribution in the vascular system is needed. For example the influence of the nutrient concentration in the xylem and the influence of the vitality of the plant on these two processes are indicated as origins of differences in the number of *V. dahliae* particles in the stems of plants, but the effect has never been proven in experiments.

In susceptible and tolerant cotton cultivars, colonisation was equal up to the point where the pathogen passed the endodermis and reached the xylem (Garber 1973). The number of vessels invaded appears to be a good measure of the severity of wilt disease, because the number of invaded vessels was related to the number of systemic infections and to the number of hyphae that progressed through the cortex from the points of colonisation.

As long as petioles of potato plants are green, distribution of *V. dahliae* in the plant is limited to the xylem, but in plants with severe disease symptoms the pith, cambium, and cortex are invaded (Garber 1973). In the leaves, infection may be confined to a single pinna, or may involve the entire leaf (Garber 1973).

Plant reactions and yield reductions (Figure 1: f)

In potato, symptoms of *V. dahliae* are difficult to distinguish from normal senescence and may initially involve only reduced growth (Street and Cooper 1984; Haverkort et al. 1990). Early foliar symptoms may appear as unilateral chlorosis of lower leaves on a few plants. Later some wilting of whole leaflets or leaves may occur, but the unilateral death of lower leaves is more typical (Isaac and Harrison 1968).

The activities of the fungus stimulate the plant to produce a suberin-like coating, tyloses and gels inside the vascular elements. Tyloses are extensions from parenchyma cells into the xylem (Newcombe and Robb 1988). Gels arise from perforation plates, end-walls and pit membranes by a process of distension of primary wall and middle lamella constituents (Van der Molen et al. 1977). This can lead to occlusion of the vessels in the vascular system immediately above primary infection sites (Van der Molen et al. 1977; Harrison and Beckman 1982; Newcombe and Robb 1988). A light brown vascular discoloration is often visible at the stem base when sliced (Isaac and Harrison 1968). The accumulation of specific chemicals such as terpenoid aldehydes in the vessels accompanies or follows the presence of the pathogen and reduces the viability of the fungal propagules (Harrison and Beckman 1982). As colonisation of the xylem proceeds, the vessels may become plugged by hyphae (Garber 1973). In potato plants the plugging has been traced from the root tip up to the top of the stem.

Inoculated potato plants produced no symptoms until tuberisation commenced (Busch and Edgington 1967), suggesting that before symptom expression of Verticillium wilt becomes evident, the host must be in an advanced stage of development (Busch et al. 1978). By altering the photoperiod to prevent tuberisation, few or no symptoms developed (Busch and Edgington 1967). These observations are consistent with other observations of an association between lateness of cultivar and resistance to *V. dahliae* (Busch et al. 1978).

Kotcon et al. (1985) associated disease incidence of *V. dahliae* with reduced root growth, foliar weight, and tuber yield. Infected plants exhibited lower specific leaf areas (area produced/dry weight of leaf tissue), higher leaf weight ratios (dry weight of the leaf system/dry weight of the whole plant) and higher leaf area ratios (area of the leaves produced/dry weight of the whole plant), and, under dry conditions, lower relative growth rates and lower leaf growth rates (increase in dry weight/unit leaf area/week) (Harrison and Isaac 1969).

Bowden et al. (1990) and Haverkort et al. (1990) showed that the initial decrease in photosynthesis caused by *V. dahliae* was caused by stomatal closure. The low stomatal conductance was correlated with low leaf water potential (Bowden et al. 1990). In potato, *V. dahliae* caused a reduction in the light conversion efficiency, but a stronger reduction of stomatal conductance, resulting in decreased internal/external CO_2 ratios and in a higher net photosynthesis at similar values of stomatal conductance (Haverkort et al. 1990). The reduction of net photosynthesis does not seem to be responsible for more than 10% of the reduction of dry matter production. In areas where the disease causes an early and rapid senescence of the leaf canopy, reduction of intercepted radiation may be a more important component of damage than reduction of photosynthesis (Haverkort et al. 1990).

Although there are interconnections between the vascular bundles in the stem, these are absent in the petiole. As a consequence blockage of petiole bundles can be more damaging than a proportional blockage of stem bundles (Garber 1973).

A lesser root length of potato plants due to *V. dahliae* may decrease the water supply and cause the development of foliar symptoms (Kotcon et al. 1984). Also root surface areas and volumes were reported to be affected negatively by *V. dahliae*.

Reproduction of V. dahliae (Figure 1: g)

As infected plants become senescent, the fungus permeates the surrounding tissues and forms MS within dead tissue (Powelson 1970). There is no evidence of other than transient increases in inoculum from sources other than plant debris.

Potato stems colonised by *V. dahliae* will be filled with MS. A 1 cm segment of stem may contain 8,000–20,000 viable MS and populations up to 1,000 MS per gram of soil have been reported in fields repeatedly cropped to potato, which roughly equals 50 million MS in the soil volume occupied by the roots of one plant (Menzies 1970). Differences in production of MS from 7,000–9,000 propagules per gram of stem tissue have been reported in several potato cultivars (Slattery 1981).

In most plant species infected with *V. dahliae*, water is required for formation of MS (Powelson 1970). However, in the temperate zones, humidity is usually sufficiently high during the senescence of the potato haulms to ensure abundant MS formation without rain.

External factors can have a large influence on the formation of microsclerotia. Autumn-grown crops of potato in the Negev area of Israel had approximately 100 times higher microsclerotial production than the spring-grown crops (Ben-Yephet and Szmulewich 1985). Either the cool and moist conditions in the autumn-winter season enabled the plants to dry slowly, favouring formation of MS, or the cooler weather allowed better survival of the fungus in the plant tissue.

Ioannou et al. (1977) examined the formation of MS in tomato debris in soil subjected to different irrigation and flooding regimes under field conditions. Few, if any MS were produced during the flooding treatment. This inhibition was attributed to decreased O_2 and increased CO_2 concentrations in the flooded soil. Upon drainage, the concentrations of O_2 and CO_2 returned rapidly to normal atmospheric levels and formation of MS was resumed. The numbers of MS eventually produced following 10-, 20-, and 40-day flooding treatments were 90, 44, and 46% respectively, of the average numbers in the non-flooded treatments.

The major part of the new inoculum is produced in the aerial parts of the potato plant (Ben-Yephet and Szmulewich 1985; Mol 1994). A direct way for control of *V. dahliae* is to interfere with the formation and dispersal of MS. In practice this is accomplished by various sanitation measures such as removal or destruction of diseased plant material before it enters the soil or before it releases the inoculum otherwise. This process of field sanitation is not widely practised because of expense, lack of equipment, and through farmers' aversion to destroying organic matter. Another possibility for the control of the reproduction is the use of mechanical haulm killing techniques. Compared to haulm killing by the application of a herbicide, the reproduction in the aerial parts of the potato plant was reduced by up to 80% following mechanical haulm killing (Mol 1994).

Production of MS has been reported in plant roots without systemic infection. Microsclerotia were formed in large numbers in only few sections of wheat roots (Krikun and Bernier 1990). Taking into account total root length of wheat and the number of MS found (up to 100 per root fragment of 1.3 × 0.4 mm), the contribution to the inoculum in soil may be significant (Krikun and Bernier 1990).

Weed control is important to limit multiplication of *V. dahliae* (Woolliams 1966; Johnson et al. 1980). Host species include many common weeds and native plants. As in crop plants, symptoms are not always apparent in infected weed plants.

Host specificity

Adaptation of *V. dahliae*, leading to host specificity, may confuse the search for possible resistance or tolerance mechanisms. Zilberstein et al. (1983) reported that germination on agar and pathogenicity of MS of *V. dahliae* to egg-plant, potato and tomato was affected by growth medium and host origin. The virulence of an isolate of *V. dahliae* depends on the host species, and geographical origin, susceptibility of the host cultivar/genotype, and the organ on the plant (Zilberstein et al. 1983; Michail 1989). Isolates obtained from susceptible cultivars attacked these cultivars only. Isolates from resistant cultivars attacked both susceptible and moderately resistant cultivars (Michail 1989).

Production of MS variants might permit the fungus to adapt to new host species and varieties or to new environmental conditions (Tolmsoff 1973). Mint isolates were originally not pathogenic to tomato. After one passage through tomato, however, the isolates became more pathogenic to tomato and lost pathogenicity to mint. The isolates from a region of continuous cropping to one crop tend to be similar and all display high virulence against that particular crop but generally a weak virulence against other species which they can infect nevertheless (Vigoroux 1971). Thus a notion of 'preferential' and 'occasional' hosts is formed. It is known that great variability is possible in *Verticillium* species (Vigoroux 1971).

Puhalla and Hummel (1983) found 16 different vegetative compatibility groups in *V. dahliae*. Using another method, Joaquim and Rowe (1990; 1991) were able to reduce the number of vegetative compatibility groups to four. Isolates from two different vegetative compatibility groups showed significant differences in pathogenicity towards potato (Joaquim and Rowe 1991).

Thus, in soils, populations of individuals with different properties can be built up. It appears that the kind of crop constitutes a determinant factor for the quantitative and qualitative composition of the population of *V. dahliae* in cultivated soils (Vigoroux 1971; Tjamos 1981). So, the choice of cultivar, crop rotation, and cultural practices will not be sufficient to keep the crop from infection, but they will still be key factors in the controlling the severity of the disease.

Acknowledgements

We thank Dr. K. Scholte, Professor P.C. Struik, and Dr. J. Vos for critically reading the manuscript.

References

Bell A A (1973) Nature of disease resistance. Pages 47–62 in Verticillium Wilt of Cotton. Proceedings of a Work Conference, National Cotton Pathology Research Laboratory College Station USDA Agricultural Research Service, Texas, USA.

Ben-Yephet Y, Szmulewich Y (1985) Inoculum levels of *Verticillium dahliae* in the soils of the hot semi-arid Negev region of Israel. Phytoparasitica 13:193–200.

Bloomfield B J, Alexander M (1967) Melanin and resistance of fungi to lysis. Journal of Bacteriology 93:1276–1280.

Bowden R L, Rouse D I, Sharkey T D (1990) Mechanism of photosynthesis decrease by *Verticillium dahliae* in potato. Plant Physiology 94:1048–1055.

Busch L V, Edgington L V (1967) Correlation of photoperiod with tuberization and susceptibility of potato to *Verticillium albo-atrum*. Canadian Journal of Botany 45:691–693.

Busch L V, Smith E A, Njoh-Elango F (1978) The effect of weeds on the value of rotation as a practical control for Verticillium wilt of potato. Canandian Plant Disease Survey 58:61–64.

Coley-Smith J R, Cooke R C (1971) Survival and germination of fungal sclerotia. Annual Review Phytopathology 9:65–92.

Curl E A, Truelove B (1986) The Rhizosphere. Advanced Series in Agricultural Sciences 15. Springer-Verlag, Berlin, Germany.

Emmatty D A, Green R J (1969) Fungistasis and the behaviour of the microsclerotia of *Verticillium albo-atrum* in soil. Phytopathology 59:1590–1595.

Evans G, Gleeson A C (1973) Observations on the origin and nature of *Verticillium dahliae* colonizing plant roots. Australian Journal of Biological Sciences 26:151–161.

Evans G, Wilhelm S, Snyder W C (1967) Quantitative studies by plate counts of propagules of the Verticillium wilt fungus in cotton field soils. Phytopathology 57:1250–1255.

Farley J D, Wilhelm S, Snyder W C (1971) Repeated germination and sporulation of microsclerotia of *Verticillium albo-atrum* in soil. Phytopathology 61:260–264.

Fitzell R, Evans G, Fahy P C (1980) Studies on the colonization of plant roots by *Verticillium-dahliae* with use of immuno fluorescent staining. Australian Journal of Botany 28:357–368.

Garber R H (1973) Fungus penetration and development. Pages 69–77 in Verticillium Wilt of Cotton. Proceedings of a Work Conference, National Cotton Pathology Research Laboratory College Station USDA Agricultural Research Service, Texas, USA.

Gerik J S, Huisman O C (1988) Study of field-grown cotton roots infected with *Verticillium dahliae* using an immunoenzymatic staining technique. Phytopathology 78:1174–1178.

Green R J (1981) Fungal wilt diseases of plants. Pages 1–24 in Mace E, Bell A A, Beckman C H (Eds.) Fungal Wilt Diseases of Plants. Academic Press, New York, USA.

Griffiths D A (1973) An electron microscopic study of host reaction in roots following invasion by *Verticillium dahliae*. Skokubutsu Byogai Kenkyu 8:147–154.

Hancock J G, Benham G S (1980) Fungal decay of buried cotton *Gossypium-hirsutum* stems. Soil Biological Biochemistry 12:35–42.

Harrison M D (1976) The effect of barley straw on the survival of *Verticillium albo-atrum* in naturally infested field soil. American Potato Journal 53:385–394.

Harrison N A, Beckman C H (1982) Time/space relationships of colonization and host response in wilt-resistant and wilt susceptible cotton (*Gossypium*) cultivars inoculated with *Verticillium dahliae* and *Fusarium oxysporum* f. sp. *vasinfectum*. Physiological Plant Pathology 21:193–207.

Harrison J A C, Isaac I (1969) Host/parasite relations up to the time of tuber initiation in potato plants infected with *Verticillium* spp. Annals of Applied Biology 64:469–482.

Haverkort A J, Rouse D I, Turkensteen L J (1990) The influence of *Verticillium dahliae* and drought on potato crop growth. 1. Effects on gas exchange and stomatal behaviour of individual leaves and crop canopies. Netherlands Journal of Plant Pathology 96:273–289.

Howell C R (1973) Pathogenicity and host-parasite relationships. Pages 42–46 in Verticillium Wilt of Cotton. Proceedings of a Work Conference, National Cotton Pathology Research Laboratory College Station USDA Agricultural Research Service, Texas, USA.

Huisman O C, Ashworth L J (1976) Influence of crop rotation on survival of *Verticillium albo-atrum* in soils. Phytopathology 66:978–981.

Ioannou N, Schneider R W, Grogan R G (1977) Effect of flooding on the soil gas composition and the production of microsclerotia by *Verticillium dahliae* in the field. Phytopathology 67:651–656.

Isaac I, Harrison J A C (1968) The symptoms and causal agents of early-dying disease (verticillium wilt) of potatoes. Annals of Applied Biology 61:231–244.

Itoh S, Komoda H, Monma T, Amano T (1989) Development of field diagnosis system (FDS) for preventing continuous cropping injury of crop. 12. Study of factors related to the development of a prediction model of Verticillium yellows in Chinese cabbage. Bulletin of the National Agriculture Research Center, Japan 16:33–53.

Joaquim T R, Rowe R C (1990) Reassessment of vegetative compatibility relationships among strains of *Verticillium dahliae* using nitrate-nonutilizing mutants. Phytopathology 80:1160–1166.

Joaquim T R, Rowe R C (1991) Vegetative compatibility and virulence of strains of *Verticillium dahliae* from soil and potato plants. Phytopathology 81:552–558.

Johnson K B (1992) Evaluation of a mechanistic model that describes potato crop losses caused by multiple pests. Phytopathology 82:363–369.

Johnson W M, Johnson E K, Brinkerhoff L A (1980) Symptomatology and formation of micro sclerotia in weeds inoculated with *Verticillium-dahliae* from cotton *Gossypium-hirsutum*. Phytopathology 70:31–35.

Jordan V W L, Sneh B, Eddy B P (1972) Influence of organic soil amendments on *Verticillium dahliae* and on the microbial composition of the strawberry rhizosphere. Annals of Applied Biology 70:139–148.

Kotcon J B, Rouse D I, Mitchell J E (1984) Dynamics of root growth in potato fields affected by the early dying syndrome. Phytopathology 74:462–467.

Kotcon J B, Rouse D I, Mitchell J E (1985) Interactions of *Verticillium dahliae, Colletotrichum coccodes, Rhizoctonia solani*, and *Pratylenchus penetrans* in the early dying syndrome of Russet Burbank potatoes. Phytopathology 75:68–74.

Krikun J, Bernier C C (1990) Morphology of microsclerotia of *Verticillium dahliae* in roots of gramineous plants. Canadian Journal of Plant Pathology 12:439–441.

Lacy M L, Horner C E (1966) Behaviour of *Verticillium dahliae* in the rhizosphere and on roots of plants susceptible, resistant, and immune to wilt. Phytopathology 56:427–430.

Martinson C A, Horner C E (1962) Importance of non-hosts in maintaining the inoculum potential of *Verticillium*. Phytopathology 52:742.

Menzies J D (1970) Factors affecting plant pathogen population in soil. Pages 16–21 in Toussoun T A, Bega R V, Nelson P E (Eds.) Root Diseases and Soil-Borne Pathogens. University of California Press, Berkeley, USA.

Menzies J D, Griebel G E (1967) Survival and saprophytic growth of *Verticillium dahliae* in uncropped soil. Phytopathology 57:703–709.

Michail S H (1989) Fusarium and Verticillium wilts of cotton. Pages 199–217 in Agrihotri V P, Singh N, Chaube H S, Singh U S, Dwivedi T S (Eds.) Perspectives in Phytopathology. Today and Tomorrow's Printers & Publishers, New Delhi, India.

Mol L (1994) Control of *Verticillium dahliae* by catch-crops and haulm killing techniques. Pages 223–228 in Struik P C, Vredenberg W J, Renkema J A, Parlevliet J E (Eds.) Plant Production on the Threshold of a New Century. Kluwer Academic Publishers, Dordrecht, The Netherlands.

Newcombe G, Robb J (1988) The function and relative importance of the vascular coating response in highly resistant, moderately resistant and susceptible alfalfa infected by *Verticillium albo-atrum*. Physiological and Molecular Plant Pathology 33:47–58.

Pegg G F (1974) Verticillium diseases. Review of Plant Pathology 53:156–182.

Pegg G F (1984) The impact of Verticillium diseases in agriculture. Phytopathologia Mediterranea 23:176–192.

Powelson R L (1970) Significance of population level of Verticillium in soil. Pages 31–33 in Toussoun T A, Bega R V, Nelson P E (Eds.) Root Diseases and Soil-Borne Pathogens. University of California Press, Berkeley, USA.

Puhalla J E, Hummel M (1983) Vegetative compatibility groups within *Verticillium dahliae*. Phytopathology 73:1305–1308.

Rouse D I (1985) Some approaches to prediction of potato early dying disease severity. American Potato Journal 62:187–193.

Schnathorst W C (1981) Life cycle and epidemiology of Verticillium. Pages 81–111 in Mace M E, Bell A A, Beckman C H (Eds.) Fungal Wilt Diseases of Plants. Academic Press, New York, USA.

Scholte K (1989) Effects of crop rotation and granular nematicides on the incidence of *Verticillium dahliae* Kleb. and *Colletotrichum coccodes* (Wallr.) Hughes, in potato. Potato Research 32:377–385.

Scholte K, s'Jacob J J (1989) Synergistic interactions between *Rhizoctonia solani* Kuhn, *Verticillium dahliae* Kleb., *Meloidogyne* spp. and *Pratylenchus neglectus* (Rensch) Chitwood & Oteifa, in potato. Potato Research 32:387–395.

Scholte K, Veenbaas-Rijks J W, Labruyère R E (1985) Potato growing in short rotations and the effect of *Streptomyces* spp., *Colletotrichum coccodes*, *Fusarium tabacinum* and *Verticilium dahliae* on plant growth and tuber yield. Potato Research 28:331–348.

Schreiber L R, Green R J (1963) Effect of root exudates on germination of conidia and microsclerotia of *Verticillium albo-atrum* inhibited by soil fungistatic principle. Phytopathology 53:260–264.

Slattery R J (1981) Inoculum potential of *Verticillium* – infested potato cultivars. American Potato Journal 58:135–142.

Street P F S, Cooper R M (1984) Quantitative measurement of vascular flow in petioles of healthy and *Verticillium*-infected tomato. Plant Pathology 33:483–492.

Tjamos E C (1981) Virulence of *Verticillium dahliae* and *Verticillium albo-atrum* isolates in tomato seedlings in relation to their host origin and the applied cropping system. Phytopathology 71:98–100.

Tolmsoff J (1973) Life cycles of *Verticillium* species. Pages 20–38 in Verticillium Wilt of Cotton. Proceedings of a Work Conference, National Cotton Pathology Research Laboratory College Station USDA Agricultural Research Service, Texas, USA.

Tolmsoff W J, Young R A (1959) The influence of crop residues and fertilizer on the development and severity of Verticillium wilt of potatoes. Phytopathology 49:114.

Van der Molen G E, Beckman C H, Rodehorst E (1977) Vascular gelation: a general response phenomenon following infection. Physiological Plant Pathology 11:95–100.

Vigoroux A (1971) Hypothesis to explain the anomalous pathological behaviour of some *Verticillium* isolates. Page 31 in Proceedings of the International Verticillium Symposium. Wye College, London, UK.

Wheeler T A, Madden L V, Rowe R C, Riedel R M (1992) Modelling of yield loss in potato early dying caused by *Pratylenchus penetrans* and *Verticillium dahliae*. Journal of Nematology 24:99–102.

Wilhelm S (1955) Longevity of the Verticillium wilt fungus in the laboratory and field. Phytopathology 45:180–181.

Woolliams G E (1966) Host range and symptomatology of *Verticillium dahliae* in economic, weed, and native plants in interior British Columbia. Canadian Journal of Plant Science 46:661–669.

Zilberstein Y, Chet I, Henis Y (1983) Influence of microsclerotia source of *Verticillium dahliae* on inoculum quality. Transactions of the British Mycological Society 81:613–617.

17. Modelling the dynamics of *Verticillium dahliae*

A.J. TERMORSHUIZEN[1] and L. MOL[2]

[1] *Department of Phytopathology, Wageningen Agricultural University, P.O. Box 8025, 6700 EE Wageningen, The Netherlands*

[2] *Department of Agronomy, Wageningen Agricultural University, Haarweg 333, 6709 RZ Wageningen, The Netherlands*

Abstract. Models dealing with *Verticillium dahliae* focus either on simple models which relate inoculum density to disease incidence and yield loss or on complex models handling the whole life cycle of *V. dahliae*. From the published literature on the relation between inoculum density and disease incidence no general model appears, which may have its origin in the various methodological approaches. Five complex models have been published: two crop loss models, one mechanistic model, and two decision models. Many complex models have limited practical value because of problems with calibration. They should be considered as quantitative summaries of the present knowledge. Coupling the effects of biotic stress factors on radiation interception and radiation use efficiency has shown to be a simple way to introduce multiple biotic stress factors in one model.

Introduction

Researchers dealing with *Verticillium dahliae* share the experience that results of experiments are often inconsistent, apparently as a consequence of the fact that the behaviour of this pathogen is insufficiently understood. This may account for the limited progress in modelling the dynamics of *V. dahliae* and the diseases it causes. This review investigates the progress in quantitative studies on *V. dahliae* and attempts to investigate the methodologies used and the way they may have influenced the results.

Many types of models have been recognized, including empirical, mechanistic, and decision-supporting. Since little progress has been made in quantitative modelling of *V. dahliae*, these subdivisions are not used here but rather models are grouped only as simple and complex. Simple models describe relations between one or a few independent and dependent variables. In complex models, multiple variables are involved.

Simple models related to inoculum density, disease incidence and yield loss

Many attempts have been made to relate soil inoculum density to disease incidence, in order to predict yield loss from pre-plant inoculum density.

A.J. Haverkort and D.K.L. MacKerron (eds.), Potato Ecology and Modelling of Crops under Conditions Limiting Growth, 265–280.

Inconsistent results have been obtained, ranging from absence of correlations to positive correlations. Negative correlations have not been reported.

DeVay et al. (1974) were unable to demonstrate a significant correlation between inoculum density in field soils and percentage of diseased cotton plants. Two fields with inoculum densities greater than 120 microsclerotia g^{-1} soil showed no diseased plants, whereas four fields containing 30 microsclerotia g^{-1} soil had more than 50% diseased plants. They suggested that the density of microsclerotia in soil is not the determining factor in disease development. Similar results were obtained by Ashworth et al. (1972).

Several other reports seemed to indicate a clear relation between inoculum density and disease incidence. Significant disease incidences have been shown to occur at inoculum densities < 1–5 microsclerotia g^{-1} soil (Ashworth et al. 1979a,b; Grogan et al. 1979; Nicot and Rouse 1987; Paplomatas et al. 1992). The logistic model fitted best to data of inoculum density and disease incidence (Nicot and Rouse 1987). Due to the high variability in their results, more research is needed to confirm that the logistic model holds, since a monomolecular model would be expected if other factors would remain constant. One factor which may influence the model is the effect of inoculum density on root growth (Kotcon et al. 1984), which may in turn affect disease progress (Jeger 1987).

In some studies, the relation between inoculum density of *V. dahliae* and yield loss was obtained by including the interaction with root-infecting nematodes (Wheeler et al. 1992).

The ultimate aim of these studies, i.e. to use the relation between inoculum density and disease incidence or yield loss for advisory systems, has been achieved in only a few cases (Nnodu and Harrison 1979). Failure to achieve this goal may be due to inconsistent relationships between inoculum density and disease incidence or yield loss among different fields (DeVay et al. 1974; Keinath and Fravel 1992) and also to the influence of weather on disease progress (e.g. Francl et al. 1990). Since most control measures can be implemented only prior to planting (site selection, choice of crop or cultivar, soil fumigation, flooding), yield loss is unpredictable because long-term weather forecasts are not reliable. The factors which may influence the relation between inoculum density and disease incidence are presented in more detail in the following sections.

Instead of focusing on the effects of pre-plant inoculum density on the current crop, a wider view includes the whole cropping sequence as a measure of control (Bollen et al. 1989; Davis et al. 1994) and the future dynamics of new microsclerotia as target. These studies include the inoculum density as a variable instead of a constant, which is usually assumed for single growing-season studies.

Design of experiments. Empirical relationships between inoculum density and disease incidence or yield loss have been studied in either field or pot experiments. The term microplot is used for experiments with large sized pots on containers with open bottoms that are dug into the field (Francl et al. 1988).

Field studies have been conducted either by collecting data from naturally infested fields or by adding different amounts of inoculum to soils.

Addition of known amounts of microsclerotia can be accomplished using either naturally produced microsclerotia collected from the field (Ashworth et al. 1979a,b; Huisman 1988b) or microsclerotia produced artificially on susceptible crops in the greenhouse, on sterilized plant material like rye seed (MacGuidwin and Rouse 1990b), on autoclaved potato stems (Mol, current work), or on an artificial medium (Francl et al. 1988)). The addition of known amounts of inoculum to soil (with a known germination rate) avoids potential errors related to detection and quantification. On the other hand, the process of adding microsclerotia to field soil influences soil structure, and a 'natural' distribution over the soil profile will not be obtained. However, it may be questioned whether there is such a thing as a 'natural' distribution, since all distributions are due to soil cultivation. Pot experiments may be useful in standardizing the environment and the inoculum density, but they may also impose artifacts such as limited soil volume, atypical soil temperatures and unnatural soil structure.

Zilberstein et al. (1983) obtained evidence that the germinability and pathogenicity of microsclerotia depended on their origin (i.e. different plant species or growth medium). Mol (unpublished) noticed that the mortality rate of microsclerotia produced on autoclaved potato stems was greater than those collected in the field and stored in the same way. Then again, the use of microsclerotia collected from infested field material may provoke unwanted effects where the stems have been colonized by other fungi, e.g. *Colletotrichum coccodes*, *V. albo-atrum* or *V. tricorpus*, or by antagonists. Further, the genetic variation in collected field inoculum may lead to more variable results.

Ashworth et al. (1979a) and Pullman and DeVay (1982a) suggested that the absence of relationships between inoculum density and disease incidence as reported by Ashworth et al. (1972) and DeVay et al. (1974) was due to the fact that they collected data from different fields. Ashworth et al. (1979a) reported a positive correlation between inoculum density, created by addition of soil using tomato shoots containing microsclerotia of *V. dahliae*, and disease incidence. However, in the same paper the authors reported that they failed to find such a correlation when the inoculum densities and disease incidences from different fields were compared. Similarly, Nnodu and Harrison (1979) found a positive correlation between inoculum density and yield loss at one location, but not at another. Additional experiments provided substantial evidence that the suppression of disease at one location was caused by suboptimal temperatures for wilt development caused by *V. dahliae* (Nnodu and Harrison 1979).

In several cases, however, significant relations between inoculum density and disease incidence have been obtained by collecting data from different fields (Grogan et al. 1979; Harris and Yang 1990 [except two cases]; Paplomatas et al. 1992). Significant correlations between disease incidence and inoculum density were obtained by Nicot and Rouse (1987), who based the inoculum density of *V. dahliae* on soil samples taken from the vicinity of small groups of plants on which the disease incidence was based. When this was done, data between different locations yielded consistent results.

Variation in results due to different methodologies between researchers makes comparisons highly problematic (Conroy et al. 1972; Harris et al. 1993). Variable results may also be obtained when comparing samples from one field collected at different periods of time (Ashworth et al. 1976).

Measurement of inoculum density. Inoculum density of *V. dahliae* in soil is based on the density of microsclerotia. Although sporulation in soil has been reported (Farley et al. 1971), conidia are short-lived (Schreiber and Green 1962; Green 1969). Linear relations between the density of microsclerotia in soil, the density of colonies isolated from roots, and the random distribution of colonies on the root system further suggest that conidia do not play a significant role in the process of root infection (Evans and Gleeson 1973).

In most investigations, inoculum density is expressed as the number of microsclerotia per unit dry weight of soil. However, some researchers express the inoculum density per unit volume of fresh soil (e.g. the research groups at Wooster, Ohio and Berkeley, California). The latter is biologically more correct, since distances between entities in soil are volume-based and not weight-based. Since soil densities among different soil textures may vary by up to 50%, assessments of inoculum densities based on soil weight may be biased by an equal amount.

Since there is no single method for estimating the inoculum density of microsclerotia in soil, researchers use various methods, making their results difficult to compare. Methods of analysis vary in composition of isolation medium and handling of soil samples. In addition, other variables include duration of air drying and storage of soil samples (Smith and Rowe 1984), duration of incubation and amount of soil suspension per agar plate. A concise review on this subject was recently provided by Harris et al. (1993). At present, the most commonly applied methods involve dilution plating of air-dried, ground and sieved (20–c. 130 μm) samples on a polypectate-based semi-selective medium. Many kinds of the polypectate-based medium exist (cf. Harris et al. 1993), and it seems that every laboratory is claiming the optimal nutrient medium for isolation of *V. dahliae* from soil samples. For the matter of comparability between laboratories, standardization in composition of the polypectate-based nutrient medium seems to be a minimal requirement. The Anderson sampler method is used by several researchers for samples with high inoculum densities (e.g. Davis et al. 1994). However, dilution plating seems to result in a lower variability in the results (Smith and Rowe 1984).

Inoculum densities in experiments using microsclerotia produced artificially are based on enumeration of the microsclerotia either directly, before addition to the soil, or afterwards by soil dilution plating. Care must be taken when interpreting inoculum densities from dilution plating, since results may be influenced by soil type. After addition of a given amount of inoculum to two different field soils, Ashworth et al. (1979a) recovered by soil dilution plating 23.9 and 80.0 microsclerotia per gram of soil, respectively.

Problems in identification of *V. dahliae* on soil dilution plates are rarely mentioned. Some confusion may arise with *V. tricorpus*, which is distinguished by its larger microsclerotia and, attached to the microsclerotia, short, black, hyphae. The size of microsclerotia of both *V. dahliae* and *V. tricorpus* can be influenced by the nutrient medium used (Termorshuizen, unpublished). *Verticillium tricorpus* often produces a yellow pigment on certain nutrient media, but this is inconsistent. The yellow pigmentation only rarely occurs on soil dilution plates and the same is true for cultures on ethanol agar (Termorshuizen, unpublished). Huisman (1988a), citing Taylor (1969), adjusted his polypectate medium by adding cellophane-extract to increase the production of yellow pigment by *V. tricorpus. Verticillium dahliae* is easily distinguished from *V. albo-atrum* by the presence of microsclerotia and the absence of dark hyphae. Previously, some authors considered *V. dahliae* as a microsclerotial form of *V. albo-atrum*. However, in addition to the morphological differences, recent research on RFLP-patterns and DNA-sequences clearly indicate the distinction between these two species (Hu et al. 1993; Nazar et al. 1991). Problems with identification may arise on soil dilution plates when small colonies of only few microsclerotia are formed. In this case, colonies should be subcultured on potato dextrose agar.

Inoculum density (= the number of microsclerotia per unit volume of soil) is not necessarily the same as inoculum potential (= the ability to cause disease), which can be measured only by means of bio-assays. Evans et al. (1974) provided a bio-assay for inoculum potential by measuring the density of colonies on root systems of susceptible plants. This assay has not been generally accepted, presumably because of the high variability obtained experimentally (Huisman 1988b). In other words, the factors influencing the bio-assays are insufficiently understood and investigated. Ben-Yephet and Pinkas (1977) obtained evidence that the potential of microsclerotia to germinate, and thus their inoculum potential, was dependent on their size. If consistent size differences in microsclerotia among locations occur, various relations between results of bio-assays and measurements of inoculum densities will be found.

Soil sampling to estimate inoculum density. Several studies have examined spatial aggregation of microsclerotia of *V. dahliae* in soil. Mean densities of microsclerotia in an Ohio potato field appeared to follow the negative binomial distribution when analyzed in a 15.2 m grid pattern (Smith and Rowe 1984), indicating spatial aggregation of inoculum. Johnson et al. (1988) arrived at the same conclusion using plots set out on a 10 m grid pattern in three potato fields in Oregon. Nicot (1985) examined the distribution of microsclerotia at three grid sizes: 54 m, 6.3 m and 0.125 m in four potato fields. Aggregation was shown for the 54 m grids in three fields but a random pattern in the fourth field. At the other grid sizes, most patterns were random. The variation among mean inoculum densities were on average smaller among proximate areas than among more remote areas. Nicot (1985) pointed out that random patterns of distribution of inoculum density might indicate the planting of *V. dahliae*-infected seed tubers, as has been reported several times (e.g. Beckman 1973).

From the few studies on spatial distributions of inoculum in soil, it may be concluded that aggregation of inoculum can be found at some scale so that a composite sample is a biased estimator of the inoculum potential of the whole field. Smith and Rowe (1984) estimated that for one field 107 soil samples would have been necessary to reduce the sampling error to 10%. However, due to the amount of work involved in handling the samples, composite samples over larger areas are the rule rather than the exception. Nicot (1985) argued that due to the very different patterns of distributions between fields, a sampling strategy adapted to one particular field may not be appropriate for others. In contrast to most studies performed, that by Nicot and Rouse (1987) related inoculum densities of samples taken within 0.125 m squares to the three nearby potato plants. This produced positive correlations between inoculum density and disease incidence within as well as between the plots. This is a strong argument that spatial heterogeneity of inoculum in soil (and associated sampling problems) may well explain some of the conflicting reports of relationships between estimates of inoculum density and disease incidence.

The few data available on the vertical distribution of microsclerotia in soil show that the majority of inoculum is present in the surface layer of 0.30 m (Green 1957; Wilhelm 1950; Wilhelm et al. 1967; Huisman 1988b). However, viable propagules have been found at depths greater than 1 m (Wilhelm et al. 1967; Huisman 1988b). Sampling is usually done in the upper 0–15 or 0–20 cm of soil, but occasionally deeper, e.g. 10–20 cm below soil surface (DeVay et al. 1974).

Measurement of disease incidence. Root infection is usually determined by plating out washed root segments on a semi-selective medium. Evans and Gleeson (1973) advised against the use of disinfectants as they negatively affect the superficial colonies of *V. dahliae*. As there is a chance of cutting one colony into two pieces, Evans and Gleeson (1973) and Huisman (1988a) provided calculations to express the number of colonies of *V. dahliae* per unit length of the root.

Disease incidence in the shoot is usually determined either by isolation of fungal tissue from the stem base or by assessing wilt symptoms (wilting of leaves and vascular browning), accompanied with incidental stem isolations. Care must be taken when evaluating disease incidence on the basis of wilt symptoms alone (e.g. Ashworth et al. 1979a), since this method will not reveal symptomless infections. In addition, other diseases may cause wilt symptoms as well. Assessments of propagule densities in the xylem can provide quantitative information about the prevalence of the pathogen (Hoyos et al. 1991). However, results can be influenced by the host tissue itself, which may affect germination of the conidia or growth of the mycelium (Pegg 1978). For potato, Hoyos et al. (1991) reported a significant positive correlation between colony-forming units of *V. dahliae* in the stem base and the amount of wilting (expressed as a disease index). Harrison and Beckman (1982) described the build-up of inoculum of *V. dahliae* in the stem base (inoculated by addition of conidia through severed tap-roots) by an exponential model.

Yield loss is a less accurate estimator for disease incidence than direct isolation of the pathogen, but is more directly related to practice. It requires the estimation of two variables, yield of the control and that of the treatment, leading to an increase in error and variation. Yield loss, but not necessarily disease incidence, is usually positively related to infection by *V. dahliae* under dry growing conditions (Francl et al. 1990). On the other hand, isolations of the pathogen during the growing season are indispensable for distinguishing among treatments because disease incidences may converge towards the end of the growing season for different treatments, especially in late-maturing crops (Scholte 1989). However, such convergence has not been observed in cotton and potato production areas in the United States (Ashworth et al. 1979; Pullman and DeVay 1982a; Nicot and Rouse 1987). Ashworth et al. (1979a) showed that relationships between disease incidence and yield loss may vary among different cotton cultivars. From the point of view of the farmer, economical yield loss is a more sensible parameter than disease incidence. Harris and Yang (1990) established their threshold levels for strawberries on the maximum permissible incidence of Verticillium wilt allowable for certification.

Influence of environment. The biotic and abiotic environment have a profound effect on disease progress and yield loss. The biotic environment may be synergistic or antagonistic. Root-infecting nematodes generally increase the reaction of the plant to infection by *V. dahliae*. Their effect is dependent on the species of nematode involved and generally results in steeper disease progress curves and steeper regression curves between inoculum density and incidence of *V. dahliae*. Regression equations on the interaction of root infecting nematodes and *V. dahliae* have been published by the Ohio group (Francl et al. 1988; 1990; Wheeler et al. 1992).

Up to the present the effects of antagonists, including hyperparasites, on *V. dahliae* have not been modeled. This may be due in part to the variable results obtained with biocontrol agents. The effects of fungistasis which may explain different recoveries of microsclerotia (Ashworth et al. 1979a) and different rates of disease progress (Azad et al. 1985; 1987; Keinath and Fravel 1992) on different soil types urgently need to be clarified and to be quantified.

In tritrophic or tripartite systems, the densities of the organisms involved are usually mutually affected. In contrast to entomology, modelling of density-dependent relations in fungal pathogens has barely been attempted. Reasons for this may be the complexity of such studies and the lack of understanding of mechanisms of interaction. For example, MacGuidwin and Rouse (1990b) and Martin et al. (1982) obtained evidence that the effect of *V. dahliae* on potato was dependent on the density of *Pratylenchus penetrans*, in contrast to results obtained by Wheeler et al. (1992) on the same system and MacGuidwin and Rouse (1990a) with *Meloidogyne hapla*. Conflicting results have also been reported on the effect of *V. dahliae* on the density of *P. penetrans* in soil (cited by Martin et al. 1982).

One aspect which makes predictions about yield loss prior to planting precarious is the influence of weather on disease progress and the unpredictability of weather. Although the general effects of temperature and moisture are known (e.g. Nnodu and Harrison 1979; Ioannou et al. 1977; see also Mol and Termorshuizen 1995), modelling on this topic has been performed only rarely. Francl et al. (1990) showed that yield loss of *V. dahliae*-infected plants was correlated with temperature and stated that prediction of yield loss may be possible based on weather prediction in advance of symptom appearance. However, information on inoculum density or disease incidence also is then needed. In addition to climatic conditions, also irrigation and fertilization have been shown to influence disease incidence significantly (Davis and Everson 1986; Cappaert et al. 1992).

Relating inoculum density to disease incidence. Although not investigated, the time of assessing inoculum density in soil in relation to the time of assessing disease incidence is likely to influence the results. Assessment of inoculum density usually takes place around planting or sowing date. However, Grogan et al. (1979) assessed inoculum density at the time that disease incidence was measured. Inoculum density can vary considerably during the growing season, as indicated by Evans et al. (1967) and Huisman (1988a). Factors involved in this variation are: germination of microsclerotia, root growth, release of clustered microsclerotia from decomposing debris (Huisman 1988a), change in fungistasis and death of microsclerotia.

Disease incidence is often assessed only at harvest time. Since disease progress is influenced clearly by temperature, soil moisture and plant nutrition, the slope of the relation between inoculum density and disease incidence is likely to be affected by this. A more proper method would be to analyze disease progress in time in relation to the dynamics of inoculum density.

In this regard, either normal time or thermal time (Kelvin days) could be used as a time-base. Thermal time may reflect crop growth and fungal development more realistically (Pullman and DeVay 1982b). However, one should be aware that temperature dependence may vary between plant and fungus, and temperature relations may be consistent only over a limited range.

Conclusions. Knowledge regarding relationships among inoculum densities, disease incidence and yield loss does not seem to be generally applicable. The variation has been attributed to non-specified differences among research sites, but these 'explanations' need still to be converted into quantifiable variables. One possibility, disease suppressiveness, has been investigated in only a few cases (Keinath and Fravel 1992). The only mechanistic explanation for disease suppressiveness was provided by Ashworth et al. (1976) who attributed the development of a soil fungistasis to presence of high concentrations of Cu^{2+} in soil. The extent of this phenomenon is however not known.

Presumably, the main problem in understanding the relationships between inoculum density and disease incidence or yield loss is that too little attention has been paid to the study of intermediate processes. Much more insight is needed on the effects of general microbial activity on the germination of microsclerotia and subsequent root infection, the fraction of cortical colonies that actually penetrate the xylem, the biomass of *V. dahliae* in the stem as a function of the number of xylem infections, the root growth and root architecture, and the effects of other biotic and abiotic parameters on these processes. The influence of abiotic factors on disease progress has been studied surprisingly little under controlled conditions. For example, when referring to the effect of temperature on growth of *V. dahliae* in vitro, references are most often made to the review by Schnathorst (1981), but original contributions are rare and incomplete (Kendrick and Middleton 1959). As has been discussed in previous sections, the methodology employed may cause artifacts which lead to different conclusions (e.g. Pegg 1978). It may be questioned if all intermediate processes can be analyzed quantitatively in full detail. For example, the time of root infection and the production of conidia in the xylem are probably the results of complex stochastic processes, which may be impossible to approach deterministically. Beckman et al. (1962; 1972) provided some interesting quantification of the dynamics of *Fusarium oxysporum* in xylem vessels. A tentative, semi-quantitative model was later published on this subject (Beckman 1987).

In conclusion, it seems to be far too simple to aim to relate inoculum density in soil to subsequent disease incidence or yield loss because many intermediate processes occur over a relatively long time interval. Ways to handle this effectively are quantitative analysis of intermediate processes and continuous critical evaluation of research methods. Which intermediate processes are crucial to explain the relation between inoculum density and disease incidence, can only be determined after a thorough conceptual and experimental analysis of the intermediate processes has been carried out.

Complex models

Crop loss models. Johnson (1988; 1992) constructed two models on yield loss in potato, the earlier with *V. dahliae* only and the later with multiple biotic stress factors, including *V. dahliae*.

Incidence of *V. dahliae* in stems was expressed by a Gompertz model as a function of disease onset (revealed by isolations of the fungus) and infection rate. These parameters were fitted to the data of field experiments in 1985, 1986 and 1987. In 1987, treatments included supplemental irrigation and two planting dates about 6 weeks apart. Supplemental irrigation in that experiment gave lower infection rates, and the late planting resulted in a higher infection rate.

It was assumed that increased incidence of *V. dahliae* in the stems resulted in accelerated senescence of leaf tissue, which was assumed to be positively and linearly correlated with daily mean temperature above 17 °C. Accelerated senescence of leaf tissue resulted in reduced interception of radiation and lower radiation use efficiency.

The simulations showed a good fit with the field data, confirming the importance of temperature in symptom expression. The importance of soil matric potential and transpiration stream in the dynamics of *V. dahliae* within the plant was considered by Johnson (1988), but not implemented in the model.

In a subsequent model, Johnson (1992) integrated the simplified versions of submodels developed earlier that simulated crop loss caused by potato leafhopper (*Empoasca fabae*), *Alternaria solani* and *V. dahliae*. All three biotic stress factors were coupled to the potato growth model through modification of radiation interception and radiation use efficiency. Both the field data and the simulation results suggested that total crop loss in the presence of all biotic stress factors was less than the sum of losses from each stress factor acting alone. Johnson (1992) concluded that competition between the diseases and potato leafhopper for photosynthates was the dominant interaction in the system.

The approach of Johnson (1992) differs from many others which are based on linear, empirical relations between crop yield and disease severity. This relation is, according to Johnson (1992), not necessarily linear, and the situation becomes more complex with multiple biotic stress factors, which is the common field situation. The advantage of Johnson's model is its simplicity and its partly mechanistic approach, being based on physiological processes. Another advantage is its flexibility in being able to incorporate other stress factors which affect radiation interception or radiation use efficiency.

Gutierrez et al. (1983) presented a model for cotton which describes the effects of inoculum density, virulence and the timing of foliar symptom appearance on cotton phenology, dry matter accumulation and lint yield. The relations between inoculum density and disease incidence, plant infection and symptom appearance were derived from published literature. Pathotype of *V. dahliae* was expressed as a scalar variable Z, where $Z = 1$ corresponds to absence of symptoms and $Z = 0$ to maximum symptoms. In the model, values for Z were fitted to data for a defoliating pathotype ($Z = 0.4$) and for a non-defoliating pathotype ($Z = 0.5$). Effects of *V. dahliae* on the plant were modeled as a decrease in photosynthesis rate. The model also simulated the effects of shading and compensation for diseased plants, assuming random distribution of these plants. Field data derived from published literature on effects of stem-injected conidial suspensions of *V. dahliae* at different times fitted reasonably well with part of the model. The authors admitted that their assumptions about effects of different pathotypes, expressed in one scalar variable, were probably too simplistic. In addition, they did not validate the relation between inoculum density and disease incidence, since the starting point in their validation efforts was the injected stems. The model was presented as a concise summary of what is known about the interaction of cotton and *V. dahliae*.

Mechanistic model. Termorshuizen and Rouse (1993) constructed a model for the development of *V. dahliae* in potato by coupling a Verticillium submodel to an existing potato crop growth model, using mainly biologically meaningful and measurable parameters. This submodel describes disease development in single plants. Root infection was calculated by applying an adapted pathozone concept as developed by Gilligan (1985). The number of stem base infections was calculated by assuming that only a fraction of root infections would reach the stem base. Stem base infection was modeled as a stochastic process. In the model, *V. dahliae* affected maximum photosynthetic rate as a function of the abundance of conidia in the stem xylem. Development of *V. dahliae* in the plant was assumed to be influenced by air temperature. Due to the stochastically modeled stem base infection, a set of reruns of the model resulted in averages and standard deviations.

One of the problems in parameter-rich models is to find optimal parameter fits. This problem was approached in three ways. First, parameters to which the potato growth model appeared to be insensitive were fixed to their nominal values, thus reducing the number of potato crop growth parameters to 11. Second, the parameters of the potato crop growth model were calibrated in advance of those of the Verticillium subroutine. By doing so, corrections could be made for lack of fit in the potato growth model, prior to evaluation of the Verticillium submodel. Third, a computerized calibration procedure was incorporated, based on work of Klepper and Rouse (1991). For this, biologically plausible ranges in parameter values were determined or derived from published literature. By chosing parameter values using an optimizing algorithm (cf. Klepper and Rouse 1991) and rerunning the model many times, optimal sets of parameters were found.

Calibration was performed using two years of experimental field data at two levels of microsclerotial density in soil. The model did not include water stress, since the experiments were conducted on irrigated fields. Simulation of root infection and incidence of stem colonization by *V. dahliae* correlated well with experimental data. However, deviation from published data at inoculum densities which were not used in the calibration procedure indicated that further calibration had to be performed using additional field data and incorporating parameters describing fungal dynamics. Runs of the model correlated most optimal with published data at pathozones (Gilligan 1985) of c. 30 μm, which may indicate that infections occur only when a microsclerotium is located very close to the root.

Decision model. A theoretical decision model was published for cotton by Regev et al. (1990). It integrates simple economic parameters (costs of control, yield) with changes in inoculum density and virulence of *V. dahliae*. The virulence index was developed by Gutierrez et al. (1983). Four decision variables were defined: no treatment, soil solarization, fumigation and growth of an alternative crop. The optimal decision was based on the income of the farmer (being proportional to crop yield minus yield loss due to wilt injury and costs of control). Subsequently, a risk element was built in, assuming a relation between variability in yield and control decision. The integration of economic parameters, disease and acceptation of control measures by the farmer is a relatively unexplored field. Future goals, using the approach of Regev et al. (1990) as a starting point, will be the incorporation of multiple biotic stress factors in a multi-year agro-ecosystem model that considers a continuously changing economic market.

A model which was published in the Japanese language (Itoh et al. 1989) included among its parameters the number of successful infections, fraction of infected plants, disease intensity, number of root infections, production of microsclerotia, effects of inoculum density, temperature, organic matter and density of *Pratylenchus penetrans*. The program evaluated preceding management of a field and predicted the disease severity assuming that certain types of management were implemented, but it was not validated successfully.

Conclusions

The complex models discussed here have different aims. There is a tendency in phytopathology to shift from simply relating disease incidence to inoculum density towards the development of more realistic, detailed, and applicable models (Johnson 1992; Termorshuizen and Rouse 1993; Regev et al. 1990). In general, the density of microsclerotia in soil is the starting point, but Johnson (1988) used the day of onset of the disease as input. His idea that time of infection is important in explaining crop loss is very valuable, but it needs to be coupled to inoculum density. However, as we have pointed out, models relating inoculum density to crop parameters or disease incidence have poor ability to generalize. Therefore, more data is needed on the basis of relationships between inoculum density and disease incidence. Key factors may be temperature, soil moisture potential and microbial activity in soil. These factors are generally lacking in the models published so far, although their importance has been stressed several times. The best way to limit the number of new parameters implied by including these factors is to search for simple relations, as in the approach of Johnson (1992). For example, general microbial activity may be measured as soil respiration or density of propagules, and may be expressed as a single factor influencing the pathozone (Gilligan 1985) surrounding the root. Detailed experiments are needed to study whether simplifying statements can be constructed. The incorporation of effects of those soilborne pathogens that do not directly affect radiation interception or radiation use efficiency (e.g. *Rhizoctonia solani*) is another challenge.

A final point to be considered is the presentation of models in a readable and understandable form. Due to their complexity, authors should present their models in a way which will be clear to those who do not normally work with models. Tables need to be included with explanations of mathematical notations, lists of variables and their units, and the assumptions made should be stated clearly. Only when non-modelers are able to understand and use the results of modelers, will the usefulness of modelling to the development of sustainable agriculture become generally evident.

Acknowledgements

We are grateful to Dr. R.C. Rowe for critically reading the manuscript.

References

Ashworth L J, McCutcheon O D, George A G (1972) *Verticillium albo-atrum*: the quantitative relationship between inoculum density and infection of cotton. Phytopathology 62:901–903.

Ashworth L J, Huisman O C, Grogan R G, Harper D M (1976) Copper induced fungistasis of microsclerotia of *Verticillium albo-atrum* and its influence on infection of cotton in the field. Phytopathology 66:970–977.

Ashworth L J, Huisman O C, Harper D M, Stromberg L K, Bassett D M (1979a) Verticillium wilt disease of cotton: Influence of inoculum density in the field. Phytopathology 69:483–489.

Ashworth L J, Huisman O C, Harper D M, Stromberg L K (1979b) Verticillium wilt of tomato: influence of inoculum density and root extension upon disease severity. Phytopathology 69:490–492.

Azad H R, Davis J R, Schnathorst W C, Kado C I (1985) Relationships between rhizoplane and rhizosphere bacteria and Verticillium wilt resistance in potato. Archives of Microbiology 140:347–351.

Azad H R, Davis J R, Schnathorst W C, Kado C I (1987) Influence of Verticillium wilt resistant and susceptible potato genotypes on populations of antagonistic rhizosphere and rhizoplane bacteria and free nitrogen fixers. Applied Microbiology and Biotechnology 26:99–104.

Beckman C H, Halmos S, Mace M E (1962) The interaction of host, pathogen, and soil temperature in relation to susceptibility to Fusarium wilt of bananas. Phytopathology 52:134–140.

Beckman C H, Elgersma D M, MacHardy W E (1972) The localization of fusarial infection in vascular tissue of single-dominant-gene resistant tomatoes. Phytopathology 62:1256–1260.

Beckman C H (1973) The incidence of *Verticillium* species in soils, vines and tubers of Rhode Island-grown potatoes. Plant Disease Reporter 57:928–932.

Beckman C H (1987) The Nature of Wilt Diseases of Plants. The American Phytopathological Society, St. Paul, Minnesota, USA.

Ben-Yephet Y, Pinkas Y (1977) Germination of individual microsclerotia of *Verticillium dahliae*. Phytoparasitica 5:159–166.

Bollen G J, Hoekstra O, Scholte K, Hofman T W, Celetti M J, Schirring A (1989) Incidence of soilborne pathogens in potato related to the frequency of potato growing on a clay loam. Pages 203–222 in Vos J et al. (Eds.) Effects of Crop Rotation on Potato Production in the Temperate Zones. Kluwer Academic Publishers, Dordrecht, The Netherlands.

Cappaert M R, Powelson M L, Christensen N W, Crowe F J (1992) Influence of irrigation on severity of potato early dying and tuber yield. Phytopathology 82:1448–1453.

Conroy J J, Green R J, Ferris J M (1972) Interaction of *Verticillium albo-atrum* and the root lesion nematode, *Pratylenchus penetrans*, in tomato roots at controlled inoculum densities. Phytopathology 62:362–366.

Davis J R, Everson D O (1986) Relation of *Verticillium dahliae* in soil and potato tissue, irrigation method, and N-fertility to Verticillium wilt of potato. Phytopathology 76:730–736.

Davis J R, Pavek J J, Corsini D L, Sorensen L H, Schneider A T, Everson D O, Westermann D T,

Huisman O C (1994) Influence of continuous cropping of several potato clones on the epidemiology of Verticillium wilt of potato. Phytopathology 84:207–214.

DeVay J E, Forester L L, Garber R H, Butterfield E J (1974) Characteristics and concentration of propagules of *Verticillium dahliae* in air-dried field soils in relation to prevalence of *Verticillium*. Phytopathology 64:22–29.

Evans G, Wilhelm S, Snyder W C (1967) Quantitative studies by plate counts of propagules of the Verticillium wilt fungus in cotton field soils. Phytopathology 57:1250–1255.

Evans G, Gleeson A C (1973) Observations on the origin and nature of *Verticillium dahliae* colonizing plant roots. Australian Journal of Biological Sciences 26:151–161.

Evans G, McKeen C D, Gleeson A C (1974) A quantitative bioassay for determining low numbers of microsclerotia of *Verticillium dahliae* in field soils. Canadian Journal of Microbiology 20:119–124.

Farley J D, Wilhelm S, Snyder W C (1971) Repeated germination and sporulation of microsclerotia of *Verticillium albo-atrum* in soil. Phytopathology 61:260–264.

Francl L J, Rowe R C, Riedel R M, Madden L V (1988) Effects of three soil types on potato early dying disease and associated yield reduction. Phytopathology 78:159–166.

Francl L J, Madden L V, Rowe R C, Riedel R M (1990) Correlation of growing season environmental variables and the effect of early dying on potato yield. Phytopathology 80:425–432.

Gilligan C A (1985) Probability models for host infection by soilborne fungi. Phytopathology 76:61–67.

Green R J (1957) The vertical distribution of *Verticillium albo-atrum* in muck soils and its control. Phytopathology 47:522 (Abstract).

Green R J (1969) Survival and inoculum potential of conidia and microsclerotia of *Verticillium albo-atrum* in soil. Phytopathology 59:874–876.

Grogan R G, Ioannou N, Schneider R W, Sall M A, Kimble K A (1979) Verticillium wilt on resistant tomato cultivars in California: virulence of isolates from plants and soil and relationship of inoculum density to disease incidence. Phytopathology 69:1176–1180.

Gutierrez A P, DeVay J E, Pullman G S, Friebertshauser G E (1983) A model of Verticillium wilt in relation to cotton growth and development. Phytopathology 73:89–95.

Harris D C, Yang J R (1990) Pre-planting prediction of strawberry wilt (*Verticillium dahliae*) risk as an aid in disease management. Proceedings of the Brighton Crop Protection Conference on Pests and Diseases 1:117–122.

Harris D C, Yang J R, Ridout M S (1993) The detection and estimation of *Verticillium dahliae* in naturally infested soil. Plant Pathology 42:238–250.

Harrison N A, Beckman C H (1982) Time/space relationships of colonization and host response in wilt-resistant and wilt-susceptible cotton (*Gossypium*) cultivars inoculated with *Verticillium dahliae* and *Fusarium oxysporum* f.sp. *vasinfectum*. Physiological Plant Pathology 21:193–207.

Hoyos G P, Zambino P J, Anderson N A (1991) An assay to quantify vascular colonization of potato by *Verticillium dahliae*. American Potato Journal 68:727–742.

Hu X, Nazar R N, Robb J (1993) Quantification of *Verticillium* biomass in wilt disease development. Physiological and Molecular Plant Pathology 42:23–36.

Huisman O C (1988a) Seasonal colonization of roots of field-grown cotton by *Verticillium dahliae* and *V. tricorpus*. Phytopathology 78:708–716.

Huisman O C (1988b) Colonization of field-grown cotton roots by pathogenic and saprophytic soilborne fungi. Phytopathology 78:716–722.

Ioannou N, Schneider R W, Grogan R G, Duniway J M (1977) Effect of water potential and temperature on growth, sporulation and production of microsclerotia by *Verticillium dahliae*. Phytopathology 67:637–644.

Itoh S, Komada H, Momma I, Amano I (1989) Development of field diagnosis systems (FDS) for preventing continuous cropping injury of crop. 13. Development of a model for prediction of severity of Verticillium yellows of Chinese cabbage. Bulletin of the National Agricultural Research Center Japan 17:1–43 (in Japanese).

Jeger M J (1987) The influence of root growth and inoculum density on the dynamics of root disease epidemics: theoretical analysis. New Phytologist 107:459–478.

Johnson K B (1988) Modelling the influences of plant infection rate and temperature on potato foliage and yield losses caused by *Verticillium dahliae*. Phytopathology 78:1198–1205.

Johnson K B, Apple J D, Powelson M L (1988) Spatial patterns of *Verticillium dahliae* propagules in potato field soils of Oregon's Columbia basin. Plant Disease 72:484–488.

Johnson K B (1992) Evaluation of a mechanistic model that describes potato crop lossed caused by multiple pests. Phytopathology 82:363–369.

Keinath A P, Fravel D R (1992) Induction of soil suppressiveness to Verticillium wilt of potato by successive croppings. American Potato Journal 69:503–513.

Kendrick J B, Middleton J T (1959) Influence of soil temperature and of strains of the pathogen on severity of Verticillium wilt of pepper. Phytopathology 49:23–28.

Klepper O, Rouse D I (1991) A procedure to reduce parameter uncertainty for complex models by comparison with real system output illustrated on a potato growth model. Agricultural Systems 36:375–395.

Kotcon J B, Rouse D I, Mitchell J E (1984) Dynamics of root growth in potato fields affected by the early dying syndrome. Phytopathology 74:462–467.

MacGuidwin A E, Rouse D I (1990a) Effect of *Meloidogyne hapla*, alone and in combination with subthreshold populations of *Verticillium dahliae*, on disease symptomatology and yield of potato. Phytopathology 80:482–486.

MacGuidwin A E, Rouse D I (1990b) Role of *Pratylenchus penetrans* in the potato early dying disease of Russet Burbank potato. Phytopathology 80:1077–1082.

Martin M J, Riedel R M, Rowe R C (1982) *Verticillium dahliae* and *Pratylenchus penetrans*: Interactions in the early dying complex of potato in Ohio. Phytopathology 72:640–644.

Mol L, Termorshuizen A J (1995) Life cycle and ecology of *Verticillium dahliae* in potato. This volume. in Haverkort A J, MacKerron D K L (Eds.) Potato Ecology and Modelling of Crops under Conditions Limiting Growth. Kluwer Academic Publishers, Dordrecht, The Netherlands.

Nazar R N, Hu X, Schmidt J, Culham D, Robb J (1991) Potential use of PCR-amplified detection and differentiation of Verticillium wilt pathogens. Physiological and Molecular Plant Pathololgy 33:47–58.

Nicot P C (1985) Epidemiology of Verticillium Wilt of Potato: Spatial Patterns of Soilborne *Verticillium dahliae* and Incidence of Stem Colonization. Ph.D. Thesis, University of Wisconsin, Madison, USA.

Nicot P C, Rouse D I (1987) Relationship between soil inoculum density of *Verticillium dahliae* and systemic colonization of potato stems in commercial fields over time. Phytopathology 77:1346–1355.

Nnodu E C, Harrison M D (1979) Relationship between *Verticillium albo-atrum* inoculum density and potato yield. American Potato Journal 56:11–26.

Paplomatas E J, Bassett D M, Broome J C, DeVay J E (1992) Incidence of Verticillium wilt and yield losses of cotton cultivars (*Gossypium hirsutum*) based on soil inoculum density of *Verticillium dahliae*. Phytopathology 82:1417–1420.

Pegg G F (1978) Effect of host substrate on germination and growth of *Verticillium albo-atrum* and *V. dahliae* conidia and mycelia. Transactions of the British mycological Society 71:483–489.

Pullman G S, DeVay J E (1982a) Epidemiology of Verticillium wilt of cotton: a relationship between inoculum density and disease progression. Phytopathology 72:549–554.

Pullman G S, DeVay J E (1982b) Epidemiology of Verticillium wilt of cotton: effects of disease development on plant phenology and lint yield. Phytopathology 72:554–559.

Regev U, Gutierrez A P, DeVay J E, Ellis C K (1990) Optimal strategies for management of Verticillium wilt. Agricultural Systems 33:139–152.

Schnathorst W C (1981) Life cycle and epidemiology of *Verticillium*. Pages 81–111 in Mace M E, Bell A A, Beckman C H (Eds.) Fungal Wilt Diseases of Plants. Academic Press, New York, USA.

Scholte K (1989) Effects of crop rotation and granular nematicides on the incidence of *Verticillium dahliae* and *Colletotrichum coccodes* in potato. Potato Research 32:377–385.

Schreiber L R, Green R J (1962) Comparative survival of mycelium, conidia, and microsclerotia of *Verticillium albo-atrum* in mineral soil. Phytopathology 52:288–289.

Smith V L, Rowe R C (1984) Characteristics and distribution of propagules of *Verticillium dahliae* in Ohio potato field soils and assessment of two assay methods. Phytopathology 74:553–556.

Taylor J B (1969) Fluctuations in natural populations of *Verticillium tricorpus*. Canadian Journal of Botany 47:737–740.

Termorshuizen A J, Rouse D I (1993) Towards a mechanistic model for the *Verticillium dahliae*-potato system. Netherlands Journal Plant Pathology 99 (Suppl. 3):201–218.

Wheeler T A, Madden L V, Rowe R C, Riedel R M (1992) Modelling of yield loss in potato early dying caused by *Pratylenchus penetrans* and *Verticillium dahliae*. Journal of Nematology 24:99–102.

Wilhelm S (1950) Vertical distribution of *Verticillium albo-atrum* in soils. Phytopathology 40:368–376.

Wilhelm S, Sagen J E, Tietz H, George A (1967) Vertical distribution of fungus suggests... rigolen plowing for control of Verticillium wilt in cotton. California Agriculture 21:2–4.

Zilberstein Y, Chet I, Henis Y (1983) Influence of microsclerotia source of *Verticillium dahliae* on inoculum quality. Transactions of the British Mycological Society 81:613–617.

18. Use of a crop-growth model coupled to an epidemic model to forecast yield and virus infection in seed potatoes

THOMAS NEMECEK[1], JACQUES O. DERRON[1], ANDREAS FISCHLIN[2] and OLIVIER ROTH[3]

[1] *Federal Agricultural Research Station of Changins, 1260 Nyon, Switzerland*
[2] *Systems Ecology, Institute of Terrestrial Ecology, Swiss Federal Institute of Technology, Grabenstr. 3, 8952 Schlieren, Switzerland*
[3] *Widenstr. 3, 8302 Kloten, Switzerland*

Abstract. Pathosystems of vector-borne plant viruses consist of the elements viruses, vectors and plants, influenced by man and the natural environment. Most virus epidemic models emphasize the importance of vectors and viruses, but do not consider the role of the plants.

The model 'EPOVIR' is the first virus epidemic model coupled to a crop-growth submodel which, for its part, is coupled to a soil water-balance submodel. The two submodels are used in three ways: tuber yield and tuber size are calculated, the physiological age of the leaves and the drought stress are used to estimate the susceptibility to virus infections ('age resistance') and finally, the fraction of soil covered by the canopy is needed to calculate landing rates of the vectors in the potato field. Since the rate of virus spread is a function of plant physiology and phenology as simulated by the crop submodel, the epidemic should react appropriately to changes in plant growth, caused by man or the environment. As an example we show how planting density influences virus infection.

The model EPOVIR is integrated in the decision-support system 'TuberPro', which forecasts tuber yields graded by size and the infection of the tubers by PVY and PLRV. It supports optimization of haulm destruction dates in seed potato production. The system calculates expected seed yield and the probability that virus infection remains below the tolerance limit used in seed certification. The combination of the two factors gives the expected average certified seed yield for a production region, which can subsequently be optimized.

Introduction

Pathosystems of vector-borne plant viruses consist of the elements virus, vectors and plants, all influenced by man and the natural environment (Robinson 1976). Most forecasting systems and virus epidemic models focus on vectors and viruses and their relations (e.g. Marcus and Raccah 1986; Kisimoto and Yamada 1986; Miyai et al. 1986; Madden et al. 1990; Kendall et al. 1992). The role of the plants in virus infection and translocation and their influence on the vector populations is not included, or at best, described by a few constant parameters or functions (e.g. Sigvald 1986; Ruesink and Irwin 1986; Van der Werf et al. 1989). However, to describe appropriately the effects of plant physiology and phenology on the epidemic process and the influence of the environment on these relations, a crop-growth submodel is necessary. Crop models have been used in plant pathology in combination with fungal epidemic models (Rouse 1988), but to our knowledge, crop models have never been coupled to virus epidemic models.

A.J. Haverkort and D.K.L. MacKerron (eds.), Potato Ecology and Modelling of Crops under Conditions Limiting Growth, 281–290.

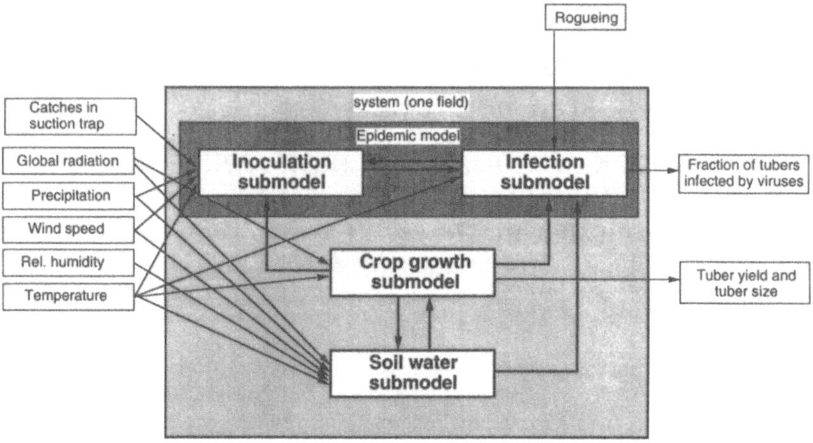

Figure 1. Structure of the model EPOVIR (= *e*pidemiology of *po*tato *vir*uses).

We present a project, where a potato crop submodel has been coupled to a virus epidemic model to predict yield and virus infection in seed potatoes. The crop submodel for its part is coupled to a soil water-balance submodel, which accounts for suboptimal water supply conditions. These two submodels influence the epidemic model, but are independent of it. The reason why we did not study the yield reduction due to viruses is that we are mainly interested in seed potato production, where the yield loss caused by virus infection is very small. Nevertheless, the economic impact of virus infection can be very high, since a seed lot can be rejected, leading to almost no revenue to the farmer. We have developed a decision-support system for seed potatoes that allows us to forecast virus infection of tubers and, particularly, the risk that virus infection will exceed the tolerance limits for a certain quality class. We also forecast tuber yields graded by size. The system, called 'TuberPro' (= *Solanum* ***tuber****osum* **pro**gnosis), allows us to optimize haulm destruction dates in seed potato production (Nemecek et al. 1994).

The aim of this paper is to show, why and how the crop submodel is used in the virus epidemic model and within 'TuberPro'.

Model outline

The model 'EPOVIR' (= *e*pidemiology of *po*tato *vir*uses) forms the core of TuberPro. EPOVIR simulates crop growth and virus epidemic from crop emergence to haulm destruction in a single field. A detailed description of a first version is given by Nemecek (1993). It consists of four submodels (Figure 1):

- An *inoculation submodel*, which calculates vector intensity (after Irwin and Ruesink 1986) from vector abundance in the field, vector propensity and vector behaviour. Vector abundance is estimated from catches in a suction trap. Wingless aphid vectors are considered for PLRV transmission only. The submodel represents the role of the vectors and their relations with the viruses and the crop.

- An *infection submodel*, which determines the infection of plants and tubers by PVY and PLRV, respectively. Further, it estimates the fraction of plants serving as infection sources by accounting for the latent period and the age resistance of the plants to virus infections. The submodel represents mainly the role of the viruses and their relations with the crop.
- A *crop-growth submodel*, which calculates the dry mass of leaves, stems, roots, tubers and assimilates, the physiological state of the canopy and graded tuber yields.
- A *soil water-balance submodel*, which calculates the water content of the soil and the water stress of the potato plants from the ratio of actual and potential transpiration.

The crop model has been developed by Johnson et al. (1986; 1987). The original model used measurements of soil moisture, which were not available in our experiments so we replaced them by a soil water-balance submodel, based on work of Van Keulen and Wolf (1986). A complete description of the coupled crop-growth and soil water-balance submodels and validation results for cv. Maris Piper are given by Roth et al. (1995). The crop submodel was adapted to varieties cultivated in Switzerland and extended by a tuber-size submodel. These changes and the corresponding validation results are given by Nemecek and Derron (1994). The crop model was capable of reproducing experimental data collected under wet and dry conditions with sufficient reliability.

EPOVIR is implemented in the programming language Modula-2 (Wirth 1985) by using the simulation environment ModelWorks (Fischlin et al. 1994) on Apple™ Macintosh™ computers.

Use of the crop-growth submodel

Following our definition of a plant-vector-virus pathosystem, the submodels for crop growth and soil water-balance are used in three ways in the model EPOVIR: to calculate yield, to describe the influence of plant physiology on virus multiplication and translocation in plants and to describe the effect of plant phenology on vector behaviour.

Plants

Forecasting tuber yield and tuber size are, of course, highly important in the decision-support system. The crop submodel calculates total tuber yield and the fraction being between arbitrarily chosen size limits (see example in Figure 4 below).

Plant-virus interaction

Potato tubers are less readily infected by viruses as plants grow older (Beemster 1987). This phenomenon is called 'age resistance' or 'mature-plant resistance'. Its causes are unknown, but it seems to be related to the physiological activity of the leaves (Venekamp et al. 1980). The submodel keeps track of the leaf mass produced per day as well as its physiological age (Johnson et al. 1987; Roth et al. 1995). We use the fraction l_y of leaf mass younger than a physiological age threshold ($0 \leq l_y \leq 1$) to provide an estimate of the age dependent susceptibility to virus infections *AS*. The susceptibility to virus infections depends also on drought: Wislocka (1982) demonstrated that drought (suboptimal water supply) can favour tuber infection, apparently by partially breaking the age resistance. The effect of water stress *wStress* ($0 \leq wStress \leq 1$, 1 = optimal growing conditions, 0 = no growth, see Roth et al. 1995) is also included in the equation for *AS*:

$$AS = \begin{cases} 1-(1-ly)(1-0.5(1-wStress)) & \text{if drought} \\ l_y & \text{otherwise} \end{cases}$$

The constant 0.5 is a weighting factor that describes the effect of drought (soil water content below critical soil moisture, Roth et al. 1995) on the susceptibility. Figure 2 shows a typical evolution of *AS*. In young plants, *AS* equals 1, irrespective of drought. As the plants grow older, l_y decreases gradually to 0 and consequently *AS* decreases also. Drought can increase the susceptibility at

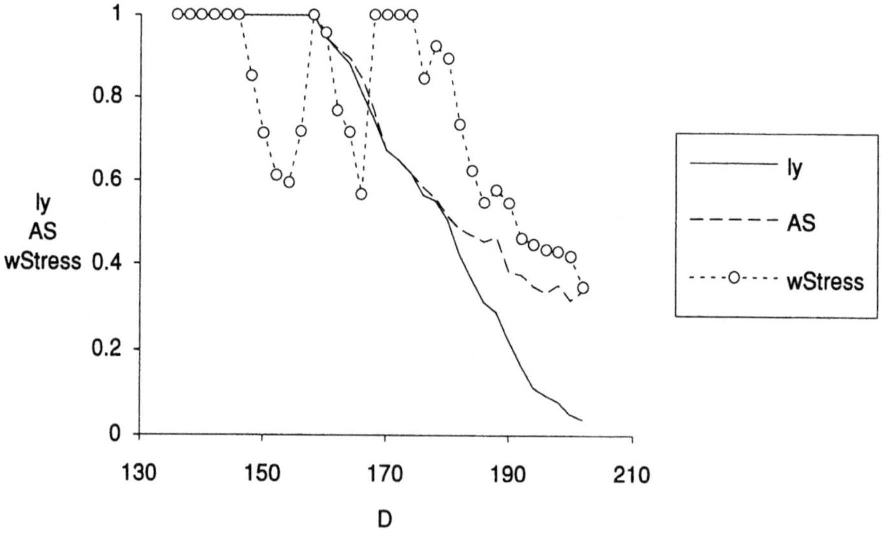

Figure 2. Age dependent susceptibility to virus infections *AS*, which is a function of the fraction l_y of leaf mass younger than a physiological age threshold and the water stress factor *wStress*. *D* = day of the year.

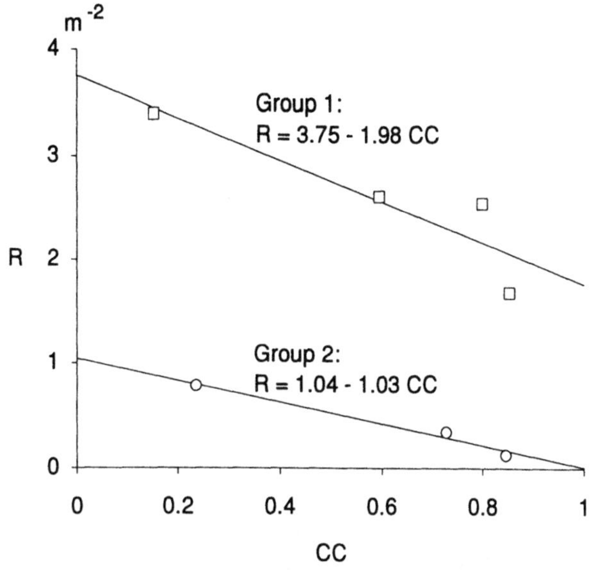

Figure 3. Relative landing rate of aphids: R = ratio between landing rate [# m^{-2} d^{-1}] in a potato field and catches in a suction trap [#/d] as functions of the canopy cover CC (= fraction of soil area covered by the canopy). The following species are contained in group 1: *Aphis spp.*, *Myzus persicae* Sulz. and *Macrosiphum euphorbiae* Thomas; group 2: *Acyrthosiphon pisum* Harris, *Brachycaudus helichrysi* Kalt., *Phorodon humuli* Schrk. and *Rhopalosiphum padi* L.

this stage, i.e. the plants become more susceptible to virus infection compared with a situation, where water supply is optimal. Without drought, AS would be equal to l_y.

Plant-vector interaction

The plants influence not only virus infection, multiplication and translocation, but also the development and behaviour of aphid populations, and therefore indirectly virus spread (Irwin and Kampmeier 1989; Nemecek 1993). The host suitability strongly influences vector behaviour and virus spread. Nemecek (1993) has shown that vector species that do not colonize potato plants, more often show sequences of behaviour that increase their efficiency as vectors. Simulation studies revealed that this attribute will lead to about twice as efficient transmission of nonpersistent viruses, compared to species that colonize potato plants.

Plant phenology influences the behaviour of vector populations. The relative landing rate R (number of aphids landing per m^2 for each aphid captured in the suction trap (Taylor and Palmer 1972) in the same time interval) decreases with increasing canopy cover (Figure 3). The landing rate was calculated indirectly from catches in a fisherline trap (Derron and Goy 1993) and their relation to the landing rates on potato plants (Derron et al. 1989). The canopy cover is simulated by the crop submodel. Regressions between the average R and canopy

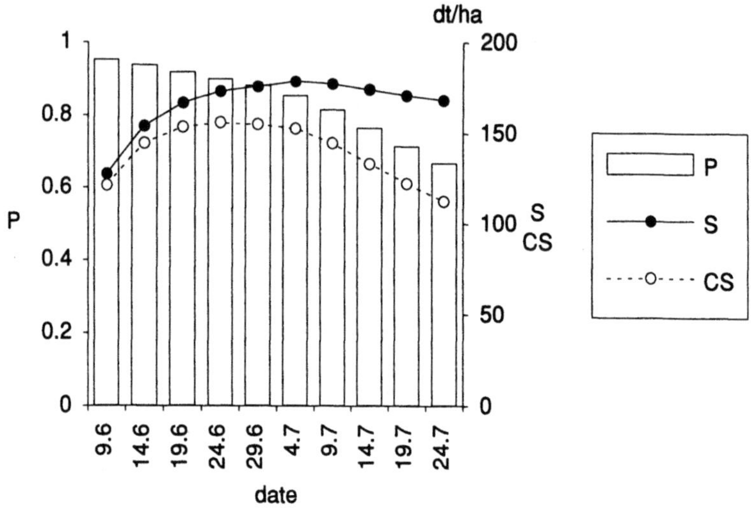

Figure 4. Development of the probability *P* that seed will be certified, the expected yield in seed grade *S* [dt/ha] and the expected average certified yield in seed grade *CS* [dt/ha]. The results are actually averages of 30 simulation runs: three situations (favourable, mean and unfavourable) were combined with the input data of the last 10 years after 14 June (up to this date the data of the current year have been used).

cover were calculated for 3, 4, 5 and 10 groups with approximately the same number of data points. The regression with the highest r^2, i.e. the best linear relation, was used in the model (Figure 3). Since the canopy cover is a variable calculated by the model, it was easy to include this relation. The landing rate and consequently virus transmissions will vary with canopy development. Further we can distinguish two groups of vector species: group 1 had a relative landing rate that was on average 8.6 times higher than that of group 2.

Model application

We show two examples of model application (see Nemecek et al. 1994 for further examples). The principal goal of 'TuberPro' is to forecast the following variables for decision making:

- the expected tuber infection by PVY or PLRV and its variance;
- the expected tuber yield of a certain size grade and its variance.

Logit transformed forecasts of virus infection are approximately normally distributed. The mean and variance describe the probability distribution of the forecasts. They include the uncertainty caused by unknown future development of input time series (weather variables and aphid abundance), as well as the error in the estimates of certain parameters. Unknown future values of inputs are replaced by data from recent years. The scope of the simulations is always one field. However, it is also possible to define average situations that are representative for regional or national production. Such an example is given in

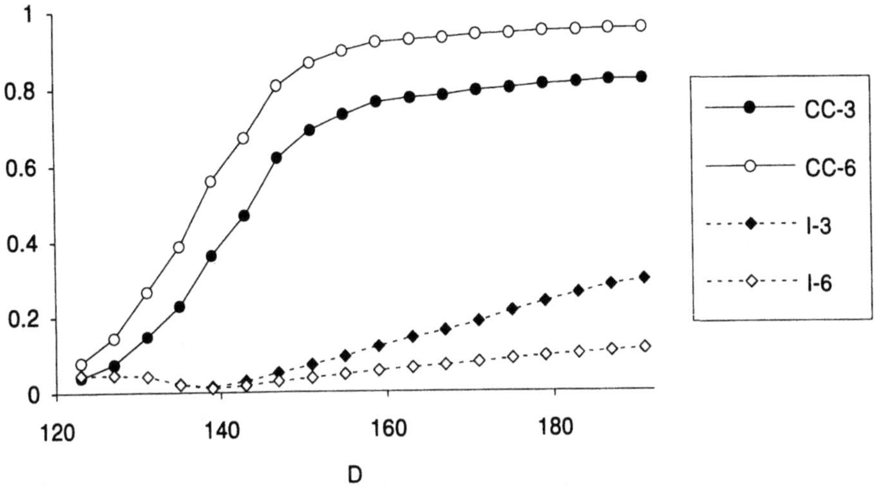

Figure 5. Effect of planting density on virus spread: I = fraction of tubers infected by PVY, CC = fraction of soil covered by the canopy, D = day of the year. The numbers 3 and 6 refer to the planting density of 3 respectively 6 seed pieces/m^2.

Figure 4. In Switzerland, a seed lot with an infection with up to 10% by PVY or PLRV can be certified as class A seed. The probability, P, that a lot will be certified is therefore the probability that the virus infection of tubers is $\leq$ 10%, which is also the fraction of lots that are expected to be certified. The maximum expected yield in seed grade, S, is reached on 4 July. P and S allow calculation of the expected average certified yield in the seed grade, CS, which is simply the product of the two variables ($CS = S \times P$). The maximum of CS is reached as early as on 24 June. So, 24 June would be the optimal haulm destruction date for production of a maximum of certified seed.

Another strategy would be to maximize the farmer's income. The farmer is also interested in the larger tubers, since they can often be sold for human consumption. For this reason maximum expected income will usually be reached later than maximum expected certified seed. The user of TuberPro must define the strategy to achieve his objectives.

The model can also be used to evaluate cultural practices. Simulation studies show that virus infection can be substantially reduced by increasing planting density (Figure 5). This is due to increased canopy cover (see Figure 3) and a dilution effect (Power 1990; 1992). Both factors lead to a lower number of vectors per plant, which will lower the proportion of plants infected. The high density planting reduced the average number of vectors per plant by 58% and the virus infection by 62% compared with the low density planting, i.e. virus infection was reduced even more than vector density. This difference is due to the fact that vector abundance was decreased mainly early in the growing season, when virus transmission is most dangerous (Nemecek 1993).

Discussion

The crop reacts to influences exerted by the natural environment and by man. As the multiplication of potato viruses depends completely on the host plant metabolism, changes in it will also affect the epidemic process. For example if leaf growth continues longer, this will most likely favour virus spread by delaying age resistance. The vectors have a very close relationship to the crop as well. Such indirect influences on the epidemic process are best described by a virus epidemic model coupled to a crop-growth model.

Since well-validated models are now available for many crops, we think that crop models should be more widely used in modelling virus epidemics. The use of a crop submodel makes the entire model more flexible and adaptable to other conditions, e.g. in other countries with different production schemes. By including the plant-virus and plant-vector relations, the model also becomes more explicit and offers better possibilities for testing hypotheses and evaluating cultural measures (Nemecek 1993). A later unpublished version of the model of Ruesink and Irwin (1986) also considers the effect of the canopy cover on the landing rate. However, the authors use measured time series as inputs and so, new experiments will be required each time growth parameters are changed. That is not necessary in our approach. Our last argument for the use of crop submodels is perhaps the most obvious one: the crop submodel calculates tuber yield and so allows seed potato production to be optimized according to different criteria (see Figure 4). It allows also evaluation of the effect of virus control measures on yield.

Coupling a crop submodel to a virus epidemic model has also some drawbacks. The complexity of the whole model increases considerably and a lot more parameters are needed. As these parameters will not be available in all cases, default or average values must be used instead. By doing so, one loses some of the advantages of the crop submodel. Finally, the development and coupling of models is laborious and time consuming, especially where no suitable crop model is already available. The decision to use a crop submodel will depend on the objectives of the research, the availability of a suitable model and the available resources.

TuberPro is a powerful system management tool for seed potato production. It allows production to be optimized according to various strategies which are chosen by the decision maker. A farmer will probably try to assure a certain income. A seed growers' organization will perhaps try to increase its profits and the national service is mainly preoccupied with maintaining a high quality. TuberPro calculates the variables necessary to find optimal solutions at the different levels, especially to determine the haulm destruction dates according to the specified goal. Moreover, it has proved to be useful for the instruction of farmers and experts and for the evaluation of cultural and virus control measures (Nemecek et al. 1994). TuberPro is now being tested in practice in Switzerland. As it is very flexible, it should be possible to adapt it to other conditions with a reasonable amount of effort.

Acknowledgements

We thank R. Schwärzel for his valuable contributions. We are also grateful to the Swiss Alcohol Board and the Swiss National Science Foundation (grant no. 31–8766.86) for their financial support.

References

Beemster A B R (1987) Virus translocation and mature-plant resistance in potato plants. Pages 116–125 in De Bokx J A, Van der Want J P H (Eds.) Viruses of Potatoes and Seed-Potato Production, 2nd Ed. Pudoc, Wageningen, The Netherlands.

Derron J O, Goy G (1993) Description et mode d'emploi d'un piège pour l'étude du vol des pucerons vecteurs de virus. Revue Suisse d'Agriculture 25:135–137.

Derron J O, Goy G, Genthon M (1989) Le piégeage des pucerons ailés: potentialités et limites de différents types de pièges. Pages 71–82 in Cavalloro R (Ed.) Euraphid Network: Trapping and Aphid Prognosis. Proceedings of a Meeting of the EC-Experts' Group, Catania, Italy, 7–9 November 1988. ECSC-EEC-EAEC, Brussels, Belgium.

Fischlin A, Roth O, Gyalistras D, Ulrich M, Thöny J, Nemecek T, Bugmann H, Thommen F (1994) ModelWorks 2.2: An Interactive Simulation Environment for Personal Computers and Workstations. Internal Report No. 14. Systems Ecology. ETH, Zürich, Switzerland.

Irwin M E, Kampmeier G E (1989) Vector behaviour, environmental stimuli and the dynamics of plant virus epidemics. Pages 14–39 in Jeger M J (Eds.) Spatial Components of Plant Disease Epidemics. Prentice Hall, Englewood Cliffs, NJ, USA.

Irwin M E, Ruesink W G (1986) Vector intensity: a product of propensity and activity. Pages 13–33 in McLean G D, Garett R G, Ruesink W G (Eds.) Plant Virus Epidemics (Monitoring, Modelling, and Predicting Outbreaks). Academic Press, Sydney, Australia.

Johnson K B, Johnson S B, Teng P S (1986) Development of a simple potato growth model for use in crop-pest management. Agricultural Systems 19:189–209.

Johnson K B, Teng P S, Radcliffe E B (1987) Coupling feeding effects of potato leafhopper, *Empoasca fabae* (Homoptera: Cicadellidae), nymphs to a model of potato growth. Environmental Entomology 16:250–258.

Kendall D A, Brain P, Chinn N E (1992) A simulation model of the epidemiology of barley yellow dwarf virus in winter sown cereals and its application to forecasting. Journal of Applied Ecology 29:414–426.

Kisimoto R, Yamada Y (1986) A planthopper-rice virus epidemiology model: rice stripe and small brown planthopper *Laodelphax striatellus* Fallén. Pages 327–344 in McLean G D, Garett R G, Ruesink W G (Eds.) Plant Virus Epidemics (Monitoring, Modelling, and Predicting Outbreaks). Academic Press, Sydney, Australia.

Madden L V, Raccah B, Pirone T P (1990) Modelling plant disease increase as a function of vector numbers: nonpersistent viruses. Research in Population Ecology 32:47–65.

Marcus R, Raccah B (1986) Model for the spread of non-persistent virus diseases. Journal of Applied Statistics 13:167–175.

Miyai S, Kiritani K, Nakasuji F (1986) Models of epidemics of rice dwarf. Pages 459–480 in McLean G D, Garett R G, Ruesink W G (Eds.) Plant Virus Epidemics (Monitoring, Modelling, and Predicting Outbreaks). Academic Press, Sydney, Australia.

Nemecek T (1993) The Role of Aphid Behaviour in the Epidemiology of Potato Virus Y: A Simulation Study. Ph.D. Thesis. No. 10086. ETH, Zürich, Switzerland.

Nemecek T, Derron J O (1994) Validation et application d'un modèle de croissance de la pomme de terre. Revue Suisse d'Agriculture 26:311–315.

Nemecek T, Derron J O, Schwärzel R, Fischlin A, Roth O (1994) Un modèle de simulation au service des producteurs de plants de pommes de terre. Revue Suisse d'Agriculture 26:17–20.

Power A G (1990) Cropping systems, insect movement and the spread of insect transmitted diseases in crops. Pages 47–69 in Gliessman S R (Ed.) Agroecology (Ecological Studies 78). Springer, New York, USA.

Power A G (1992) Host plant dispersion, leafhopper movement and disease transmission. Ecological Entomology 17:63–68.

Robinson R A (1976) Plant Pathosystems. Springer, Berlin, Germany.

Roth O, Derron J, Fischlin A, Nemecek T, Ulrich M (1995). Implementation and parameter adaptation of a potato crop simulation model combined with a soil water subsystem. In: Kabat P, Van den Broek B J, Marshall B, Vos J, Van Keulen H (Eds.) Modelling and Parametrization of the Soil-Plant-Atmosphere System – A Comparison of Potato Growth Models. Wageningen Pers, Wageningen, The Netherlands (in press).

Rouse D I (1988) Use of crop growth-models to predict the effects of disease. Annual Review of Phytopathology 26:183–201.

Ruesink W G, Irwin M E (1986) Soybean mosaic virus epidemiology: a model and some implications. Pages 295–313 in McLean G D, Garett R G, Ruesink W G (Eds.) Plant Virus Epidemics (Monitoring, Modelling, and Predicting Outbreaks). Academic Press, Sydney, Australia.

Sigvald R (1986) Forecasting the incidence of potato virus Y^{O}. Pages 419–441 in McLean G D, Garett R G, Ruesink W G (Eds.) Plant Virus Epidemics (Monitoring, Modelling, and Predicting Outbreaks). Academic Press, Sydney, Australia.

Taylor L R, Palmer J M P (1972) Aerial sampling. Pages 189–234 in Van Emden H F (Eds.) Aphid Technology. Academic Press, London, UK.

Van der Werf W, Rossing W A H, Rabbinge R, De Jong M D, Mols P J M (1989) Approaches to modelling the spatial dynamics of pests and diseases. Pages 89–119 in Cavalloro R, Delucchi V (Eds.) Parasitis 88. Proceedings of a Scientific Congress, 25–28 October 1988, Barcelona, Spain.

Van Keulen H, Wolf J (Eds.) (1986) Modelling of Agricultural Production: Weather, Soils and Crops. Pudoc, Wageningen, The Netherlands.

Venekamp J H, Schepers A, Bus C B (1980) Mature plant resistance of potato against virus diseases. III: Mature plant resistance against potato virus Y^{N}, indicated by decrease in ribosome-content in ageing potato plants under field conditions. Netherlands Journal of Plant Pathology 86:301–309.

Wirth N (1985) Programming in Modula-2, 3rd Ed. Springer-Verlag, Berlin a.o., Germany.

Wislocka M (1982) Einfluss der Trockenheit vor und zu verschiedenen Zeitpunkten nach Inokulation auf den Knollenbefall der Kartoffelsorte 'Uran' mit Kartoffelvirus Y. Potato Research 25:293–298.

19. Prescriptive crop and pest management software for farming systems involving potatoes

W.R. STEVENSON[1], J.A. WYMAN[2], K.A. KELLING[3], L.K. BINNING[4], D. CURWEN[4], T.R. CONNELL[4] and D.J. HEIDER[4]

Departments of [1] Plant Pathology, [2] Entomology, [3] Soil Science, [4] Horticulture, University of Wisconsin, Madison, WI 53706, U.S.A.

Abstract. Societal concerns about pesticide and fertilizer use and the intensive management of cropping systems involving potatoes have focused attention on the choice, timing and application rates of agricultural inputs. An integrated pest management (IPM) program was developed for the Wisconsin potato industry because 1) growers have traditionally been progressive and interested in IPM tactics, 2) the crop is intensively managed with multiple inputs of pesticide, fertilizer and irrigation, 3) potato is a high value/high risk crop, 4) the crop is often grown in environmentally sensitive areas, and 5) a strong research base existed to support the IPM effort. An important aspect of the Wisconsin IPM program was the development of specialized computer software to assist growers in management decisions. Computer software has provided an effective on-farm tool for analyzing complex environmental and crop information and providing specific management recommendations. The Potato Crop Management (PCM) software was developed at the University of Wisconsin through a multidisciplinary team effort. The program contains modules for predicting crop emergence, scheduling irrigation, managing pests (diseases, insects and weeds) and assessing storage ventilation needs. Since its release to growers and IPM consultants in 1989, use of the PCM software has continued to increase. Growers and consultants now use the software on approximately 28,300 ha of potato production. Savings related to reduced use of pesticides and irrigation are estimated to exceed US $ 5,890,000 annually when compared with inputs prior to use of this software. The current software, however, has application to only a few of the many decisions confronting growers. Software enhancements are underway that focus on potato canopy development, crop nutrition, seedpiece decay and farm record-keeping. The PCM program is also being converted from an MS-DOS® environment to a Microsoft Windows™ platform. The new software entitled WISDOM will allow easier data input and exchange, graphical presentation of data and comparison of environmental and crop/pest data between fields and years. The WISDOM software will expand opportunities for managing the farming enterprise including crops often grown in rotation with potato. Modules are being developed for management of snap beans (irrigation scheduling, insect management, and risk assessment for white mold) and sweet corn (irrigation scheduling and insect management). Program modularity will facilitate the future addition of new potato and rotational crop modules as new information becomes available from field and laboratory research.

Introduction

Potato production for processing, fresh market and seed is an important asset to Wisconsin's economy. Grown on approximately 27,500 ha, the crop is valued at $ 120,000,000. Wisconsin is ranked fourth in the United States in total potato

A.J. Haverkort and D.K.L. MacKerron (eds.), Potato Ecology and Modelling of Crops under Conditions Limiting Growth, 291–304.

production (Pratt 1993). The total value to Wisconsin in jobs, processing and allied agribusiness approaches $ 350,000,000 per year. Production costs for irrigated potatoes are typical of national costs and currently average $ 4,199 per ha for long season russet potatoes while gross returns often exceed $ 4,900 per ha.

Midwestern US weather conditions not only favour the production of cool weather crops such as potato, but they also contribute to the development of a broad range of important pest problems. Production of potatoes in Wisconsin and throughout the North Central region typically includes the intensive use of pesticides to manage diseases, insects and weeds; relatively high rates of fertilizer, particularly on irrigated sandy soils; and irrigation water for optimal production. Fifteen years ago, typical Wisconsin crops of Russet Burbank on a loamy sand soil were treated with fungicide dust on the seedpieces, up to 12 applications of foliar fungicides for control of early and late blight, a systemic insecticide at planting and up to four sprays of foliar insecticides, one to three herbicide sprays for control of grass and broadleaf weeds, one to two applications of vine desiccant, irrigation consisting of three applications per week totaling 5 cm per week and fertilizer treatments totaling 280–392, 134–179 and 336–448 kg/ha of N, P_2O_5 and K_2O, respectively.

During the early 1980s, these intensive inputs began to induce concerns in citizens and growers. Their concerns focused on environmental risks, grower and consumer safety and the potential for developing pest resistance to commonly used pesticides. The reliance on intensive use of pesticides and other inputs as well as environmental concerns associated with potato production were the driving forces behind the development of a comprehensive integrated crop and pest management program for Wisconsin growers. This integrated program was designed in partnership with growers, processors and allied industries to maintain a competitive potato industry that optimizes inputs and attempts to minimize or eliminate adverse environmental inputs associated with potato production.

The foundation of this crop and pest management program was a long history of research in areas such as disease management (judicious use of fungicides in conjunction with disease forecasting), insect management (targeting vulnerable stages of pest development and economic thresholds), weed management (prediction of crop emergence, plant canopy development and the timing of herbicide applications), fertility practices (rates, placement, source and timing of required nutrients) and irrigation management (matching applied irrigation with crop needs). Research in these and other critical areas created an extensive database that proved useful in making recommendations for solving specific production problems. During the mid-1980s, however, it became apparent to both the potato industry and University of Wisconsin researchers that to be most useful, this information must take on a more integrated and accessible format. The computer, in its early evolution for on-farm use, turned out to be a convenient vehicle for this integration and improved accessibility.

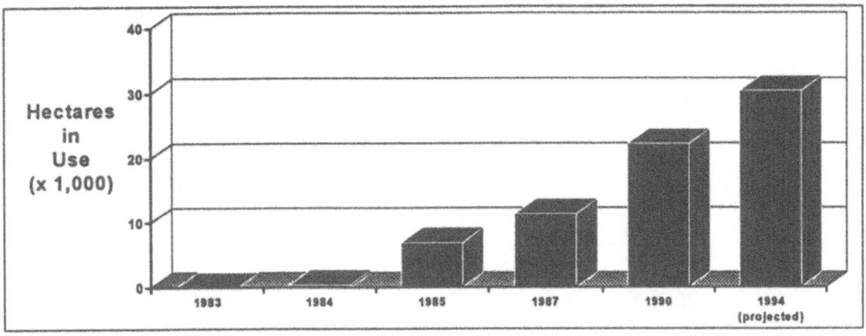

Figure 1. Adoption of the PDM/PCM software by the potato industry since 1983.

Initial development of computer aplications

Initial development of computer software included the Potato Disease Management (PDM) and Wisconsin Irrigation Scheduling (WISP) programs as stand-alone software. The PDM program addressed the prediction and control of early blight (an annual problem in Wisconsin) (Pscheidt and Stevenson 1986) and late blight (a sporadic problem occurring as a function of inoculum and weather) using a modified BLITECAST approach (Krause et al. 1975; Stevenson 1983). Growers experiencing losses to these diseases in the early 80s were receptive to computerized information processing that enabled them to predict the appearance of these diseases and then to improve disease control through optimized fungicide application scheduling and selection of fungicide application rate. The PDM program provided growers with prescriptive pest management software using weather and crop information collected in individual fields. Because of the success achieved in disease control, PDM served as a successful cornerstone for future development of more comprehensive computer software. Subsequent to the introduction and adoption of the PDM program, improved computer technology enabled our IPM computer programmers to combine the PDM and WISP programs into a more comprehensive Potato Crop Management (PCM) program. This software included modules for predicting and controlling disease, scheduling irrigation (Curwen and Massie 1984), predicting crop emergence, predicting development and managing harmful insects through improved targeting of insecticides and use of economic thresholds (Walgenbach and Wyman 1984a,b) and calculating the ventilation needs for storing potatoes. Each of these areas was requested by growers and represented areas of intensive research. The PCM program facilitated integration between modules and allowed growers to use data collected in their fields for multiple management decisions.

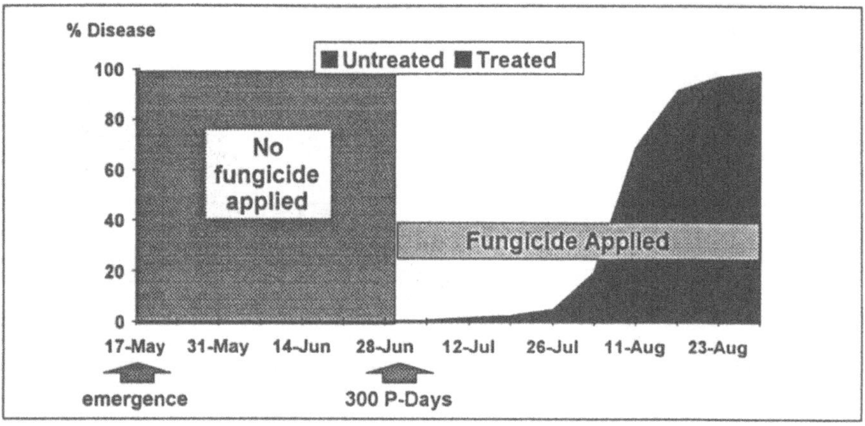

Figure 2. Delayed treatment with fungicide for early blight control using a treatment threshold of 300 P-Days. After 300 P-Days, fungicide is applied at 5- to 10-day intervals, depending on environmental conditions, for the duration of the growing season. Fungicide treatment slows disease progress.

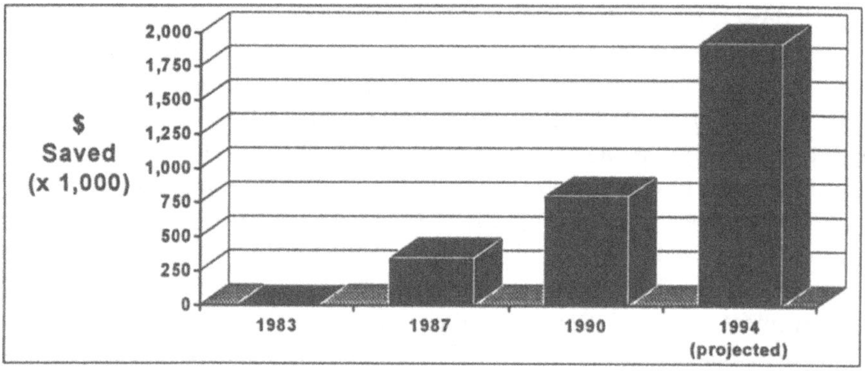

Figure 3. Estimated savings from reduced fungicide inputs related to use of the PDM/PCM software.

Value related to software use

Growers and IPM consultants report use of the PCM software on approximately 16,200 of Wisconsin's 27,500 ha of potatoes. The program is also used in neighbouring states where growers are using the software on an additional 12,100 ha (Figure 1). Use of the software saves growers an estimated $ 49 per ha (Connell et al. 1991) by eliminating 2–3 fungicide sprays early in the growing season before the threshold accumulation of 18 severity values (Krause et al. 1975) and 300 P-Days (Sands et al. 1979; Pscheidt and Stevenson 1986) (Figure 2). The estimated savings related to the delayed initiation of early blight and late blight controls for Midwestern potato growers exceed $ 1.4 million per

Table 1. Origins of reduced inputs for production of long season 'Russet Burbank' potatoes using the Potato Crop Management software and associated technology

Description of Reduced Inputs	Input Reduction/Hectare	Input Value/Hectare
Elimination of two sprays using EBDC fungicide each 1.7 kg ai/ha plus application costs ($ 12.35/ha)	3.4 kg ai	$ 49.40
Reduction in fungicide rate for remaining sprays – e.g. EBDC – 18% reduction from historical use pattern of 13.4 kg/ha	2.52 kg ai	$ 18.53
Herbicide use reduction (50%) – Eliminate one spray post emergence; Reduce rates of preemergence spray	metolachlor – 1.7 kg ai; metribuzin – 0.56 kg ai	$ 56.81
Insecticide use reduction (40%) – Eliminate systemic at planting and 1–2 foliar sprays	2.8 kg ai	$ 49.40
Reduction in irrigation and energy costs – application of water to match crop needs	13.2 cm	$ 11.12
Nitrogen fertilizer reduction (20%)	56 kg	$ 22.72
Total Input Reduction	10.9 kg ai pesticide, 56 kg N, 13.2 cm water	$ 207.97

year (Figure 3). Further reductions (15–20%) in the amount of fungicide applied during the growing season are possible when growers adjust fungicide rates according to environmental conditions and the accumulation of P-Days. Additional benefits from using PCM and associated technology include: (i) reducing herbicide rates (up to 50%) and improving the timing of herbicide applications; (ii) reducing insecticide treatment(40–50% reduction) by using economic thresholds and by targeting vulnerable life stages of economically damaging insects; (iii) reducing irrigation inputs 10–15% by matching irrigation inputs to crop demand; and (iv) reducing nitrogen inputs by 20% (Table 1 and Figure 4). Potential total savings from using the PCM software and associated technology are approximately $ 208/ha. Growers and consultants who have used the software for several years appear to achieve the greatest benefits. It should be noted, however, that these savings are partially offset by the costs of soil and plant tissue analysis, scouting services and environmental monitoring. These costs appear to range from $ 37–$ 49/ha in Wisconsin, but can be higher in other regions of the US. Wisconsin growers and allied industry are currently being surveyed to determine the full extent, value and costs of using integrated pest and crop management technology. The survey will also assist in identifying barriers to further adoption of management technology.

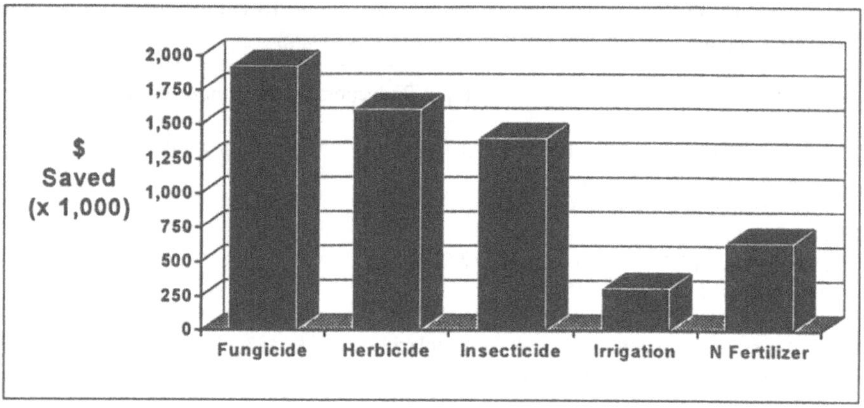

Figure 4. Projected savings from reduced inputs related to use of PCM software and associated technology on 28,300 ha in 1994.

Opportunities for enhanced use of prescriptive pest management software

Disease management

Early and late blight control. Foliar application of fungicides is directed at the control of both early (*Alternaria solani*) and late (*Phytophthora infestans*) blights. The appearance of a second mating type of *P. infestans* (A2) in many states in the US and the appearance of several strains insensitive to metalaxyl fungicide has placed added emphasis on multiple applications of protectant fungicide when weather conditions favor disease development. Potato cultivars currently grown in Wisconsin and other potato production areas of the US and Canada are susceptible to both early blight and late blight. Field evaluation of a broad spectrum of potato cultivars and breeding lines indicate potential for reducing fungicide inputs for early blight control on some cultivars and breeding lines, but the presently available cultivars do not have worthwhile levels of resistance to late blight. Field trials in Wisconsin have demonstrated the benefits of cultivar resistance to early blight for reducing disease progress in both sprayed and unsprayed blocks (Figure 5). Combining disease forecasting technology with information on cultivar susceptibility to foliar diseases would enable the grower to make informed decisions on the timing and rate of fungicide application.

Seedpiece handling. Seedpiece decay, stand loss, and loss of plant vigor pose major production risks to Wisconsin growers. The seed handling and planting aspects of potato production are among the most critical decisions faced by growers. Failure to recognize potential problems related to seedpiece decay and failure to take corrective action can result in reduced or delayed emergence. This in turn may require replanting or result in retarded growth which then affects

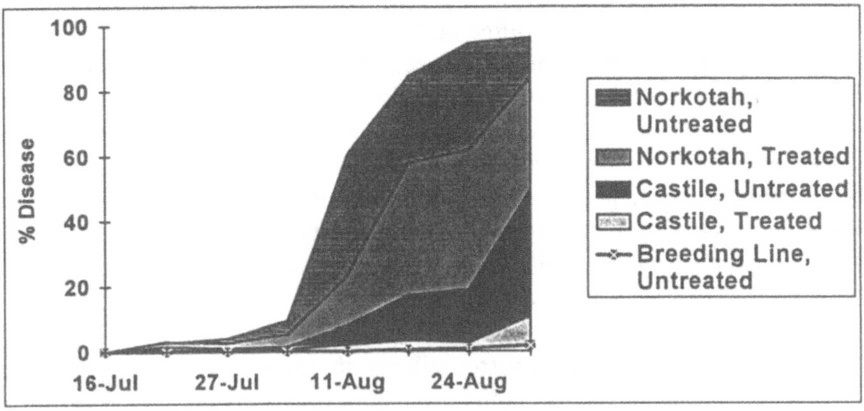

Figure 5. Changing the progress of early blight with combinations of field resistance and fungicide.

harvest scheduling, yield, marketability and crop profitability. To offset the risk of seedpiece decay, growers have traditionally coated cut-seedpieces with a fungicide dust before planting. Planting 2 metric tonnes of seedtubers/ha and applying a dust treatment at the rate of 1 kg/100 kg of seed results in each planted hectare receiving 20 kg of fungicide dust. Research at the University of Wisconsin has shown that seedpiece decay in Wisconsin is generally caused by bacteria such as *Erwinia carotovora atroseptica* and a fungicide dust treatment does little to protect the seedpiece from decay. Research has also shown that seedpiece decay is directly related to how the seedpiece is handled before and during planting and to the temperature and moisture at the time of planting and during the week after planting. Seedpiece decay is, therefore, not a random event, but rather a situation that is under the grower's control. Field and laboratory data provide the basis for development of an expert system designed to analyze information on seed handling and weather conditions and to provide site-specific recommendations for reducing the potential for seedpiece decay with minimized fungicide inputs. The addition of this expert system to the PCM program will provide an additional avenue for adoption of IPM technology and for further reduction in fungicide inputs.

Insect management. The increased interactive capacity provided by the new PCM software presents opportunities for growers to improve decisions on insect management by using real-time scouting data cross referenced with other fields and historical data. Differences in cultivar tolerance to key insect pests could be accounted for in control decisions by adjusting existing treatment thresholds to reflect higher or lower tolerance. Temporal changes in susceptibility that occur through the growth cycle could also be addressed in a similar fashion. The refinement of treatment thresholds to reflect known cultivar differences would reduce pesticide use on resistant varieties. As insecticidal inputs are decreased, the potential for integration of biological control into the general insect management program will be enhanced. Graphical presentation of real-time information, generated by field scouts, on populations of pests and beneficial insects would allow growers to assess the potential for biological regulation in individual fields and significantly increase the possibility of success. Insect resistance to chemicals continues to be a major threat to the effective management of insect pests such as the Colorado potato beetle and the green peach aphid. On-line availability of historical data on pesticide use will allow growers to follow resistance management strategies more effectively for individual species while ensuring that control decisions for other pests do not compromise such strategies.

Weed management. Weed control efforts in most years consist of pre-emergence herbicide sprays, cultivation and occasionally, post emergence sprays. The reliance on chemical controls for weed management in potatoes has directed research efforts to methods that reduce the potential environmental impacts of these applications. Recent research has focused on modelling the growth of the potato crop and evaluating the effect of crop shading on weed growth (Connell and Binning 1991; Raby and Binning 1986). A potato cultivar such as 'Russet Burbank' forms a dense canopy and provides approximately 95% shading of the soil surface seven to nine weeks after emergence. Using this knowledge, we have been able to adjust the timing and reduce the rate of herbicide applications such that only the required seven to nine weeks of control are provided from a single herbicide application. Other experiments have determined that, at these reduced rates, the potential for herbicide movement through the soil and into the groundwater is greatly reduced or is eliminated. To make this information useful for a wide range of geographical areas and different cultivars (which vary in the amount of competition they provide), we are focusing on modelling the weed growth under a variety of shading conditions. By evaluating the growth of different weed species under differing light conditions, we will be able to predict weed problems before they occur, based on historical weed geography and the amount of shading that different cultivars provide. This information can then be combined into a total weed management program to guide the grower in making environmentally and economically sound decisions.

Irrigation management. Improved management of irrigation has the potential to reduce the impact of some plant diseases, improve tuber quality and reduce the potential for leaching of pesticides and nitrogen. Over irrigation from emergence to tuberization appears to increase root infection by *Verticillium dahliae* but avoiding water stress during tuber bulking may decrease disease severity (Powelson et al. 1993). Since the activity of the common scab pathogen (*Streptomyces scabies*) is inhibited in moist soil, disease severity can be limited by maintaining 80–90% of available soil moisture during tuber initiation and bulking (Powelson et al. 1993). Soil moisture levels that deviate from the optimum during plant and tuber growth also affect the development of hollow heart, sugar ends and tuber malformation. The fine-tuning of irrigation by carefully adjusting water inputs during important phases of crop growth offers opportunities for reducing water inputs and perhaps more importantly, for improving the quality and marketability of the harvested tubers.

Nitrogen fertilizer management. Historically nitrogen fertilizer has been applied at rates totaling 280–392 kg N/ha. In most cases, yield and quality have been optimized with rates closer to 224 kg N/ha (Fixen and Kelling 1981). Recent research has shown that, through the use of a calibrated petiole nitrate-N test, the amount of early-season applied N can be reduced and later-season applications made only where the need exists. This approach results in fewer instances of excessive N being applied that can then leach to groundwater. The calibrations have been completed for several potato cultivars (Kelling and Wolkowski 1992). Subsequent research has also shown that these tests help growers to accurately account for on-farm N sources when potatoes are preceded by a forage legume (Kelling et al. 1993). This information is being incorporated into a nitrogen fertilizer management module which will help the grower to select the best rate and timing of N application for the cultivar being grown and the management options available for a given field. It will also help growers interpret in-season petiole tests for selected cultivars and, where appropriate, will recommend rates of N to apply.

Pest and crop interactions. There are often complex interactions between the crop, pests and cropping inputs which require integrated solutions. For example, plants deficient in nitrogen are more susceptible to early blight than well nourished plants. Early blight progresses rapidly on nitrogen deficient plants and prematurely defoliates infected plants even with intensive treatment with multiple fungicide sprays. Excessive irrigation promotes the leaching of nitrogen below the root zone of the potato and leads to nitrogen deficiency in the plant. Frequent and excessive irrigation also keeps the foliage wet for extended periods and thus favors spread of disease. As the nitrogen deficient plants begin to senesce prematurely, foliage is lost to early blight infection and the canopy density and shading is reduced. Light penetration stimulates weed growth which leads to weed competition for light, water and nutrients and may cause harvesting difficulties. Insects may also be attracted to stressed potato plants leading to additional defoliation. The use of prescriptive software provides opportunities to improve the management of complex problems with timely and effective remedies.

Enhancement of computer applications

The Wisconsin potato industry has requested additional program functions such as modules for soil fertility and farm record keeping that require expanded software capabilities not possible within the current PCM program in an MS-

DOS® environment. Currently, the PCM program is being converted to a Microsoft Windows™ application entitled WISDOM. This conversion allows greater flexibility in programming and it also provides many new features to the program user including:

1. An improved spreadsheet format for easier data entry and exchange of data between files: Environmental and crop data are manually entered into a spreadsheet format on the screen. Moving between days, weeks and months of the growing season is achieved by touching left or right arrows. Some growers report that they use data collected at a single monitoring station for management decisions in several nearby fields planted with different cultivars, planting dates, growth rates, etc. With the new program, data can be copied easily from one file into other files. The copy routine can be set to automatically copy newly entered data into several designated files, thus reducing data entry time for the user.
2. Presentation of data summaries in a graphical format: Information needed for grower decisions is presented as line graphs to show how various parameters (e.g. P-Days, Severity Values, Degree Days, Rainfall/Irrigation, days after emergence) change over time. Program users can be alerted when these units exceed specific thresholds so that key management decisions can proceed in a timely manner.
3. Graphical comparison of data between years: Historical data for a specific field or region can be stored in a separate file and used for comparison of production parameters with the current growing season. By using multiple comparisons per graph, program users will be able to further refine their crop and pest management activities.
4. Crop and pest record keeping: Records of field activities such as pesticide and fertilizer applications and pest scouting information can be stored for future reference. These records will play a greater significance in the future as they are linked to various program modules for tailoring specific recommendations to the occurrence of specific pests and implementation of specific management decisions. Records of pesticide application will also play useful roles in implementing strategies for managing or preventing problems of pest resistance.
5. A pictorial database of pest and crop problems: Scanned slides of insects, diseases, weeds, deficiencies, toxicities and physiological problems, stored on disk or CD-ROM, are being linked with existing software. The stored photographs, complete with descriptive information and recommendations for control can then be selected by the software user for review. This feature will help in the diagnosis of problems and in the application of the correct cultural, biological or chemical remedies.

Expanding the horizons of PCM to include the entire farming enterprise

The PCM program is comprehensive, but by design, is targeted at only a portion of the total farming enterprise, the potato crop. A need still exists to incorporate IPM information for the crops commonly used in rotation with potatoes so that growers have a guide for the entire farming system. Potato production in Wisconsin usually includes two- to three-year rotational sequences involving marketable commodities (snap beans, sweet corn, field corn, peas) and non-harvested cover crops (alfalfa, red clover, sorghum-sudan, rye). Both types of rotational crops have advantages to the whole farming enterprise. Cover crops may serve to hold the soil and nutrients in place, act as trap crops for insect pests or reservoirs for natural enemies, and have allelopathic effects on weeds and diseases. Many cover crops do not provide short-term economic returns to growers, but may promote long-term benefits to the overall enterprise which may be of equal or greater value. In contrast, marketable rotational crops provide short-term income to the grower and may also provide some additional long-term benefits. Some rotational crops, however, may contribute to the survival and increase of important pest problems and so would represent risks to the potato component of the farming enterprise.

Table 2. Six specific two- and three-year rotations involving potatoes as the crop of primary focus

Rotation No.	Year of Rotation 1	2	3	4	5	6
1	Potatoes	Snap Beans	Potatoes	Snap Beans	Potatoes	Snap Beans
2	Potatoes	Sorghum-Sudan	Potatoes	Sorghum-Sudan	Potatoes	Sorghum-Sudan
3	Potatoes	Sweet Corn	Potatoes	Sweet Corn	Potatoes	Sweet Corn
4	Potatoes	Snap Beans	Sorghum-Sudan	Potatoes	Snap Beans	Sorghum-Sudan
5	Potatoes	Snap Beans	Sweet Corn	Potatoes	Snap Beans	Sweet Corn
6	Potatoes	Snap Beans	Red Clover	Potatoes	Snap Beans	Red Clover

An integrated research project was initiated in 1991 to investigate the short- and long-term benefits and risks of six specific rotational sequences involving marketed and non-marketed crops grown in rotation with potatoes. The project was designed as a long-term (six-year) experiment to investigate six specific two- and three-year rotations involving potatoes (Table 2). Factors are being identified that contribute to or interfere with the preventive management of pests in the potato crop. Information developed from this research will be integrated with existing preventive and therapeutic strategies currently used in the PCM software. Expansion of the PCM software which will include modules for the management of both snap beans and sweet corn is underway. Management modules for snap beans include irrigation scheduling, a degree day calculator for predicting insect development and crop maturity and assessment of the potential the development of white mold, a disease affecting both potatoes and snap beans. Modules for sweet corn include irrigation scheduling and insect management. All of these modules will be included in an enterprise management program that addresses the improved management of potato and the crops grown in rotation with potato.

Summary

Growers and associated industry have been quick to adopt new IPM practices that enhance productivity, reduce crop inputs, and maintain their competitiveness with other major production areas. Computer technology has provided a convenient vehicle for delivering new IPM technology to the farm for implementation. Feedback from growers indicates many needs and opportunities for further adoption of IPM technology. This feedback has spawned many component and integrated research projects that will lead to additional solutions to current problems. The complexity of the PCM program has increased greatly since its first release in 1989. As we move from a single crop focus to the broader perspective of an entire farming enterprise, we are keenly aware of the risks and opportunities of this endeavor.

Acknowledgements

The authors gratefully acknowledge the support of the many undergraduate and graduate students, postdoctoral trainees and research specialists who worked with program development, growers who contributed time and facilities for program testing, staff of the UW Agricultural Research Stations where much of the component research was conducted, County Extension agents in key areas of Wisconsin potato production and independent crop consultants who continue to provide input. The authors are especially grateful to George Rice and Paul Kaarakka and Roger Schmidt, computer programming specialists, who have been instrumental in expanding the capabilities of the PCM software and who are rewriting this software in a Windows™ format.

References

Connell T R, Binning L K (1991) Canopy development model for weed management in potatoes. American Potato Journal 68:602 (Abstract).

Connell T R, Koenig J P, Stevenson W R, Kelling K A, Curwen D, Wyman J A, Binning L K (1991) An integrated systems approach to potato crop management. Journal of Production Agriculture 4:453–460.

Curwen D, Massie L R (1984) Potato irrigation scheduling in Wisconsin. American Potato Journal 61:235–241.

Fixen P E, Kelling K A (1981) Potato Fertility Requirements and Recommendations: Nitrogen.

Wisconsin Potato Production. 81-DF-B, University of Wisconsin Extension Publication, Wisconsin, USA.

Kelling K A, Wolkowski R P (1992) Interaction of potato variety and N rate on yield, quality and petiole nitrate-nitrogen levels. Proceedings of Wisconsin's Annual Potato Meetings 5:23–29.

Kelling K A, Hero D, Grau C R, Rouse D I, MacGuidwin A (1993) Potato responses to nitrogen following various legumes. Proceedings of Wisconsin's Annual Potato Meetings 6:93–104.

Krause R A, Massie L B, Hyre R A (1975) BLITECAST: A computerized forecast of potato late blight. Plant Disease Reporter 59:95–98.

Powelson M L, Johnson K B, Rowe R C (1993) Management of diseases caused by soilborne pathogens. Pages 149–158 in Rowe R C (Ed.) Potato Health Management. APS Press, St. Paul, Minnesota, USA.

Pratt L H (1993) Wisconsin Agricultural Statistics. Wisconsin Department of Agriculture, Trade and Consumer Protection, Madison, WI, USA.

Pscheidt J W, Stevenson W R (1986) Comparison of forecasting methods for control of potato early blight in Wisconsin. Plant Disease 70:915–920.

Raby B J, Binning L K (1986) Weed competition study in 'Russet Burbank' and 'Superior' potato (*Solanum tuberosum*) varieties with different management practices. Proceedings of the North Central Weed Control Conference 41:28–29.

Sands P J, Hackett C, Nix H A (1979) A model of the development and bulking of potatoes (*Solanum tuberosum* L.). I. Derivation from well-managed field crops. Field Crops Research 2:309–331.

Stevenson W R (1983) An integrated program for managing potato late blight. Plant Disease 67:1047–1048.

Walgenbach J F, Wyman J A (1984a) Colorado potato beetle (Coleoptera: Chrysomelidae) development in relation to temperature in Wisconsin. Annals of the Entomological Society of America 77:604–609.

Walgenbach J F, Wyman J A (1984b) Dynamic action threshold levels for the potato leafhopper (Homoptera: Cicadellidae) on potatoes in Wisconsin. Journal of Economic Entomology 77:1335–1340.

20. An advisory system for the management of potato cyst nematodes (*Globodera* spp)

T.H. BEEN[1], C.H. SCHOMAKER[1] and J.W. SEINHORST[2]
[1] *DLO Research Institute for Plant Protection (IPO-DLO), P.O. Box 9060, 6700 GW Wageningen, The Netherlands*
[2] *Middelberglaan 6, 6711 JH Ede, The Netherlands*

Abstract. An advisory system is presented for the management of potato cyst nematodes (*Globodera pallida*). It emphasizes the use of partially resistant potato cultivars, which provide the possibility of keeping population densities of potato cyst nematodes at a low level in short fixed rotations. Using stochastic models based on the population dynamics of potato cyst nematodes and the relation between pre-plant nematode densities and relative yield it is possible to calculate the probabilities of population development and the reductions in yield caused by these population densities. A simulation model is developed which integrates both models, using the frequency distributions of some of the most variable parameters relevant to a particular combination of potato cultivar and nematode population. Also, the natural decline in population density when non-hosts are grown is incorporated in the model. The model makes it possible to calculate the probability of a certain yield reduction, given a certain potato cultivar, nematode population and rotation. Therefore, it becomes feasible for a farmer to evaluate risks and the costs of different control measures in fixed rotations. The application of this model in the starch potato growing areas could lead to significant improvements in financial returns and a major reduction of the use of nematicides.

Introduction

Potatoes are among the most profitable agricultural crops in arable farming in the Netherlands. Therefore, they are grown as frequently as possible, especially in those areas, where farmers have almost no choice of other profitable crops. This frequency is limited by build up of potato cyst nematodes which, without control leads to considerable crop losses. Possible control measures other than crop rotation (growing susceptible potato cultivars in rotation with five to seven years of non-hosts) are growing highly- or partially-resistant potato cultivars and chemical control. Chemical control not only has a poor cost-to-benefit ratio, its use is also increasingly restricted by legislation. A 50% reduction by 1995 and a 80% reduction by the year 2000 in the use of the so-called fumigants are main objectives in the Multi-Year Crop Protection Plan of the Dutch government. Cultivars that are highly resistant to the pathotypes Pa 2 and Pa 3 of *Globodera pallida* are rare (industrial processing) or not available (human consumption). As crop loss is strongly associated with nematode density in the field at the time of planting, control should aim at preventing nematode

A.J. Haverkort and D.K.L. MacKerron (eds.), Potato Ecology and Modelling of Crops under Conditions Limiting Growth, 305–322.

densities from becoming too large. Once the population density in the field is reduced to an acceptable level a proper combination of relative susceptibility of a potato cultivar and crop rotation can keep it small. It is in the interest of potato growers and the environment to integrate the use of partially-resistant potato cultivars and other control measures in farming practices in such a way that a maximum return is obtained by minimizing the sum of cost of control and cost of yield reduction. This requires an advisory system based on prediction of population development and yield reduction in given rotations which emphasizes the use of partially resistant potato cultivars. As treatments with nematicides are expensive (about 1200 Dutch guilders/ha) and their effectiveness in reducing population densities is overrated, net returns will be considerably larger than with chemical control, when partially resistant potato cultivars are grown in the proper short rotation and, therefore with only minor crop losses due to the resulting small nematode densities. As a result the greater part of the reduction of the use of chemicals required in the Multi-Year Crop Protection Plan can be achieved without adverse economical consequences, if advices according to the advisory system are followed.

An advisory system has been developed based on a stochastic simulation model which can include a number of sub-models ranging from those of spatial distribution of population densities (resulting in new soil sampling methods), to economics for calculation of maximum returns. This paper is restricted to a discussion of the possibilities of using partially-resistant potato cultivars in certain crop rotations to minimize yield reductions and the cost of control. Four sub-models in the integrated simulation model will be discussed:

1. The simplified version of population dynamics of potato cyst nematodes applying at small to medium nematode densities at planting;
2. The concept of relative susceptibility ('susceptibility' = 'host status');
3. The relation between pre-plant nematode densities and relative yield (yield as fraction of the yield when potatoes are grown without nematodes);
4. Population decrease as a result of growing non-host crops.

Population dynamics

The relation between population density at planting or sowing (*Pi*) and the population density of the new generation at harvest (*Pf*) for nematode species with only one generation per season (potato-, oat-, white clover cyst nematodes, *Meloidogyne naasi*) is described by Seinhorst (1967; 1970; 1986a; 1993) (Figure 1). Seinhorst's (1993) most extended equation for this relation contains ten parameters of which the most important ones are the maximum rate of reproduction (occurring at very small initial population densities) and a theoretical maximum density, the number of eggs that would have been produced per unit weight of soil at very large initial nematode densities if the size of the plant would not have been reduced by the nematodes. Because of the reduction of the size of the plants by nematodes the actual maximum nematode

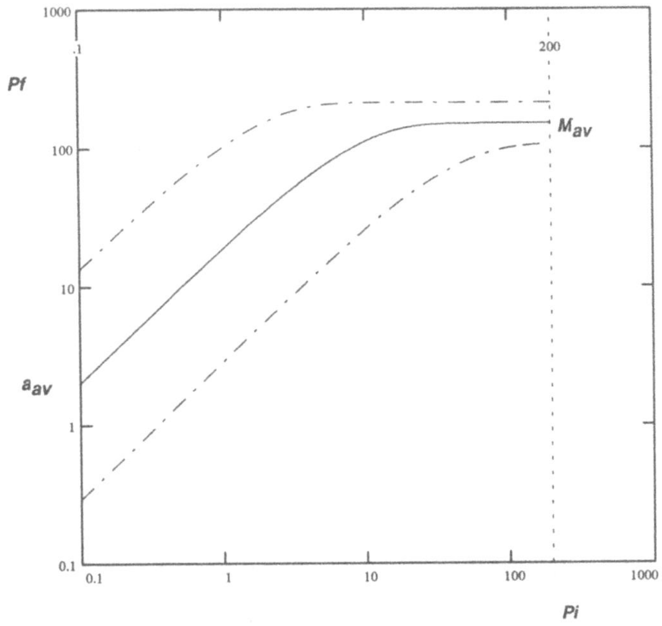

Figure 1. The relation between initial and final egg densities according to Eq. 1. Dotted lines = 95% confidence interval of the log-normal distribution of a and M (0,145a $< a_{av} <$ 6,9a and 0,71M $<$ Mg_{av} $<$ 1,41M). The scale of Pf (eggs/g soil) applies to *G. pallida* (a_{av} = 20 and M_{av} = 150 eggs/g soil on susceptible cultivars).

density after a potato crop in the field occurs at medium initial densities and also includes non hatched eggs of the parent generation (Seinhorst 1967; 1984; 1986a).

Seinhorst's (1993) equation for this relation, if extended to incorporate the part of the parent population surviving in soil that was not exploited by roots (Seinhorst 1986a), contains eight parameters; too many to be useful for the prediction of population densities and their frequency distributions after growing a cultivar with a certain degree of resistance. However, it can be used as a basis for the formulation of the constraints on the simpler equation:

$$Pf = M \times (1 - e^{-aPi/M}) \quad (1)$$

in which: Pi – initial egg densities (before planting) in eggs/g soil; Pf – final egg densities (after harvest) in eggs/g soil; a – the maximum rate of reproduction; M – a theoretical maximum final population density.

It fits to the relation between initial and final nematode densities up to the Pi value where the maximum Pf value is reached according to the extended equation. Taking this maximum as an estimate of M did not result in appreciable differences between Pf values according to Eq. 1 and observed ones in field experiments (e.g. Seinhorst 1986a), where a considerable plot-to-plot variation existed.

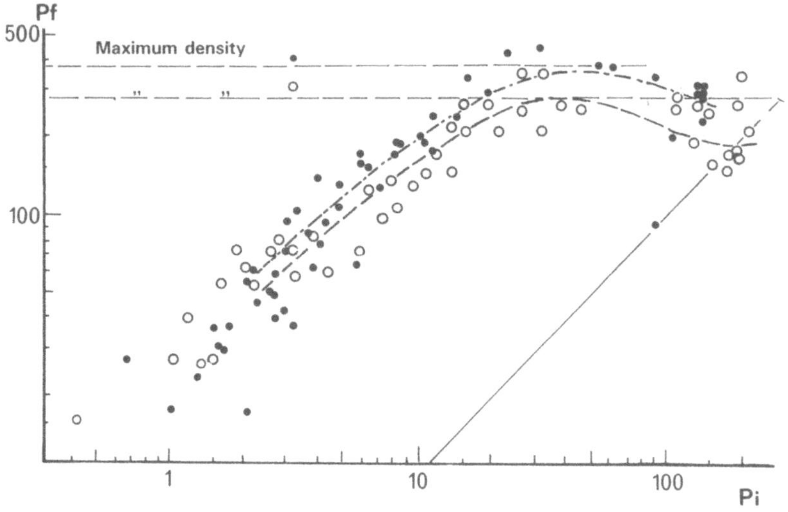

Figure 2. The relation between *Pi* and *Pf* of pathotype Ro 1 of *G. rostochiensis* on cultivar Bintje on two parts of a farmer's field (Seinhorst 1986a). *Pi* and *Pf* in eggs/g soil. The solid line indicates $Pf/Pi = 1$.

As the values of a and M are not only determined by the potato cultivar but also by external conditions, the final population density at one initial population density can vary strongly between fields and years. Both parameters not only differ between years and fields; variation can occur within a single field (Figure 2). Therefore, it is impossible to predict the development of population densities in individual fields using only average values of a and M, let alone from a only, as is common practice in contemporary advisory systems and legislation. A better approach is to establish the frequency distributions of a and M and to calculate the probability of all possible combinations of a and M and of the resulting densities *Pf* and their probabilities. According to observations on seventeen farmers' fields on several soil types in several years a for *G. rostochiensis* varied between 3 and 157 with a geometric mean of 25 and M varied between 200 and 400 eggs/g soil with a mean of 300 eggs/g soil (Seinhorst 1986c). A detailed analysis of these data will be presented elsewhere. In further calculations, the frequency distribution of both parameters is assumed to be log-normal. There are no indications that these distributions are different for partially-resistant cultivars. Therefore, the calculations for these cultivars are based on smaller average values (depending on their degree of resistance) with the same variability as for fully-susceptible cultivars.

There are not enough observations from field experiments to estimate mean values and their variances of a and M for *G. pallida*. A value of 0.8 for the ratio between a for *Globodera rostochiensis* pathotype Ro 1 and for *G. pallida* pathotype Pa 3 and a value of 0.5 for that between M for these pathotypes could be deduced from values on the susceptible cultivar Irene grown in pot experiments (Den Ouden 1974a; Seinhorst and Oostrom 1984; Seinhorst 1986b).

Therefore, it is assumed that a_{av} and M_{av} for *G. pallida* on susceptible cultivars in the field are 20 and 150 eggs/g soil respectively. These assumptions must be verified against observations from field experiments when these are available.

Measures of partial resistance and their relation

When partially-resistant potato cultivars are grown fewer females will mature than on susceptible cultivars; also the number of eggs/cyst may be smaller. Therefore, nematodes multiply less strongly on these cultivars than on susceptible ones and sustain a smaller maximum population density. However, both *a* and *M* are too variable to be suitable as a measure for partial-resistance. Therefore, the concept 'relative susceptibility', *rs*, was introduced, based on the population dynamics of the potato cyst nematode. The relative susceptibility is the ratio between the maximum multiplication rate *a* of the nematode population on the tested cultivar and on a susceptible reference cultivar or the equivalent ratio of the maximum population density *M* on these cultivars. These present two measures of partial resistance or relative susceptibility, provided that the tested cultivar and the susceptible reference are grown under the same conditions in the same experiment. Relative susceptibility is independent of external conditions which influence both *a* and *M* and contribute to their large variability. Figure 3 shows the relation between *Pi* and *Pf* of pathotype Pa 3 of *G. pallida* on the

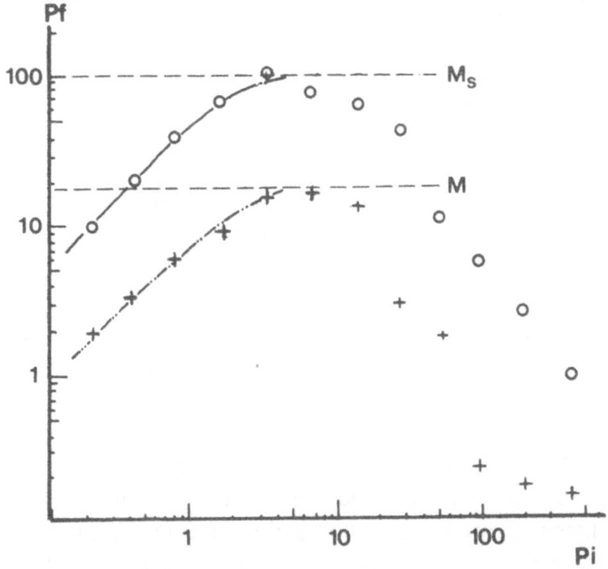

Figure 3. The relation between initial and final population density *Pi* and *Pf* of *Globodera pallida*, pathotype Pa 3 on cvs. Irene (o) and Darwina (+), (after Seinhorst and Oostrom 1984). *Pi* and *Pf* in eggs/g soil. Lines according to Eq. 1.

partially-resistant cultivar Darwina and on the susceptible cultivar Irene according to Eq. 1.

Jones et al. (1981) and Phillips (1984) assumed that the ratios ($a_{partially\ resistant}/a_{susceptible}$) × 100% and ($M_{partially\ resistant}/M_{susceptible}$) × 100% are numerically equal. According to Seinhorst (1984), Seinhorst and Oostrom (1984) and Seinhorst et al. (in press) this also applied to 9 out of 11 cultivars tested with pathotype Pa 3. However, rs_M was smaller than rs_a with cv. Activa and breeding line Karna 77/281 (Seinhorst et al. in press) and cv. Ehud (Seinhorst 1984). Seinhorst and Oostrom's 1984 data also indicate that, despite considerable differences in rates of reproduction at small initial egg densities of the same pathotype on the same cultivar in different experiments, the variation of the relative susceptibility was largely if not entirely due to experimental error, resulting from limited numbers of pots and insufficient number of cysts and eggs counted per replication.

The great similarity between rankings of cultivars according to degrees of susceptibility in pot and field experiments observed by Forrest and Holliday (1979), Phillips and Trudgill (1983; 1985), Phillips et al. (1987) supports the conclusion that relative susceptibilities do not depend on external conditions. The methods to determine the relative susceptibility of a potato cultivar with a high degree of accuracy are described in detail by Seinhorst et al. 1993.

Table 1 presents the relative susceptibilities of sixteen potato cultivars to different pathotypes of *G. pallida*, measured in 10 l pots at the IPO-DLO. The pathotypes include one of the most virulent populations found so far: Pa 3 (1) (the 'Rookmaker' population). A wide and continuous range of relative susceptibilities is apparently available in the cultivars tested.

Relation between nematode density and yield

Another essential part of the integrated simulation model is the relation between population density of eggs at the time of planting and tuber yield at harvest expressed as a proportion of the yield in the absence of nematodes. As large yield losses must be avoided, large *Pi* values must always be reduced by extra control measures to acceptable *Pi* values before a crop is grown. Therefore, the relation at small to medium nematode densities (Seinhorst 1986b) suffices:

$$y = m + (1-m) * 0.95^{(P-T)/T} \quad \text{for} \quad P > T$$

$$\text{and} \quad y = 1 \quad \text{for} \quad P \leq T \qquad (2)$$

in which: y – the yield at egg density P at planting as a proportion of that at $P \leq T$; m – the minimum relative yield (therefore a constant < 1); T – the tolerance limit, the density P up to which no yield reduction is caused.

T for susceptible cultivars varied little between experiments about an average of 2 eggs/g soil in field experiments with pathotype Ro 1 (Seinhorst 1982a; 1986b), microplots with pathotypes Ro 1 and Pa 3 (Greco et al. 1982) and pot

Table 1. Relative susceptibilities (a/a_s) of sixteen potato cultivars for pathotype Pa 1, Pa 2 and Pa 3 of *Globodera pallida* expressed in percentages of the susceptibility of the susceptible cultivar Irene for these pathotypes. Pa 3 (1), also known as the 'Rookmaker' population, is highly virulent, Pa 3b is a population with average virulence

Cultivar	Pa 1	Pa 2	Pa 3 (1)	Pa 3 (2)
Irene	100	100	100	100
Amalfi	–	3	–	39
Amera	76	63	–	98
Producent	15	3	40	34
Multa	–	12	43	40
Pansta	–	6	32	36
Promesse	–	4.5	35	21
Proton	–	1	31	18
Darwina	4.7	0.3	12	5
Santé	–	1	18	8
Atrela	–	0.7	20	9
Karna 77/270	–	–	28	–
Karna 77/281	–	–	12	–
Activa	–	–	25	–
Ellen	–	–	17	–
Seresta	–	–	2	–

– = relative susceptibility not measured. Pa 1 does not occur in the Netherlands.

experiments with pathotypes Ro 1, Ro 3 and Pa 3 (Seinhorst 1982b). Values of m varied more; 0.2 – 0.6 according to Seinhorst (1982a). But using an average of $m = 0.4$ in predictions does not result in unacceptable deviations between actual and calculated losses at densities ≤ 15 eggs/g soil. Over- or under-estimations do not exeed 5% (Figure 4). Predicting relative yield losses at larger nematode densities has no practical value as yield reductions at these densities are too high to be tolerated. Farmers will not grow a potato crop at such nematode densities without first taking control measures. In the Netherlands this implies the application of a nematicide to reduce *Pi* ánd also, as fumigation is mostly not sufficiant to attain acceptible nematode densities, the use of a systemic nematicide just before planting. The latter provide a temporal protection by delaying nematode invasion, thereby increasing the minimum yield m.

Little is known about the tolerance limit of partially-resistant cultivars except that T for pathotype Pa 3 on cv. Darwina is, also, 2 eggs per g soil and that results of field tests do not contradict such a value of T for other cultivars. The similarity of ratios M_r/M_s to a_r/a_s for the cultivars including cv Darwina, partially-resistant to pathotype Pa 3 (Seinhorst et al. in press), suggest that T for pathotype Pa 3 on these cultivars is not a multiple of the 2 eggs/g soil for cv. Darwina. An accurate assessment of the tolerance limit is desirable for two reasons. It is required for the calculation of losses to be expected at given egg population densities at the time of planting and, as the maximum population density tends to become proportional to the size of the plant when Pf approaches M, this

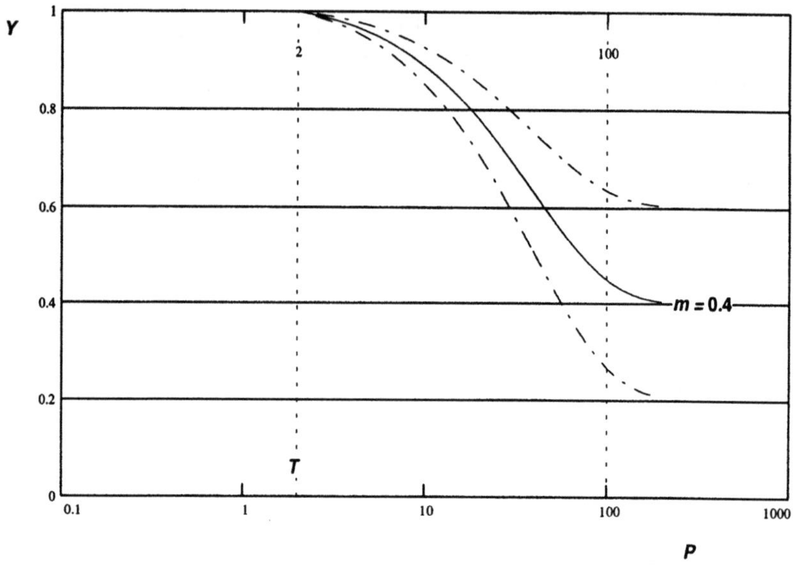

Figure 4. The relation between the initial population density (*P*) and the relative yield *y* according to the equation: $y = m + (1 - m) \times 0.95^{(P-T)/T}$ with $T = 2$ eggs/g soil and $m = 0.4$ (the mean) and $m = 0.2$ and 0.6 (the boundaries of the 95% confidence interval).

maximum will be correlated positively with the tolerance limit. However, small differences in tolerance (e.g. a ratio of tolerance limits of 1 to 2) would not have a detectable effect on M.

Population decline in the absence of a host

The nematode density at planting of potatoes is the density left by the previous potato crop multiplied by the survival rate over the period during which no potatoes were grown. This rate can be considered to be independent of nematode population density. Reduction of the population in the absence of a host crop is largely due to spontaneous hatching during a short period in spring of an apparently fixed proportion of the eggs. According to Huijsman (1961) the survival rates over a period of six years was 65% per year and this rate was (largely) independent of the age of the population as was the viability of the surviving population (Den Ouden 1963). Cole and Howard (1962) found a survival of 80 % per year during three years, and Den Ouden (1970) reported an average of 79% over 14 fields in 1969 but only 51% over six fields in 1973 (Den Ouden 1974b). However, the magnitude of field-to-field and year-to-year variation in the survival rate is unknown as, generally, sampling error was either ignored or was so large that differences could not be distinguished. However, there are indications that the survival rate during the first year after a potato crop is lower (Den Ouden 1960; Cole and Howard 1962; Andersson 1987; 1989) than in subsequent years. Therefore, in the calculations of the simulation model, a lower survival rate is used the first year after the potato crop than during later years (about 50% changing to 65%). The survival rate is used to calculate a factor for crop rotation, c, having a value of 0.5 for a 1:2 rotation; $0.5 \times 0.65 = 0.32$ for a 1:3 rotation; $0.5 \times 0.65^2 = 0.21$ for a 1:4 rotation and 0.5×0.65^3 for 1:5 rotation. But more information on survival rates in the absence of potatoes is required to adjust the model if necessary.

Simulation model

Basics

To make computations the frequency distributions of a and M were assumed to be log-normal with means 1.30103 and 2.17609 and standard deviations 0.419 and 0.07525 respectively for *Globodera pallida* pathotype Pa 3 on susceptible cultivars. Both frequency distributions were divided into 24 classes with a width of 0.25 times the standard deviation (s). Hence, a range of classes between (log a or M) + 3s and (log a or M) − 3s were used, comprising 99.7 of the frequency distribution. Each class of a and M is presented by the antilog of its class mid-mark and divided by a_{av} and M_{av} respectively and a relative frequency. Values of log a and M beyond this range are ignored. All relative frequencies are multiplied by 0.997^{-1} to obtain a total of 1 for the cumulative probability of the range considered.

Pf values following a given *Pi* for a potato cultivar with a given rs, are calculated according to Eq. 1 with every class mid-mark of the distribution of a times $rs \times a_{av}$ combined with all class mid-marks of the distribution of M times $rs \times M_{av}$. The relative frequency of each *Pf* is the product of those of the a and M used. Calculating *Pf* values for the next crop rotation as indicated above with each of these 24^2 values as *Pi* would require an unnecessarily excessive amount of work. Therefore, the *Pf* values, also, are divided into 24 classes and new mid-marks are calculated. The relative frequency of each of these class mid-marks is the sum of relative frequencies of *Pf* values belonging to that class. Now, there are 24 *Pi* values to start with, resulting in $24^3 = 13824$ *Pf* values after the second potato crop. This procedure is repeated until the required number of potato crops has been simulated.

As this paper is limited to the use of partial resistance to keep nematode densities small actual population densities as determined by soil sampling are not used as input. Instead, the almost stable frequency distribution of population densities established after five years of growing partially-resistant potatoes of the same relative susceptibility is used. These are practically independent of population densities in the first year of cropping. A more extensive description of this part of the simulation model will be presented in another paper.

Crop losses in rotations with potato cultivars of given relative susceptibility

For the assessment of relative crop losses $(1 - y)$ the relation between nematode density and relative tuber yield according to Eq. 2 is used to calculate y for the mid-marks of classes of Pf values. For the tolerance limit T a value of 2 eggs/g soil is used and for the minimum yield m 0.4. Estimates of percentages of fields with nematode densities and crop losses exceeding certain limits occurring in a given year or percentages of years in which these occur in a given field when a potato cultivar with a certain rs is grown in a certain rotation are given in Fig. 5. The probability of suffering a crop loss larger or equal than a certain value (on the x-axis) can be read off from the y-axis. From these data the average yield reduction can be calculated. When crop rotation (and the same applies to other control measuers) is used to reduce the population density between two potato cultivars with a factor c than the Pi at the next potato crop is cPf of the population density after the last potato crop. The development of population densities than depend on ca_{gem} and cM_{gem}. This implies that the frequency distributions of expected crop losses in Figure 5, which are obtained when cropping potatoes with a certain

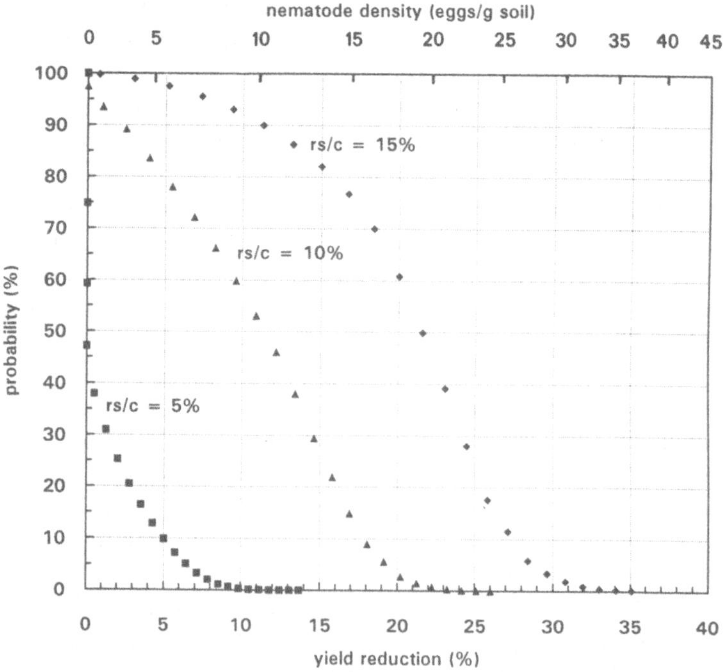

Figure 5. Probabilities of nematode densities (top x-axis) and resulting relative yield reduction $((1 - y) \times 100\%)$ (bottom x-axis) after five croppings of a partially-resistant potato cultivar with rs of $5/c$ (■); $10/c$ (▲); and $15/c$ (♦) respectively at an initial density of 5 eggs/g soil in the first year ($c = 0.5$ at 1:2 cropping frequency; 0.32 at 1:3 frequency; 0.21 at 1:4 frequency and 0.14 at 1:5 frequency). The relative yield (y) according to Eq. 2 and Figure 4 with $T = 2$ eggs/g soil and $m = 0.4$. $M_{av} = 150$ and $a_{av} = 20$ eggs/g soil.

Table 2. Relation between relative susceptibility, cropping frequency, average yield reduction and probabilities of a certain yield reductions after five croppings of a partial resistant potato cultivars at a *Pi* of 5 eggs/g soil in the first year

Relative susceptibilities (%) at cropping frequencies of 1:1	1:2	1:3	1:4	1:5	Average yield reduction (%)	Prob (%) of yield reduction larger than 0%	5%	10%	12,5%	15%	Percentage yield reduction in 95% of the fields smaller than
1	2	3	5	7	0,0	0,0	0,0	0,0	0,0	0,0	0,0
2	4	6	9	14	0,0	0,1	0,0	0,0	0,0	0,0	0,0
3	6	9	14	20	0,1	3,7	0,0	0,0	0,0	0,0	0,0
4	8	12	18	27	0,4	18,0	1,2	0,0	0,0	0,0	2,6
5	10	15	23	34	1,2	37,8	7,4	0,0	0,0	0,0	5,8
6	12	18	27	41	2,5	60,0	20,0	1,8	0,0	0,0	8,6
7	14	21	33	47	4,2	70,9	35,4	8,9	2,1	0,0	11,1
8	16	24	36	54	6,1	86,4	51,1	21,9	9,2	2,3	13,7
9	18	27	41	61	8,1	90,5	63,8	37,0	21,4	7,9	16,1
10	20	30	45	67	10,1	93,3	74,0	51,0	35,4	19,5	18,2
11	22	33	50	74	12,2	95,7	81,4	62,7	49,7	32,9	20,2
12	24	36	54	81	14,1	96,8	86,6	72,3	61,6	47,0	22,2
13	26	39	59	88	16,1	100	91,0	79,3	70,8	59,5	24,2
14	28	42	63	94	17,9	100	93,8	84,4	77,8	69,3	25,7
15	30	45	68	100	19,7	100	95,6	88,4	83,5	76,6	27,5

rs continously, are the same when a potato cultivar with *rs/c* is cropped in the corresponding rotation cycle. For a better overview concerning a range of relative susceptibilities Table 2 shows the different average yield reduction, the probabilities of more than 0%, 5%, 10%, 12.5% and 15% yield reduction and the maximum yield reduction in 95% of the fields in 5 different crop rotations and with different values of *rs*. From these figures it can be concluded that cultivars with partial resistance of ≤ 8/*c*% always control potato cyst nematodes sufficiently without the need for additional control measures. However, the level of *rs* that is still useful depends on the costs of control and the extra yield as a result of applying that control. It also depends on the financial situation of the potato grower. If he has a healthy financial reserve he can choose relative susceptibility on the basis of average yield reductions predicted. However, if no financial reserves are available and crop losses may not exceed a certain amount, risks of certain yield reductions have to be considered and the choice of relative susceptibilities becomes more limited. The percentages in Figure 5 and Table 2 can also be interpreted as chances that such densities and losses will occur in any field and year.

By applying a certain amount of control on all fields, after cropping a susceptible cultivar, calculated to prevent unacceptable yield losses even in the most heavily infested fields, too much control will be applied on a large number of fields. If the amount of control is adapted to obtain maximum average returns, yield losses in a certain proportion of the fields will be too high, while in the other fields still too much control is applied. Therefore, it would be justifiable to adjust the amount of control, after growing susceptible potatoes, to actual population densities, estimated by taking soil samples (Seinhorst 1982b; Schomaker and Been 1992). Sampling methods are then needed which provide estimates of densities of eggs and larvae with only a small error.

However, the need for soil sampling becomes obsolete when potato cultivars with high partial resistance are grown as then the small maximum population density M on these cultivars limits population increase sufficiently to densities where the cost of additional control exceeds the increase of net return obtained, whereas the required cropping frequency for potatoes is determined by other factors than crop loss caused by potato cyst nematodes. Soil sampling could then be restricted to once during a couple of rotation cycles to check for the presence of a more virulent nematode pathotype.

Sensitivity analysis

The values of ca_{av} and cM_{av} which are supposed to apply to a certain partially-resistant cultivar grown in a certain rotation in a field with a given nematode pathotype depend on several factors: The susceptibility (rs) of the cultivar, the values of a_{av} and M_{av} of the pathotype involved on the susceptible reference cultivar (of which a_{av} is the hardest to estimate in field trials), and the survival rate c of the nematode pathotype during the years without potatoes are all subject to experimental error. For practical application the indirectly derived values of a_{av} and M_{av} for pathotype Pa 3 on susceptible cultivars should be replaced, as soon as possible, by accurate averages and estimated frequency distributions derived from sufficient numbers of directly observed values of a and M from several fields and years.

Average reduction in yield (%) and the probabilities of reductions in yield > 10% were calculated using the best estimate of a_{av} of a susceptible control (20) and using the value of 25 (presuming an underestimation of the best estimate by 20%). The values for several cultivars with different rs are presented in Table 3. The underestimation of a_{av} resulted in only slightly larger calculated average crop losses than those using the best estimate of a_{av}. However, the underestimation of a_{av} resulted in a considerable underestimation of the chances of crop losses ≥ 10% in single fields. Presuming an overestimation of the best estimate of a_{av} by 20% (a_{av} then is 16) resulted in equivalent changes; calculated average crop losses were only slightly smaller, but crop losses ≥ 10% have considerable smaller probabilities than originally estimated.

Figure 6 shows the sensitivity of the relative yield to the value taken for the relative susceptibility of a cultivar. It can be seen that an over- or under-estimation of rs has a far greater effect on the average yield reduction than the same error in the estimation of a_{av}. Up to an rs/c of 4 the average yield reduction is negligible. Between $5\% < rs \times c < 15\%$ the relation becomes linear and average yield reduction increases by 2% per unit $rs \times c$. Therefore, an error of 5 pecentage points in the estimation of rs causes a deviation of $10\% \times c$ in the

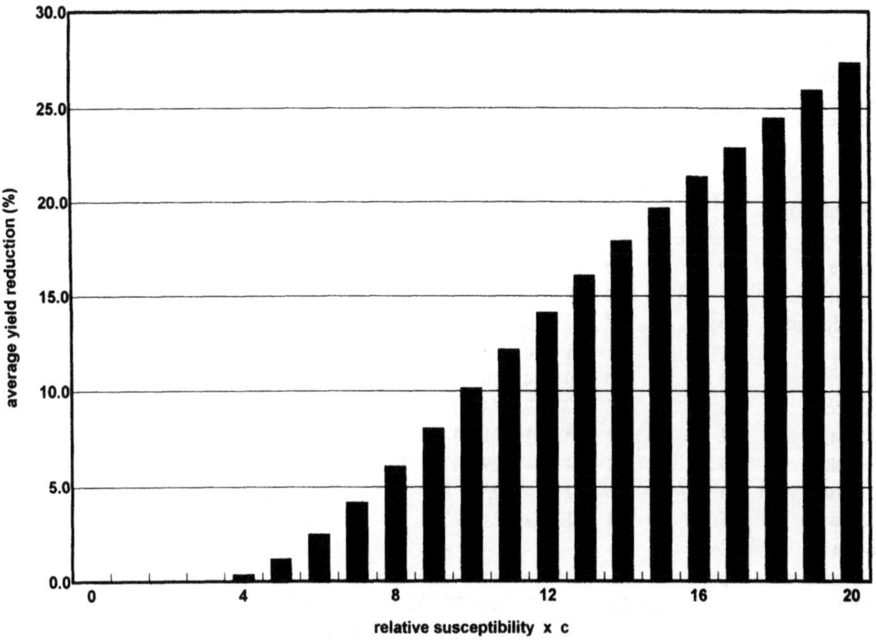

Figure 6. Sensitivity analysis of the effect of variation of estimated $rs \times c$ on the average yield reduction expressed as percentage of the expected yield when no nematodes are present after 5 croppings of a partially-resistant cultivar with the same ca_{av} and cM_{av} and Pi = 5 eggs/g soil in the first year.

predicted average yield reduction. It demonstrates the necessity of minimizing testing errors by conducting potato cultivar testing for relative susceptibility with the utmost care.

Which relative susceptibilities for which rotation schemes

The yield of potato cultivars, with the same characteristics grown in fixed rotations in fields infested with potato cyst nematodes, is negatively correlated with their relative susceptibility but there will be no difference in cost of nematode control. Therefore, the least susceptible of otherwise similar cultivars is always the best choice either because losses caused by the nematodes are smaller or because a shorter and more profitable rotation cycle can be practiced. To determine the upper limit of susceptibility that still is acceptable in a given rotation not only the average loss but also the probability of more than a certain percentage loss of a single crop is important, as, once a large nematode density has built up, the chance of larger than average population densities will be increased during the following years. Table 2 provides general information.

Table 3 presents a list of cultivars with relative susceptibilities for two different pathotypes of Pa 3 ranging from 2 to 44% with average yield losses and probabilities of more than 10% yield loss.

A farmer can decide to use this information in different ways. Assume he

Table 3. Relative susceptibilities of different potato cultivars on two populations of pathotype Pa 3, average yield reduction and probabilities of a yield reduction larger than 10% at two values of M_{av}/a_{av}. Pi = 5 eggs/g soil; – = < 0.5%. The relative susceptibilities are averages of all measurements up to now and can differ from previous publications

		Average yield reduction (%)								Prob. (%) of yield reduction > 10%							
Pa 3	% *rs*	*a* = 20				*a* = 25				*a* = 20				*a* = 25			
population 1		1:1	1:2	1:3	1:4	1:1	1:2	1:3	1:4	1:1	1:2	1:3	1:4	1:1	1:2	1:3	1:4
Irene	100	58	50	41	30	58	50	42	31	100	100	100	99	100	100	100	100
Astarte	44	48	30	19	10	48	31	21	12	100	99	87	48	100	100	94	63
Producent	40	45	27	17	8	46	29	19	10	100	98	81	35	100	99	90	50
Proton	31	39	21	11	4	40	22	13	6	100	90	55	8	100	95	70	15
Karnico	30	38	20	10	4	39	21	12	5	100	88	51	6	100	94	66	11
Santé	25	33	15	7	2	34	17	9	3	100	76	27	1	100	86	40	2
Ellen	22	30	12	5	1	31	14	7	2	98	63	13	–	100	75	22	–
Atrela	20	27	10	4	–	29	12	5	1	97	51	6	–	99	65	11	–
Karna 77/281	16	21	6	2	–	23	8	3	–	92	22	1	–	96	34	1	–
Darwina	15	20	5	1	–	21	7	2	–	88	15	–	–	94	24	–	–
Seresta	2	–	–	–	–	–	–	–	–	–	–	–	–	–	–	–	–
Pa 3	% *rs*	*a* = 20				*a* = 25				*a* = 20				*a* = 25			
population 2		1:1	1:2	1:3	1:4	1:1	1:2	1:3	1:4	1:1	1:2	1:3	1:4	1:1	1:2	1:3	1:4
Irene	100	58	50	41	30	58	50	42	31	100	100	100	99	100	100	100	100
Astarte	40	45	27	17	8	46	29	19	10	100	97	81	35	100	99	90	50
Proton	18	25	8	3	–	26	10	4	1	96	37	2	–	98	51	4	–
Atrela	9	8	1	–	–	10	1	–	–	37	–	–	–	51	–	–	–
Santé	8	6	–	–	–	8	1	–	–	22	–	–	–	34	–	–	–
Darwina	5	1	–	–	–	2	–	–	–	–	–	–	–	–	–	–	–
Seresta	2	–	–	–	–	–	–	–	–	–	–	–	–	–	–	–	–

wants to grow potatoes in a fixed rotation, for instance once in two years. He chooses which risk of yield reduction he is prepared to accept, taking into account the amount of money he could save by applying no other control measure (for instance the use of a nematicide), and chooses a cultivar with the required or better relative susceptibility that promises the largest net return because of other cultural characteristics. Another possibility is to determine in which crop rotation his favourate potato cultivar can be grown with the largest net return with or without additional control.

An example: Let's investigate whether a cultivar with a relative susceptibility of 18% yields a better net return in a 1:2 crop rotation than in a 1:3 crop rotation over a period of six years. In both cases no nematicides will be used. It is assumed that the net return of potatoes is 47% and that of a certain, extra, non-host crop is 27% of the gross return of potatoes (Kwantitatieve informatie 1990–1991, IKC-agv & PAGV). According to Table 2 the potato crop suffers an average yield reduction of 8.1% in a 1:2 rotation and of 2.5% in a 1:3 rotation. The only difference between the two rotations is that one potato crop in the 1:2 rotation is exchanged for the extra non-host crop. Therefore, the other three non-host crops can be disregarded in the calculation. Then the average net return in the remaining three years will be 3 × (47% − 8.1%)/3 = 38.9% using a 1:2 rotation and (2 × (47% − 2.5%) + 27%)/3 = 38.7% using a 1:3 rotation. So, there is still a slight advantage in growing potatoes once in two years with a cultivar with 18% relative susceptibility. However, at larger susceptibilities the balance will tip in favour of a 1:3 crop rotation.

Conclusion

By combining Eqs. 1 and 2 and using the frequency distributions of the relevant parameters, probabilities of different relative yield reductions can be calculated. As input for Eq. 1 sampling results can be used provided that these estimates of population density give a good approximation of the real density within the sampled area. When highly partially-resistant potato cultivars are grown in the proper rotation, sampling data become obsolete, as at the maximum population density M no such increase of yield can be obtained, that can balance the cost of sampling and control. The calculated frequency distribution of a and M for *G. pallida* and the rate of decrease of population density in the absence of potatoes are now verified in field experiments by the IPO-DLO on 20 farmers' fields during several years.

As the tolerance limit T is 2 eggs/g soil for most combinations of potato variety and nematode pathotype, the two important variables for Eq. 2 are $P = Pf$ of the previous potato crop (the output from Eq. 1) times c (crop rotation factor) and m. When using a value of 0.4 for m in the Netherlands, reliable predictions of yield loss can be made at economically interesting population densities. With the results of these calculations the farmer can evaluate the risks associated with the cropping of potato cultivars with a known relative susceptibility in a certain cropping frequency and choose combinations with the greatest probability of a maximum financial return based on net returns from potatoes and alternative crops in his fields.

Sensitivity analysis demonstrates that the accurate estimation of relative susceptibility of a cultivar is more important than an equally precise estimation of a and M (Eq. 1). Emphasis should be put on stabilizing experimental error when screening potatoes for partial resistance. Presently, the CPRO-DLO is testing more than 40 cultivars for their relative susceptibility against a number of populations of *G. pallida*, pathotype Pa 3, ranging in virulence from moderate to high. Field tests with some of these cultivars are being performed by the Research Station for Arable Farming and Field Production of Vegetables (PAGV).

The use of an advisory system as described above requires a mental reorientation by farmers, who still are inclined to aim at attaining maximum yields and, as a consequence, tend to opt for maximum security as actual yield losses are considered to be unpredictable. Nematicide treatments are, therefore, seen as a necessary insurance. However, in the Netherlands the use of nematicides is, at present limited to once in four years by statutory measures.

Moreover, the frequent application of nematicides causes adaptation of the microflora in soils resulting in accelerated breakdown of the fumigant (Smelt et al. 1989a,b). Therefore, chemical control cannot be used any further as an 'insurance' against losses by nematode attack in the areas producing potatoes for industrial processing, where a 1:2 cropping frequency is prevalent.

Farmers should strive to optimize returns instead of yields, not only to decrease the use of nematicides, but also to make more profit. An advisory system would provide the necessary information to apply a more profitable method of control, but whether it will be used depends on acceptance of the advice by the potato growers. The primary impulse to use the information generated by this advisory system will be the need to prevent yield reductions in those cropping years when potatoes are grown but nematicide application is not allowed.

References

Andersson S (1987) The decline of the yellow potato cyst nematode *Globodera rostochiensis* under non-host crops and under resistant potato cultivars – some preliminary results. Växtskyddsnotiser 51:145–150.

Andersson S (1989) Annual population decline of *Globodera rostochiensis* in the absence of host plants. Nematologica 34:254.

Cole C S, Howard H H (1962) The effect of growing resistant potatoes on a potato-root eelworm population – a microplot experiment. Annals of Applied Biology 50:121–127.

Den Ouden H (1960) Periodicity in spontaneous hatching of *Heterodera rostochiensis* in the soil. Report of the Fifth International Symposium in Plant Nematology. Nematologica, Supplement II:101–105.

Den Ouden (1963) Hatching and the rate of multiplication of *Heterodera rostochiensis* populations of different ages. Nematologica 9:231–236.

Den Ouden H (1970) De Afname van de Bevolkingsdichtheid van het Aardappelcysteaaltje bij Afwezigheid van Aardappelen. Jaarverslag 1969. Instituut voor Plantenziektenkundig Onderzoek, Wageningen, The Netherlands, page 133.

Den Ouden H (1974a) The multiplication of three pathotypes of the potato root eelworm on different potato varieties. Netherlands Journal of Plant Pathology 80:1–6.

Den Ouden H (1974b) Afname van de Bevolkingsdichtheid van het Aardappelcysteaaltje bij Teelt van Resistente Aardappelen en bij Braak. Jaarverslag 1973. Instituut voor Plantenziektenkundig Onderzoek, Wageningen, The Netherlands, page 96.

Forrest J M S, Holliday J M (1979) Screening for quantitative resistance to the white potato cyst nematode *Globodera pallida*. Annals of Applied Biology 91:371–374.

Greco M, Di Vito A, Brandonisio I, Giordano I (1982) The effect of *Globodera pallida* and *G. rostochiensis* on potato yield. Nematologica 28:379–386.

Huijsman C A (1961) The influence of resistant potato varieties on the soil population of *Heterodera rostochiensis* Woll. Nematologica 6:177–180.

Jones F G W, Parrott D M, Perry J N (1981) The gene-for-gene relationship and its significance for potato cyst nematodes and their Solanaceous hosts. Pages 23–36 in Zuckerman, B.M., Rohde, A.R. (Eds.) Plant Parasitic Nematodes, Vol. III. Academic Press, Inc., New York, USA.

IKC-agv & PAGV (1990) Kwantitatieve Informatie voor de Akkerbouw en de Groenteteelt in de Vollegrond, Bedrijfssynthese 1990–1991. Publicatie nr. 53.

Phillips M S (1984) The effect of initial population density on the reproduction of *Globodera pallida* on partially resistant potato clones derived from *Solamum vernei*. Nematologica 30:57–65.

Phillips M S, Trudgill D L (1983) Variations in the ability of *Globodera pallida* to produce females on potato clones bred from *Solanium vernei* or *S. tuberosum* SSP. *Andigena* CPC 2802. Nematologica 29:217–226.

Phillips M S, Trudgill D L (1985) Pot and field assessment of partial resistance of potato clones to different populations and densities of *Globodera rostochiensis*. Nematologica 31:433–422.

Phillips M S, Trudgill D L, Evans K, Lacey C N D, Mackenzie M, Turner S J (1987) The assessment of partial resistance of potato clones to cyst nematodes at six different test centres. Potato Research 30:507–515.

Schomaker C H, Been T H (1992) Sampling strategies for the detection of potato cyst nematodes: Developing and evaluating a model. Pages 182–194 in Gommers F J, Maas P W Th (Eds.) Nematology from Molecule to Ecosystem. Dekker and Huisman, Wildervank, The Netherlands.

Seinhorst J W (1967) The relationships between population increase and population density in plant parasitic nematodes. II. Sedentary nematodes. Nematologica 13:157–171.

Seinhorst J W (1970) Dynamics of populations of plant parasitic nematodes. Annual Review of Phytopathology 8:131–156.

Seinhorst J W (1982a) The relationship in field experiments between population density of *Globodera rostochiensis* before planting potatoes and the yield of potato tubers. Nematologica 28:277–284.

Seinhorst J W (1982b) Achtergronden van aaltjesbestrijding (3). Economische aspecten van vruchtwisseling en chemische bestrijding. Bedrijfsontwikkeling 13:494–500.

Seinhorst J W (1984) Relation between population density of potato cyst nematodes and measured degrees of susceptibility (resistance) of resistant potato cultivars and between this density and cyst content in the new generation. Nematologica 30:66–76.

Seinhorst J W (1986a) The development of individuals and populations of cyst nematodes on plants. Pages 101–117 in Lamberti F, Taylor C E (Eds.) Cyst Nematodes. Plenum Press, New York, USA/London, UK.

Seinhorst J W (1986b) Agronomic aspects of potato cyst nematode infestation. Pages 211–228 in Lamberti F, Taylor C E (Eds.) Cyst Nematodes. Plenum Press, New York, USA/London, UK.

Seinhorst J W (1986c) Effect of nematode attack an the growth and yield of crop plants. Pages 191–210 in Lamberti F, Taylor C E (Eds.) Cyst Nematodes. Plenum Press, New York, USA/London, UK.

Seinhorst J W (1993) The regulation of numbers of cysts and eggs per cyst produced by *Globodera rostochiensis* and *G. pallida* on potato roots at different initial egg densities. Nematologica 39:104–114.

Seinhorst J W, Oostrom A (1984) Comparison of multiplication rates of three pathotypes of potato cyst nematodes on various susceptible and resistant cultivars. Mededelingen Faculteit voor Landbouwwetenschappen, Rijksuniversiteit Gent 49 (2b):605–611.

Seinhorst J W, Oostrom A, Been T H, Schomaker C H (1995) Relative susceptibilities of eleven potato cultivars and breeders' clones to *Globodera pallida* pathotype Pa3, with a discussion of the interpretation of data from pot experiments. European Journal of Plant Pathology (in press).

Seinhorst J W, Been T H, Schomaker C H (1993) Partiële resistentie in de bestrijding van aardappelcysteaaltjes *Globodera spp.* 1: Bepaling van de graad van resistentie. Gewasbescherming 24 (1):3–11.

Smelt J H, Teunissen W, Crum S J H, Leistra M (1989a) Accelerated transformation of 1,3-dichloropropene in loamy soils. Netherlands Journal of Agricultural Sciences 37:173–183.

Smelt J H, Crum S J H, Teunissen W (1989b) Accelerated transformation of the fumigant Methylisothiocyanate in soil after repeated application of Metham-sodium. Journal of Environmental Sciences and Health, Part B. Pesticides, Food Contaminants and Agricultural Wastes 24 (5):437–455.

21. Handling qualitative and uncertain information

B. MARSHALL[1], J.W. CRAWFORD[1] and J. McNICOL[2]

[1] *Soil-Plant Dynamics Group and* [2] *Scottish Agricultural Statistics Service, Scottish Crop Research Institute, Invergowrie, Dundee DD2 5DA, U.K.*

Perhaps the most important challenge the development of computer-based systems to aid management of agriculture and the environment, is how best to utilise data and information of questionable reliability in conjunction with mathematical models that have not been extensively validated (paraphrase Smart (1992) referring to a quote of Getz and Gutierrez).

Abstract. The ability to handle uncertain knowledge improves the value of the information that can be provided. A prediction of a single value, say potato tuber yield, with unknown certainty is of no practical use to the decision-maker. In contrast, a series of possible outcomes (yield levels) with associated probabilities is clearly of benefit. Furthermore, the ability to explain how these risks are derived, what influences them and hence what may be modified, are equally important benefits. And the ability to communicate using natural language, albeit a restricted language interface, enhances the user's acceptance of decision support systems.

Knowledge

Mathematical models. Our quantitative understanding of biological systems is often expressed in mathematical models. Many such examples for potato are included in this book. Some mathematical models are used for advisory purposes. For example, a model of tuber size distribution has been developed and is being used to help decide on the optimum seed-rate or plant spacing (Marshall 1986). In practice, such models are often inflexible in the way they can be used and the information that is required and is supplied are in forms that are less than ideal. The model is a black-box within which potentially valuable knowledge is inaccessible to the user. The knowledge is not simply the value of a particular parameter e.g. average tuber size is 65.3 g, but includes equally important knowledge such as which variables affect a particular property, say number of daughter tubers produced, and in what way i.e. the structure of the model. This particular example is referred to as looking up-stream or backward propagation. Conversely, one may wish to look in the other direction (down-stream) or forward propagation – what are the consequences of changing

A.J. Haverkort and D.K.L. MacKerron (eds.), Potato Ecology and Modelling of Crops under Conditions Limiting Growth, 323–340.

number of daughter tubers on, say, average tuber size. Access to such knowledge has powerful explanatory potential. The explanation of how an answer is reached can guide the user to possible other solutions or causes. This potential is lost in the mathematical black-box or conventional simulation modelling.

Sources of uncertainty. Uncertainty, a key element in evaluating risk and decision-making, is rarely considered and when it is it is crudely handled by simulation models. Uncertainty derives from many sources: weather, inherent variability in the behaviour of systems, market fluctuations, imprecision in our knowledge (soil fertility), etc. Imprecision and variability are two aspects of uncertainty.

Imprecision refers to the degree of resolution. Mathematical models generally manipulate continuous variables with a resolution that is defined by the precision with which variables are stored in the computer. For most practical purposes this is equivalent to a continuous variable with infinite resolution. However, our knowledge of the value of a variable may be far less precise; nitrogen fertility may only be ranked on a scale from 0 to 5, the expected yield from a site may be known only to be high (say somewhere between 60 and 70 t ha^{-1}). Thus continuous variables are replaced by less precise variables with discrete ranges. The information carried by such variables is more qualitative. It is quite possible to have a hierarchy of such variables with decreasing precision i.e. progressively more qualitative.

Variability can derive from unknown sources e.g. the residual or unexplained variation of observed data about a fitted relation, or from variable, uncertain inputs e.g. future weather or future market prices. Thus a variable such as yield has a range of possible values with associated probabilities. The variable can either be quantitative (continuous or discrete values), in which case the associated uncertainty is a probability density function, or qualitative (discrete ranges) with associated probabilities or probability mass function (Figure 1). In practice, the latter case predominates.

Previously, the uncertainties associated with quantitative variables could be propagated arithmetically only in the special case that all probability density functions are normally distributed. But, in reality, this is rarely the case. Recent developments based on a Gibbs sampler, where a computer simulation of Markov processes avoids the need to integrate probability density functions, are now making it possible to handle quantitative variables with any continuous, probability density function (Cassella and George 1992). However, for the purposes of this paper it is sufficient to concentrate on qualitative variables and their associated discretised probabilities.

The challenge is to combine quantitative and qualitative knowledge with the associated uncertainties into decision support systems, ensuring that the knowledge within the systems is easily accessible.

Case studies, background. Our current understanding of factors influencing the yield of potato crops and the size distribution of the tubers has been used as

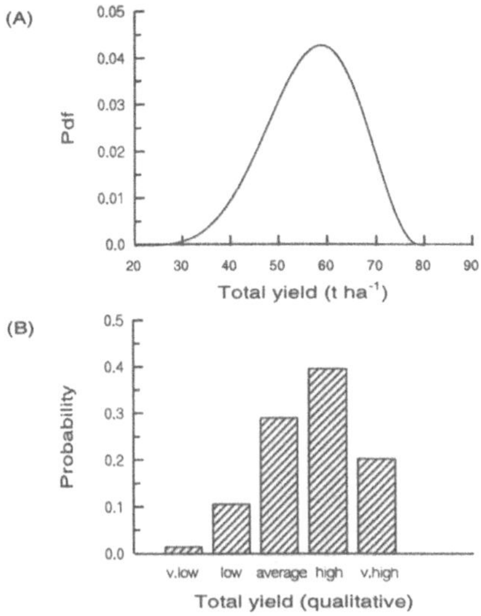

Figure 1. A distribution of potato yields that could have been observed on a farm over the last thirty years shown as (A) a probability density function (Pdf) which describes a continuous, quantitative variable and (B) a set of probabilities (P) which describe a discrete, qualitative variable.

a test case to develop and evaluate new approaches to encapsulating and accessing this knowledge. The size of a tuber is one of several important quality characteristics which determines its suitability for selected markets (pre-pack, bakers, crisps, chips, etc.). Success in achieving the optimum size distribution will increase the value of the crop. A mathematical model (seed-rate software package), referred to earlier, was developed with the Agricultural Development and Advisory Service (ADAS, UK) using traditional programming techniques (FORTRAN, Pascal, etc.). Its objective is to find the planting density for a given seed lot (given the tuber size and cultivar) which maximises the marginal price, namely market value less seed cost. Its limitation is that this is, essentially, the only problem it can solve. A large proportion of the knowledge contained within such models is inaccessible to the user. As a consequence, within the first growing season that the model was used the advisers came up with a range of equally important questions that could not be answered directly. Only an expert with computer programming skills and a detailed knowledge of the mathematical model could answer their questions.

Unlike the results produced by the mathematical model, in reality, there is variation in how potato plants perform once the tubers have been planted e.g. from one year to the next or one field to another. Later in the season the grower may have new additional information e.g. the number of stems or, even better, the number of daughter tubers actually produced. The ability to include such knowledge would considerably improve the precision of the outcome. At the end of the season the grower may wish to review the performance of the crop and identify areas of significant deviation from that intended in order to make possible adjustments in future seasons. None of that information can be fed into traditional software packages nor can consequences be examined without reprogramming.

Research in Artificial Intelligence (AI) techniques is developing rapidly. Computing power, a limitation only five years ago, has undergone major and significant advances in both hardware and software. The prospect of decision support systems of sufficient complexity and practicality is with us. There are two, complementary approaches to meeting the challenge, *rule-based* and *causal models*. We have researched and evaluated both approaches at the SCRI.

Rule-based models

Smart (1992) reviewed the potential use of rule-based models in decision support systems for agriculture. Rule-based models can encapsulate precisely the same knowledge as that contained in mathematical models. Mathematical equations are replaced by rules of the form

> *IF (planting density has the value P, seed size has the value W; stem density takes the value f[P,W])*

f[P,W] is simply the original mathematical expression quantifying the effect of plant density and seed size on the number of stems produced. A diagrammatic representation of the rule is shown in Figure 2. At first sight it would appear to be simply a more cumbersome way of implementing the original mathematical models. However there are three major advantages: model construction is automated, model definition is separated from query definition and qualitative and quantitative information can be combined.

Model construction. A set of rules, each describing one of the many relations that may have a bearing on seed cost, tuber size distribution, market value etc., are entered. At this stage only the inputs and output(s) of each rule and the relation linking them are defined as in Figure 2. Once a library of rules has been produced the rule-based model is built automatically by linking the output of each rule to the appropriate input of another rule or rules (Figure 3). At this stage the model's behaviour should be validated, if this has not already been done. All input/output values of each rule are accessible to the user. How they are accessed is determined by the query types available. Huber et al. (1990) used a similar approach in developing an integrated pest management system for apples.

Flexible querying Queries in the context of rule-based models are types of question as opposed to specific questions. A query is a sentence with a defined structure e.g. subject (variable 1) with a verb (action) and an object (variable 2). 'What is the number of daughter tubers produced if the stem density is 30

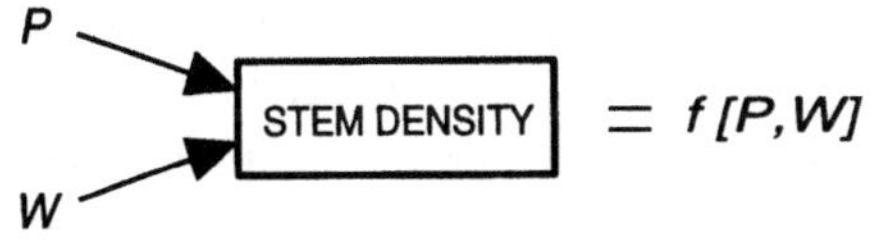

Figure 2. A diagrammatic representation of a single rule as implemented in a rule-based model. The rule takes the name of the output variable, stem density, which in this case is a mathematical function, f[*P*, *W*], of the input variables planting density, *P*, and seed size, *W*. In a complete model the input and output names are now shown in the diagrammatic representation, for clarity. Only the rule name is displayed.

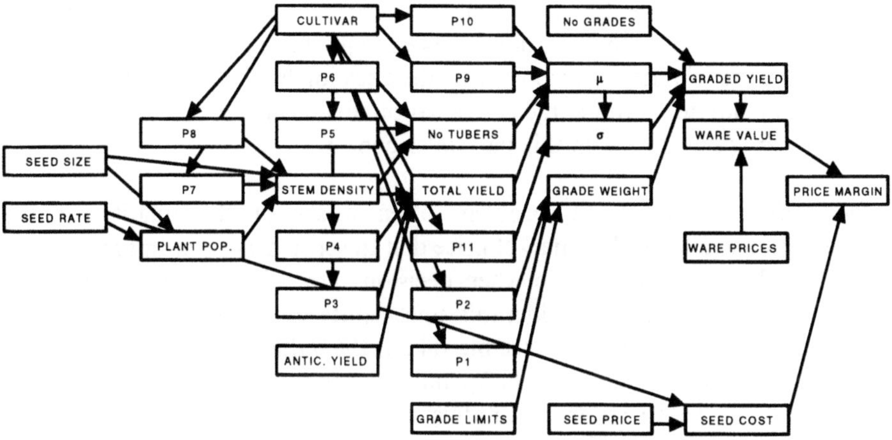

Figure 3. A rule-based model of the factors influencing tuber-size distribution, marketable yield and value, as used previously in the seed-rate software package developed jointly by the Scottish Crop Research Institute (SCRI) and the Agricultural Development and Advisory Service (ADAS), U.K.

m^{-2}?' is a specific question. The corresponding query or question type is 'What is the value of variable 1 if the value of variable 2 is X?'. While still being able to answer the specific question given here the query applies equally to any pair of variables e.g. 'What is the yield of bakers if the total yield is 65 t ha^{-1}?'. This one query is also able to answer the reverse and equally valid question, 'What is the total yield if the yield of bakers is 30 t ha^{-1}?', with the same ease. Here the rules are being exercised in the reverse direction – what input is consistent with a given output. Another example of a general query is 'What value of this variable maximises the value of that variable?'. This would answer the specific question of the original seed-rate software package, but can answer many more in addition. A further advantage is that, once a set of queries is defined for application to one particular rule-based model, the set equally applies to any rule-based model (and, indeed, queries defined elsewhere would apply to this one); without further program development being required. Thus time and effort in defining the original query is repaid many times over. This approach has considerable power and flexibility in accessing and using knowledge effectively.

A typical 'conversation' with the rule-based model of tuber size distribution is shown in Figure 4. A potato crop is part way through the growing season. The grower has first told the system the cultivar he has planted and the market (number of grades and their weight limits) he intends to grow for. The grower has set the values of three variables in the system; no other part of the model is accessed at this stage (Figure 4a). He then asks 'How does graded yield vary with total yield?'. In order to answer this more complex question the system searches back from the variable 'graded yield' to identify all those variables which influence graded yield either directly or indirectly. In this case one of the influencing variables is found to have no value set and the user is asked to enter the number of daughter tubers present in the crop (Figure 4c). Only information essential to answering the question has been requested by the system. Only that part of the model relevant to answering the question has been accessed, and the system is ready to continue the conversation answering any further questions the grower wishes to ask in this particular context.

Qualitative knowledge. A problem often arises in the quality of the information available. The market price that the crop may be ultimately sold at can only be a rough estimate when decisions are being made earlier in the season. It is subject to considerable uncertainty and is more appropriately represented by a qualitative variable taking the values e.g. 'very low', 'low', 'average', 'high', 'very high'. The quantitative form of the rule-based model must be abstracted to a more qualitative level in order for us to use this form of uncertain information. The process of abstraction requires mappings to be defined between the quantitative and qualitative levels of each variable (inputs and outputs of the rules) in the model e.g. for the variable 'Total yield' less than 30 t ha^{-1} could be classified as 'very low', 30–40 t ha^{-1} as 'low' etc. Having defined appropriate mappings for all variables the model is then exhaustively exercised to define a full set of qualitative rules. The single quantitative rule shown earlier produces a set of qualitative rules of the form.

IF (planting density is 'low', seed size is 'small'; stem density is 'low')

The limitation of this approach is the number of qualitative rules that are generated. One quantitative rule with 2 input variables, each with only 3 states (e.g. low, medium, high), and an output with say 5 states will generate 45 (3 × 3 × 5) qualitative rules! There is some possibility for reducing the final number of rules required. In particular, if changing the state of one input variable through all possible states while holding the other input(s) constant has no effect on the output then the final number of rules can be reduced. Figure 5 shows an answer to the question posed previously (Figure 4), except that this time it is using qualitative variables. This process of abstraction makes it feasible to integrate knowledge of a qualitative nature within the same system.

A rule-based approach to modelling allows full access to the knowledge contained within the model – both to the values of each variable and to the structure of the model. With abstraction it is possible to construct a hierarchy of models of decreasing precision, from the quantitative model down to the

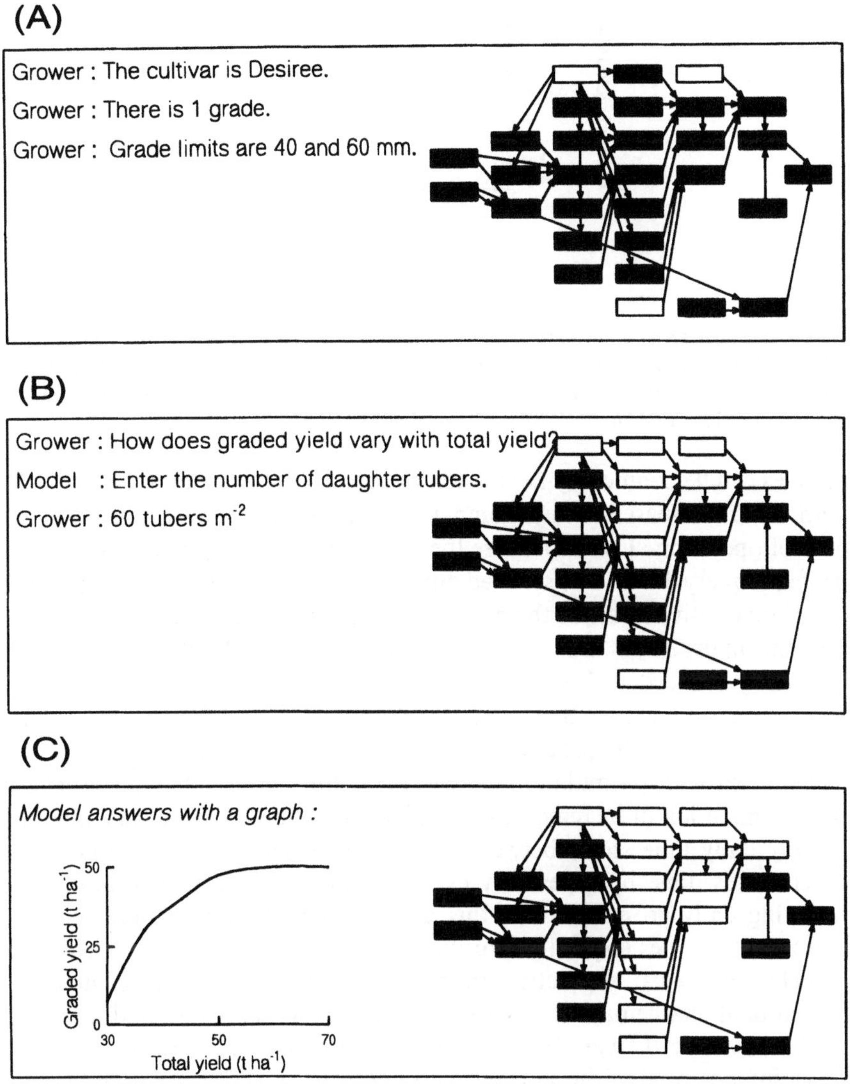

Figure 4. A typical 'conversation' between a grower and the rule-based model shown in Figure 3. The conversation is shown on the left hand side and the parts of the model accessed (open boxes) are shown on the right side. The text within the boxes has been removed for clarity.

coarsest model at the other extreme. In some cases there may not be a quantitative model. The qualitative models while being less precise, run faster and can give simpler, summarised explanations. More detailed explanations can be obtained by going to deeper levels.

The use of a restricted natural language interface allows the user more freedom and less rigour in expressing a query (Sowa 1984; Hendrix and Walter 1987; Selfridge 1989). Any ambiguities that arise are detected and the user is guided to a correct formulation of the question. Similarly, imprecise naming of

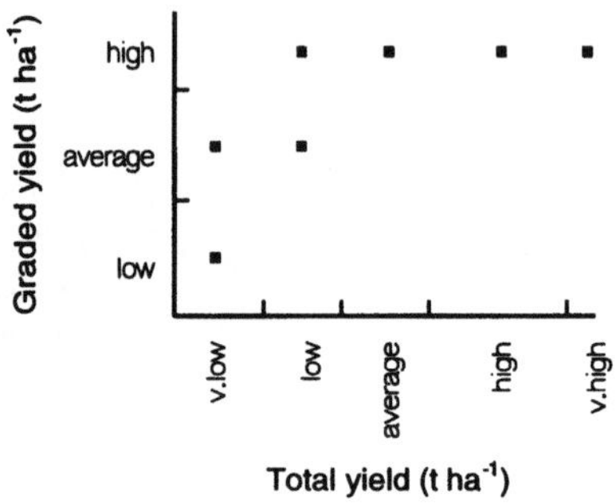

Figure 5. The final result of the same 'conversation' as in Figure 4, except that a qualitative version of the rule-based model has been used. Note the quantitative variables are replaced by qualitative ranges and the continuous line (Figure 4c) is replaced by a set of points indicating which regions of the space the relation lies in.

variables can be handled by near matches both on the name and the units. The overhead of rule-based models compared to mathematical models is the slower speed of operation. This is because the rules, which are required to answer the particular query, have to be selected during the execution of the program. In the case of qualitative levels of the model, there is considerable use of memory to store the many rules that are generated. Both these limitations are being continually reduced with each new generation of hardware and software. A more serious limitation is that program loops have to be coded explicitly. There are several reasons why program loops come about. Perhaps the most common is time. In most crop simulation models, predictions of total biomass are based on the summation of increments of simulated growth over fixed time intervals, typically daily time steps, between planting and harvest. In a mathematical model this is done by repeatedly executing the same set of subroutines for estimating daily growth. To do this in a rule-based model would require a separate set of rules (equivalent to the subroutines) for each day! Clearly this would be impractical. A practical alternative is to interface rule-based models to conventional mathematical models, but only for computational purposes. Flexible access to the variables and their structure is not available.

The full potential of this approach is only realised when the variability of or certainty in the values of each variable are attached. This can be achieved using causal models.

Causal models

Bayesian belief networks. The development and evaluation of belief networks based on Bayesian probability theory has been explored by several research groups over the last decade (Cooper 1984; Andreassen et al. 1987; Pearl 1986; Lauritzen and Spiegelhalter 1988). They are potentially a powerful tool for representing relations between entities where variability is involved, and provide a basis for manipulating uncertain beliefs. A belief network is a directed graph in which nodes represent the entities or variables and the directed links represent causal relations between the nodes (Figure 6). The nodes are usually qualitative variables such as 'seed price', 'seed rate' etc. as in the previous rule-based model. In addition, marginal probabilities are attached to the qualitative states of each node. Thus the probability that 'market price' is 'very low' may be 0.02, 'low' 0.25 and so forth, with the total probability of each state in a node adding to unity. The second difference from the rule-based model is the nature of the causal link, the connection between the nodes. Rather than a mathematical relation or a set of rules the link is a set of joint probabilities. For example the relation between seed size, plant spacing and stem density is represented by conditional probabilities of the form,

P(stem density = 'very sparse' | seed size = 'small', plant spacing = 'narrow') = 0.05

This reads as follows, the probability that stem density is 'very sparse' given (the vertical bar) the knowledge that seed size is 'small' and that plant spacing is 'narrow' is 0.05. The mapping between a qualitative state such as 'very sparse' for stem density and the corresponding quantitative range, say less than 5 stems m^{-2} in this case, is precisely analogous to the earlier abstraction process used in producing qualitative models from quantitative rule-based models. The important difference is that here the relation between these three variables has an uncertainty attached. Whereas the rule-based model may have concluded that stem density under these circumstances was either 'very sparse' or 'sparse' (Note that in qualitative models there may be more than one outcome for one set of input conditions, see Figure 5.), the causal model replaces this by a distribution of probabilities.

A simple example. A simple example can be used to illustrate the type of information that is captured in a causal model. Consider the single relation between seed size, *W* (g fresh weight), plant spacing, *P* (plants m^{-2}) and stem density, *D* (stems m^{-2}) which forms part of the original seed-rate model. The relation between these variables can be represented by the causal model shown in Figure 6, where Node A is seed size and Node B is plant density then Node C is stem density. Seed size and plant density, are assumed to be independent i.e.

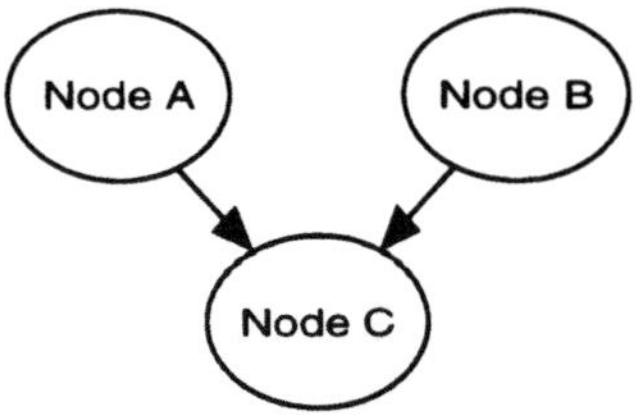

Figure 6. A directed graph showing the building blocks of a causal model. The arrows indicate which nodes (ellipses) influence or 'cause' other nodes. This information is vital in defining the probability structures of the model. In this example nodes A and B both influence node C but do not directly influence each other. Thus nodes A and B are marginally independent in this case.

changing the beliefs about the seed size does not alter the beliefs about plant density. Figure 7a shows a set of plausible data (hypothetical example) that might be collected from a series of trials on a single cultivar investigating the effect of seed tuber or seed piece and plant spacing on the number of stems produced per unit area. Each point represents say 20 tubers where W is the average fresh weight per tuber. In carrying out a trial one intentionally covers the whole spectrum of possible management practices (seed size and spacings) in order to understand and quantify the relation linking the variables. The data collected can be thought of as the training information. With a mathematical model one has to fit functions to the data. In this case the relation can be approximated by the mathematical relation.

$$D = (\mathrm{a} + \mathrm{b}W)\,P$$

The mathematical approximations for three of the plant spacings used in this hypothetical experiment are shown in Figure 7b. Clearly there is a degree of variability about the mathematical relations (lines in this case) i.e. unexplained variation, which is ignored by the mathematical model. Yet this variation is real and of significance to the grower. The causal model takes a different approach. The continuous variables, D, W and P are divided up into their respective qualitative states. For clarity of presentation consider just one instance of plant spacing, say 'medium' (Figure 7c). The proportion of data points landing in a box provides a measure of the probability that a particular combination of states will occur e.g. given plant spacing is 'medium' then there is no chance that with 'small' seed a 'very dense' stand of stems will be achieved whereas the chance is high with 'large' or 'v. large' seed given the same spacing. The precise probabilities along with the marginal probabilities are given in Table 1. The marginal probabilities indicate the distribution of individual data points over, in this case, the experimental data space. They are relatively uniformly distributed, reflecting the experiment design. The general trend, stem density increasing with seed size, is captured by the causal model. As one shifts from 'small' seed size to 'v. large' with plant spacing 'medium' so the highest probabilities of stem density shift from being 'sparse' to being 'dense' (Table 1). But, most importantly, the uncertainty surrounding the mathematical relations is also captured.

Table 1. The joint probabilities linking seed size and stem density given the fact that plant spacing is 'medium' (see Figure 7c)

	Seed size					
Stem density	v. small	small	medium	large	v. large	Marginal
v. small	0.03	0.03	0.01	0.00	0.00	0.07
sparse	0.03	0.06	0.07	0.01	0.00	0.17
medium	0.05	0.05	0.08	0.12	0.03	0.33
dense	0.04	0.04	0.06	0.10	0.18	0.42
v. dense	0.00	0.00	0.00	0.00	0.01	0.01
Marginal	0.15	0.18	0.22	0.23	0.22	

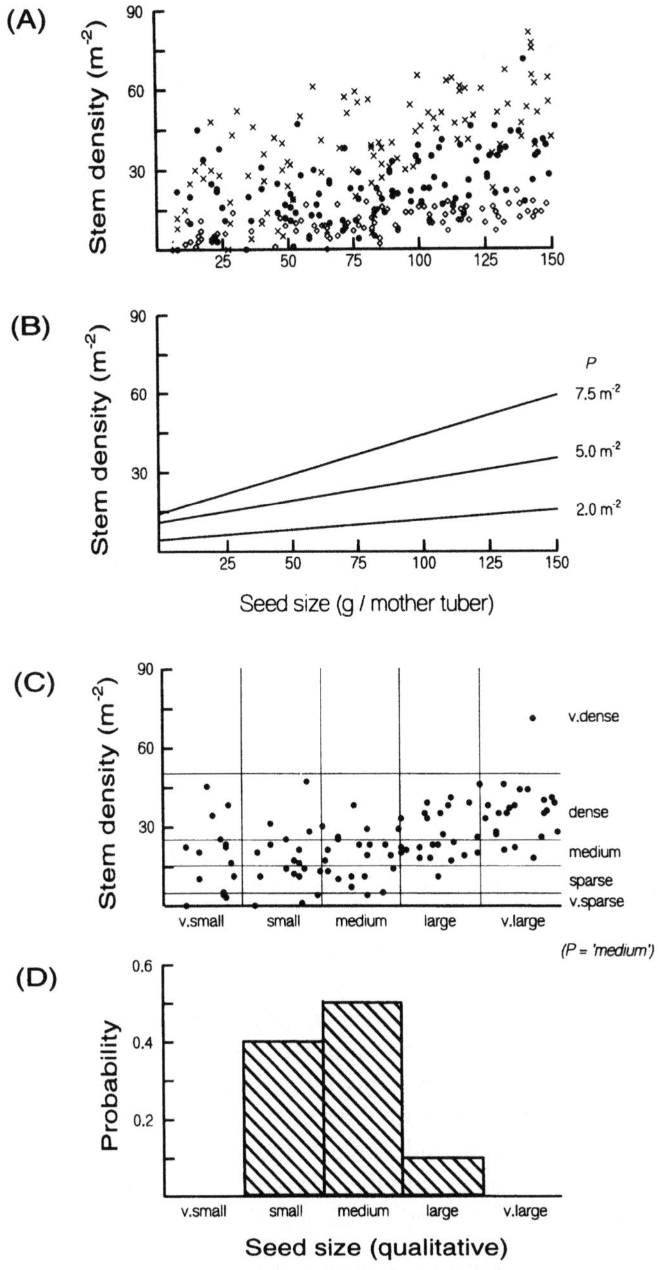

Figure 7. Data from a hypothetical experiment to study the relation between plant spacing, seed size and the number of stems produced. Three plant spacings are used: 'wide', 2 plants m^{-2} (diamonds); 'medium' 5.0 plants m^{-2} (closed circles); and 'narrow', 7.5 plants m^{-2} (crosses). Each point represents the average of 20 seed tubers planted, (A) is the raw data, (B) shows the mathematical approximations to the data, one line per plant spacing, (C) shows the qualitative partitioning of the data for 'medium' plant spacing only, and (D) shows a typical distribution of seed size from a seed lot that a grower has purchased (unlike the experimental data set the distribution of seed sizes is not uniform).

A grower now wishes to use the knowledge in the causal model to assess the chances of achieving a 'sparse' density of stems or worse with a particular seed lot he has purchased (Figure 7d). Belief in the distribution of seed sizes in the seed lot is clearly different from the uniform, marginal distribution of the experimental data set. There is a peak falling in the 'small' to 'medium' classes and the range is much narrower than that experienced in the experiments. Consequently the weighting of the data space is altered and the probabilities (both joint and marginal for stem density) change (Table 2). Whereas previously the probability of achieving a 'sparse' stem density or worse was 0.24 (marginal probabilities 0.07 + 0.17, Table 1) the chance has now increased to 0.38 (0.09 + 0.29, Table 2). The grower may consider this risk too great and reduce the plant spacing. This is an example of querying a causal model. New evidence or belief is entered and the consequences of that new belief are propagated.

Table 2. The effect on the joint probabilities linking seed size and stem density and the marginal probabilities for stem density given the fact that plant spacing is 'medium' and the marginal probability for seed size now reflects a particular seed lot (see Figure 7d)

	Seed size					
Stem density	v. small	small	medium	large	v. large	Marginal
v. small	0.00	0.07	0.02	0.00	0.00	0.09
sparse	0.00	0.13	0.16	0.00	0.00	0.29
medium	0.00	0.11	0.18	0.06	0.00	0.35
dense	0.00	0.09	0.14	0.04	0.00	0.27
v. dense	0.00	0.00	0.00	0.00	0.00	0.00
Marginal	0.00	0.40	0.50	0.10	0.00	

Constructing more complex causal models. This is best illustrated by considering a causal model describing the influence of weather (radiation, temperature and rainfall) and management (planting and harvest dates) on the yield of potatoes (Figure 8). In consultation with experts, the key nodes and major causal links are identified. Having defined the model, the probability structures are created automatically by a computer program.

The nodes, Planting date and Temp-1, are common parents to the node, Emergence date, and together with their daughter form what is referred to as a 'clique' for which a set of N conditional probabilities can be defined,

> *probability that Emergence date is 'early given Planting date is 'early' and Temp-1 is 'cold' is* p_1
>
> *probability that Emergence date is 'early' given Planting date is 'early' and Temp-1 is 'average' is* p_2
>
> ...
>
> *probability that Emergence date is 'late' given Planting date is 'late' and Temp-1 is 'hot' is* p_N

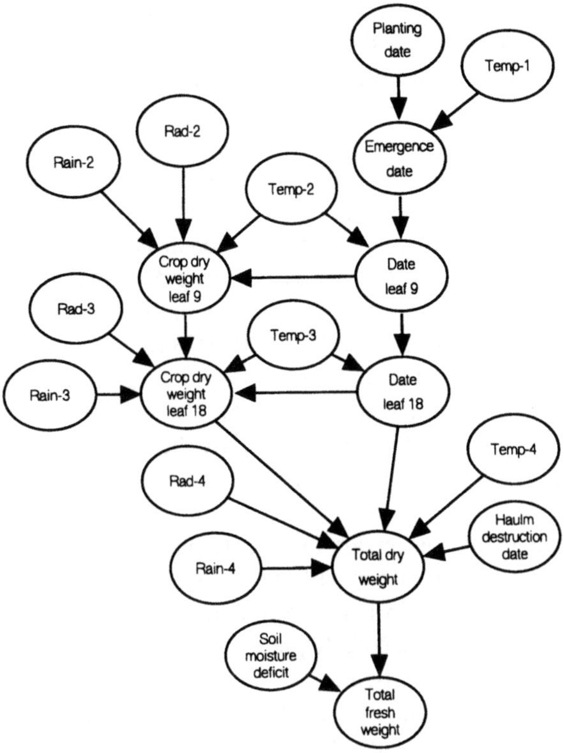

Figure 8. Causal model of the water constrained yield of potato (after Gu et al. 1994).

The formation of cliques helps to minimise the number of probability structures that need to be created (Lauritzen and Spiegelhalter 1988).

The number of probability structures can grow rapidly as more detail is added to the model (new nodes, more qualitative states per node, or more causal links). This can require a large amount of memory and time to run the model. As hardware and software continue to advance, so the level of complexity which is practical is increasing. The skill is in constructing a model which incorporates just sufficient detail to capture the important behaviour of the system in question. While one may expect that temperature and radiation during a period may be correlated no such causal link has been included in this causal model e.g. Temp-2 and Rad-2 are not linked (Figure 8). Similarly correlations between one period and another e.g. Temp-1, Temp-2, Temp-3 and Temp-4, are not included. Inclusion of such links would enable the correlations that exist between periods to be captured e.g. when a season has above average temperature. To include all such possibilities would increase the complexity of the model exponentially. In practice these correlations are weak in comparison with other effects and can be ignored – this is an example of the skill that is required in defining the structure of the model. It is analogous to the decisions made in constructing mathematical models and deciding which higher order interactions between variables, if any, to include.

Training causal models. The next stage is to enter the probabilities. With relatively simple models as shown previously it is possible to collect the necessary data, calculate the appropriate probabilities and enter them by hand (Figures 6 and 7). However, with more complex models the number of probabilities not only prohibits manual entry but it is also unlikely that there is sufficient data upon which to calculate them in the first place. A practical alternative is to use mathematical or rule-based models to provide simplified representations of the real world. This approach has been used to define the interactions between management, weather, crop development and growth (Gu et al. 1994). An exhaustive simulation of a mathematical model of water-constrained yield of potato, developed at SCRI, was carried out using all combinations of a 100-year run of current weather, planting and harvest dates (approximately 50,000 simulations). This provides sufficient information on the variability of the system with which to calculate the necessary probabilities. This is done automatically with the aid of a computer program which 'trains' the causal model.

A weather generator is necessary in order to provide an adequate 'experience' of the variability of the weather with which to train the model. Such runs of observed weather are not always available and hence weather generators are developed to produce a data set of adequate length (e.g. Peiris and McNicol 1994).

Not considered in this particular example is the variability surrounding the relations used in the mathematical model of water-constrained yield of potato. Theoretically there is no reason why this variability should not be included. Bouman and Jansen (1993) have shown one way in which this can be achieved. The variability surrounding the relations leads to a probability distribution for the estimates of parameter values. The uncertainty in the parameter estimates can be included in the simulations. For each season of weather combined with each instance of management practice (planting date and haulm destruction date), the uncertainties in the dates of developmental stages and crop weights at these various stages, due to the uncertainties in the parameter values, could be estimated. Just one season and one instance of management practice will require hundreds of simulations (maximum of 2000), and is currently limited to a maximum of 5 parameters. The parameter estimates are assumed to be independent of each other. This assumption should not be overlooked. Where correlations are known to exist, empirical relations can be derived specifically for pairs of parameters. This was done for certain parameters in the water balance model (Bouman and Jansen 1993).

Querying causal models. A simple query was shown earlier with the causal model in Figure 6 where the effects of changing the marginal distribution of seed size on stem density were examined. A similar query can be posed to the more complex causal model of potato yield. The distribution of planting dates used to train the causal model is typical of those achieved by a potato grower in NE Scotland. The grower now wants to assess the effect on yield of changing from an early to a late haulm destruction date. To present this query the probability distribution for an early haulm destruction is first fed in (probability of 1 of destroying the haulm before the 4 September). This gives a distribution of yields with a peak probability of 0.45 of obtaining a 'high' yield (in the range 51 to 64 t ha^{-1}) (Figure 9). A new set of marginal probabilities corresponding to a late haulm destruction (a probability of 1 of destroying the haulm during the third week of September) is then fed in. The effect of changing this marginal probability is then propagated throughout the model. The effect on yield is to reduce the risk of a 'low' yield – now there is only a probability of 0.1 of obtaining a yield less than 51 t ha^{-1} – and doubling the probability of a 'high' yield, in excess of 78 t ha^{-1}. The exhaustive simulation of the original mathematical model is performed once, to gain sufficient experience of the influence of weather and management on crop yield with which to train the causal model. Once trained the causal model can then be queried indefinitely.

As with rule-based models queries can also be addressed in the reverse direction. This approach was used in a recent study of the impact of climate change on a range of crops in Scottish Agriculture (Gu et al. 1994). The predictions of climate change are uncertain and are on a crude scale relative to the size of Scotland and the farm-types within it. Further, they are subject to frequent revision as understanding improves. The use of crop simulation models is extremely cumbersome and costly in time, requiring a complete repeat of previous exhaustive simulations whenever a new climate scenario is suggested. With the causal model already trained, with a set of different climates which cover the plausible range of future climates, it is a simple matter to enter the latest beliefs without further recourse to exhaustive simulations. Even more effective is to ask how much does the climate have to change for, say, the probability of high yield to double? Policy makers can then keep an eye on the latest climate predictions to check the likelihood of this critical climate scenario

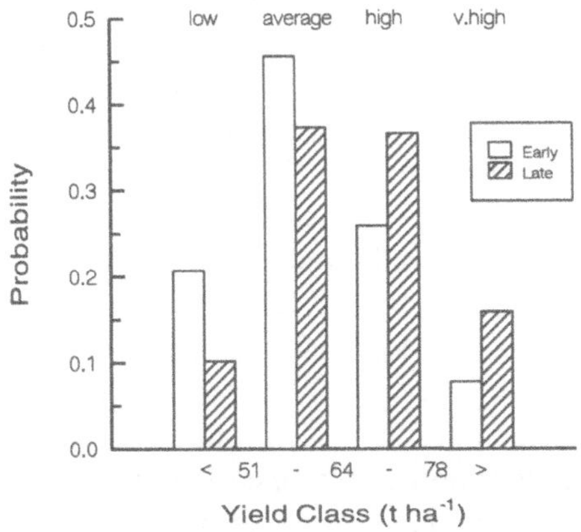

Figure 9. The effect of changing from an early to a late haulm destruction date on the probabilities of achieving various classes of total yield as estimated by the causal model of water-constrained yield of potato (Figure 8) trained for conditions prevailing in NE Scotland.

being achieved. This backwardly directed question could not be answered using the original mathematical model. Such models are restricted to pursuing pre-selected scenarios.

Applying a new belief to a node is not the same as applying a query to the rule-based model. There are some constraints on querying causal models when more than one marginal distribution or belief is to be changed. The relations in the rule-based models are unique. In the causal model one has a representation of the real world and its status after training i.e. a set of states (if you like) and the probabilities of being in each of the possible states. The model has associated probability structures that define the links between these states. A new belief e.g. that temperature is 'hot' with a probability of 0.8 and cold with '0.2', means that you believe the relative frequencies of these states has changed i.e. your belief in the world has changed. Then a second query is applied to this changed world. Thus in general the order of a set of queries is important. Only when two or more nodes are independent is it possible to update one's beliefs in these new nodes 'simultaneously'; the outcome is then independent of the order in which the new beliefs are entered. They are also reversible in this case – entering the old beliefs will restore the world to its original state. Thus nodes 'Rain-2', 'Rad-2' and 'Temp-2' can be updated simultaneously. In contrast nodes, 'Temp-2' and 'Date leaf 9' are not marginally independent and the outcome is dependent on the order in which new beliefs are entered and the system can not be restored to its original state.

Work at SCRI and elsewhere has demonstrated that there are considerable advantages to be gained in modelling by exploiting recent developments in computer science and statistics. Rule-based models allow both qualitative and quantitative knowledge to be combined, and enable several levels of precision, or abstraction, to be available within the one model. Causal models measure the strength of relations by probabilities and provide a simple means of incorporating uncertainty about the values of the model variables.

With rule-based systems the values of all the variables in the model are available at all times to the user. The addition of a stylised, natural language query interpreter enables them to answer a wide range of questions about the effect of one variable on another, in a manner accessible to a wide range of users.

Immediate and valuable progress is possible in at least two directions. Existing, simple, deterministic, mathematical models – type programmed to produce the value of a single variable only – can be improved dramatically either by incorporating the flexibility of the rule-based approach, or the uncertainty of the causal model. This work alone will increase the demand for, and use of, models – providing training, better risk assessment, ease of use and highlighting areas of ignorance which result in large uncertainties and require further research. However there is now also scope to improve modelling itself. Causal models would benefit by having the same natural language querying facility that is available for rule-based systems. The ultimate aim should be to integrate rule-based, causal and mathematical models in a single decision support environment, making the benefits of all three approaches available to the user.

The greatest challenge lies in the incorporation of new knowledge in the light of previous experiences. The standard approach, of which we are all very familiar, to interpreting the results of the latest experiment is to apply appropriate statistical analyses to the new data set while ignoring previous data. The only roles that previous knowledge play are in the design of new experiments and qualitative comparisons presented in publications that emanate from such experiments. We have the possibility to develop techniques that can incorporate *all* relevant experience both past and present in decision support systems. With the combination of advanced user interfacing (hypertext, natural language) and sophisticated modelling approaches (referred to earlier), it should be possible, when addressing new data sets (new knowledge), to ask what we already know about the system under study and identify the consistencies and apparent contradictions between past experience and current experiment. Apparent contradictions may lead to the detection of new, significant interactions. Irrespective of what is found, this approach means that new beliefs are based on past experience as well as latest findings and that a much broader base of knowledge can be brought to bear on a problem.

Acknowledgements

This work was supported by the Scottish Office Agriculture and Fisheries Department and contributed to by Dr. J. Smart, Dr. Y. Gu and Miss D.R. Peiris. The authors thank Mr. J. Fraser at the Artificial Intelligence Applications Institute (Edinburgh), Mr. A. Cairns at the University of Dundee and Professor J.B. Dent at the University of Edinburgh for their advice and guidance in the development of rule-based systems.

References

Andreassen S, Woldbye M, Falck B, Anderson S (1987) MUNIN – A causal probabilistic network for the interpretation of electromyographic. Pages 366–372 in Proceedings of the 10th International Joint Conference on Artificial Intelligence, Milan, Italy, Vol. 2, Morgan Kaufmann, Pub. Inc., San Mateo, California, USA.

Bouman B A M, Jansen M J W (1993) RIGAUS – Random Input Generator for the Analysis of Uncertainty in Simulation. Simulation Reports CABO-TT No. 34. AB-DLO, Wageningen, The Netherlands.

Cassella G, George I E (1992) Explaining the Gibbs sampler. American Statistician 46:167–174.

Cooper G F (1984) A Computer-Based Medical Diagnostic Aid that Integrates Causal and Probabilistic Kowledge. Ph.D. Thesis, Department of Medical Information Sciences, Stanford University, Stanford, USA.

Gu Y, Peiris D R, Crawford J W, McNicol J, Marshall B, Jefferies R A (1994) An application of belief networks to future crop production. In Proceedings of the 10th Conference on Artificial Intelligence for Applications, 1–4 March 1994, San Antonio, Texas, USA (in press).

Hendrix C G, Walter B A (1987) The intelligent assistant: technical considerations involved in designing Q&A's natural language interface. BYTE, December 1987.

Huber B, Nyrop J P, Wolf W, Reissig H, Angello A, Kovach J (1990) Development of a knowledge-based system supporting IPM decision making in apples. Computers and Electronics in Agriculture 4:303–313.

Lauritzen S L, Spiegelhalter D J (1988) Local computations with probabilities on graphical structures and their applications to expert systems. Journal of the Royal Statistical Society 50:157–224.

Marshall B (1986) Tuber size. Aspects of Applied Biology 13:393–396.

Pearl J (1986) Fusion, propagation, and structuring in belief networks. Artificial Intelligence 29:241–288.

Peiris D R, McNicol J W (1994) Modelling daily weather with multivariate time series. In Agriculture and Forest Meteorology, special issue based on Conference: Impact of Climate Change on Scottish Agriculture, Dundee 1993 (Accepted).

Selfridge M (1989) Towards a natural language-based causal model acquisition system. Applied Artificial Intelligence 3:191–212.

Sowa J F (1984) Conceptual Structures: Information processing in mind and machine. Addinson-Wesly, Reading, Massachusetts, 481 pp.

Smart J A C (1992) Model Abstraction: an AI Approach to Flexible Crop Modelling. Ph.D. Thesis, University of Dundee, Dundee, UK.

22. The use of models in potato breeding

D. ELLISSÈCHE[1] and J. HOOGENDOORN[2]

[1] *Institut National de la Recherche Agronomique (INRA), Ploudaniel, France*
[2] *DLO-Centre for Plant Breeding and Reproduction Research (CPRO-DLO), Wageningen, The Netherlands*

Abstract. How models can be of help for breeding activities is reviewed from literature and from some recent data. The examples as presented show that models can be used for evaluating breeding strategies and for analysing some complex genetic characteristics and identifying their most promising components for selection in breeding programmes. Modelling plant-pathogen interaction may give indications about the durability of various types of resistance. Limitations to the use of models in breeding programmes are discussed.

Introduction

Typically a potato breeding programme, from crossing parental lines to the final selection of a new cultivar, will take 10 to 15 growth cycles (Caligari, 1992). Such long-term investments of labour and facilities are only cost effective if the programme is carried out using efficient procedures. A model is a simplified representation of a limited part of reality that contains inter-related elements (Rabbinge and De Wit 1989). Simulation studies using such models are tools for forecasting and exploration, and may support decision making. In plant breeding many decisions have to be made. In a potato breeding programme parental lines for crossing have to be chosen, progenies have to be evaluated, superior genotypes need to be identified, and promising cultivars should be evaluated in the most appropriate areas. Models may help potato breeders by facilitating some of these activities, both by improving selection methodology, and by showing what positive or negative side-effects selection on one characteristic may have on other agronomically important traits.

In this paper, some possible uses of models in potato breeding are presented, mainly in the form of a critical review of the literature. The topics which are covered are tuber yield, and drought tolerance, in which the genetic variation for the interaction between crop and environmental parameters is very important. Furthermore, the use of models to optimize resistance breeding and resistance implementation strategies are discussed, where both the genetics and the physiology of the plant-pathogen interaction can be simulated to supply the breeder with improved selection strategies for parents and progeny.

A.J. Haverkort and D.K.L. MacKerron (eds.), Potato Ecology and Modelling of Crops under Conditions Limiting Growth, 341–356.

The case studies have been chosen because they represent complex traits which are economically important, but difficult to select for.

Tuber yield

Tuber yield is still the most important trait to be selected for, since there is an important need for sufficient food in the world and potato is one of the four main food crops. Further, yield remains the basis on which the grower obtains an economic return. Marketable yield increased over the last decades by technological improvement and by selection. However yield assessment takes up a lot of time and effort within a breeding programme. Therefore, more efficient selection would be welcomed, for example by means of indirect selection.

Tuber yield can be explained (Spitters 1987) in terms of cumulative light absorption (ΣPARa), light utilization efficiency or radiation use efficiency (RUE) and harvest index (HI). This was applied to analyze the variation in yield between cultivars (Spitters 1987; Van der Zaag and Doornbos 1987; Spitters et al. 1989). It is an interesting approach for the breeder to split a complex trait into more simple ones, provided genetic variation exists for these traits, and their assessment would be easier than that of tuber yield itself.

From observations in 8 cultivar trials which were conducted over 5 years at 4 locations and included 19 cultivars, Van der Zaag and Doornbos (1987) concluded that differences in tuber dry matter yield could be explained primarily by differences in ΣPARa when little or no stress occurred during the growing period. Under stress (drought or insufficient nitrogen dressing) differences in RUE were as important as differences in ΣPARa and, under high temperature conditions (Israel), all three components (ΣPARa, RUE, HI) contributed to the variation in yield.

Cultivar differences in ΣPARa were significant in 10 cultivars out of the 19. The cultivar × location interaction was not significant. The values of ΣPARa were clearly associated with the maturity classes of the cultivars, the later the maturity class, the higher the value.

Cultivar differences in RUE (not calculated using total dry matter yield but tuber yield) were significant in 8 cultivars out of 19. Again there was no significant cultivar location interaction. There was a trend for early cultivars to show higher RUE than those of later, although not late, cultivars. It was noticed that some widely grown cultivars such as Alpha, Désirée, Kennebec and Spunta had low RUE. The variations in RUE are more clearly explained by differences in the rate of photosynthesis than in foliage structure (Dwelle et al. 1981).

One of the two above-average yielding cultivars had a high ΣPARa and a low RUE while the other had both above-average ΣPARa and RUE. It was stated that in breeding programmes aiming at increasing yield, high ΣPARa should be combined with high RUE.

Cultivar differences in HI were important under high temperature conditions while, under cooler climatic conditions no genetic variation for this trait was found. The results of van der Zaag and Doornbos (1987) suggest that HI may be an important trait for adaptation to hot climates but it is unlikely that genotypic differences for this trait will be observed under the temperate climatic conditions in which most breeding companies are based at present. Screening for HI needs on-the-spot trials or glasshouse experiments.

Spitters et al. (1989), studied tuber fresh yield (Y) in an experiment covering two years and 5 cultivars differing in maturity classes, and in another one that included 16 cultivars, using the following equation:

$$Y = \Sigma PARa \times RUE \times HI/Dm,$$

where Dm stands for tuber dry matter concentration.

The variation in fresh tuber yield was mainly explained by variation in ΣPARa and in Dm. The differences found in ΣPARa were due to differences in foliage maturity. Late cultivars had a lower HI and a higher Dm. There was no relation between RUE and cultivar maturity.

In this paper it was also discussed how a breeder could easily assess these yield components in practice, since a complete growth analysis is not feasible for the great number of genotypes included in a breeding programme. It was suggested that ΣPARa could be conveniently measured by the percentage of ground covered by green foliage evaluated either visually or by viewing the canopy through a grid (Burstall and Harris 1983), such observations being made at regular time intervals. Then ΣPARa could be calculated from data on incident solar radiation obtained from a nearby weather station. More simply, genetic differences in ΣPARa could be characterized satisfactorily by the time till the ground cover by green foliage was reduced to 50%, which needs only a few observations during foliage senescence. This is in fact almost identical to scoring maturity. Radiation use efficiency is estimated by the total biomass production per unit of absorbed light. Total biomass may be estimated by adding the haulm dry weight at the haulm maximum to the tuber dry weight at final harvest. Tuber dry weight is commonly measured in breeding practice by underwater weighing. Haulm dry weight can be calculated from measured haulm fresh weight and a specific dry matter content coefficient (e.g. 0.09). The harvest index is estimated as tuber biomass divided by total biomass. It is suggested that it should be easier to estimate HI halfway the growing period at the time of haulm maximum, since the value of HI at that moment seems to be well correlated with its final value.

We carried out a similar experiment on the analysis of genetic differences in yield components in 1993 at INRA station of Ploudaniel (48° N latitude, 4°E longitude, 36 m altitude) according to usual potato crop technology. The cultivars studied were Fanette (8), Bintje (7), Claustar (7), Spunta (7), Désirée (5), Kennebec (4.5), Korrigane (4), Kaptah (2). The figures in brackets indicate the maturity of the cultivars, on a scale with 9 being very early and 1 being very late, as used in the French 'Bulletin des Variétés de Pomme de terre'" (Anonymous 1993). Periodic harvesting was done 15 times from 1 month after planting until almost complete death of the foliage. Seven plants per cultivar were separately lifted by hand at each sampling date. Leaf area indices (LAI) were calculated from leaf areas estimated by the disc method (Khurana and McLaren 1982) and determinations of dry weight. Dry weights of all plant parts were recorded after oven drying at 80 °C during 48 h. Percentages of ground covered by the foliage were determined with a 0.96 m × 0.70 m grid consisting of 100 equal rectangles. Daily incoming total radiation (Rg) was measured with a pyranometer at the agrometeorological station of Ploudaniel (Enerco 411-Cimel Electronique). Total dry weights (TDW) were related to PARa accumulated on a daily basis and to RUE according to the equation:

$$\mathrm{TDW} = \Sigma\mathrm{PARa} \times \mathrm{RUE}.$$

The daily estimation of PARa was made according to the model:

$$\mathrm{PARa} = \mathrm{Ei} \times \mathrm{PARi}.$$

PARi is the incoming photosynthetically active radiation, estimated as (Varlet-Grancher et al. 1982):

$$\mathrm{PARi} = 0.48 \times \mathrm{Rg}.$$

Ei is the interception efficiency calculated as (Varlet-Grancher and Bonhomme 1979):

$$\mathrm{Ei} = \mathrm{Emax} \times (1 - \exp(-\mathrm{K} \times \mathrm{LAI})).$$

Emax is close to 0.95 for most of the agricultural situations (Varlet-Grancher et al. 1989). We did not measure Ei, and we chose to use K = 0.8 (Bonhomme, pers. comm.) as the value for the extinction coefficient K. Daily LAI was computed assuming that it varied linearly between two successive sampling dates. A similar methodology was adopted by Damay and Le Gouis (1993) in experiments on sugar beet.

Our data showed no significant cultivar differences for RUE. The average value from all data was:

$$\mathrm{RUE} = 1.975 \pm 0.09\ \mathrm{g}\quad \mathrm{MJ}^{-1}.$$

When total dry matter (TDM) was plotted as a function of accumulated PARa, the equation of the common regression was:

$$y = 1.975x + 12.43;\ (r^2 = 0.94).$$

The values of final tuber fresh yield (Y), total accumulated PARa (ΣPARa), harvest index (HI), and tuber dry matter content (Dm) were calculated as average values from the data of the 3 last sampling dates, at which for all tested cultivars ground cover by green foliage was reduced to less than 50%. These average values were respectively:

$$\mathrm{Y} = 6002\ \mathrm{g}\quad \mathrm{m}^{-2} \qquad \Sigma\mathrm{PARa} = 641\ \mathrm{MJ}\quad \mathrm{m}^{-2}$$
$$\mathrm{HI} = 0.90 \qquad \mathrm{Dm} = 0.21.$$

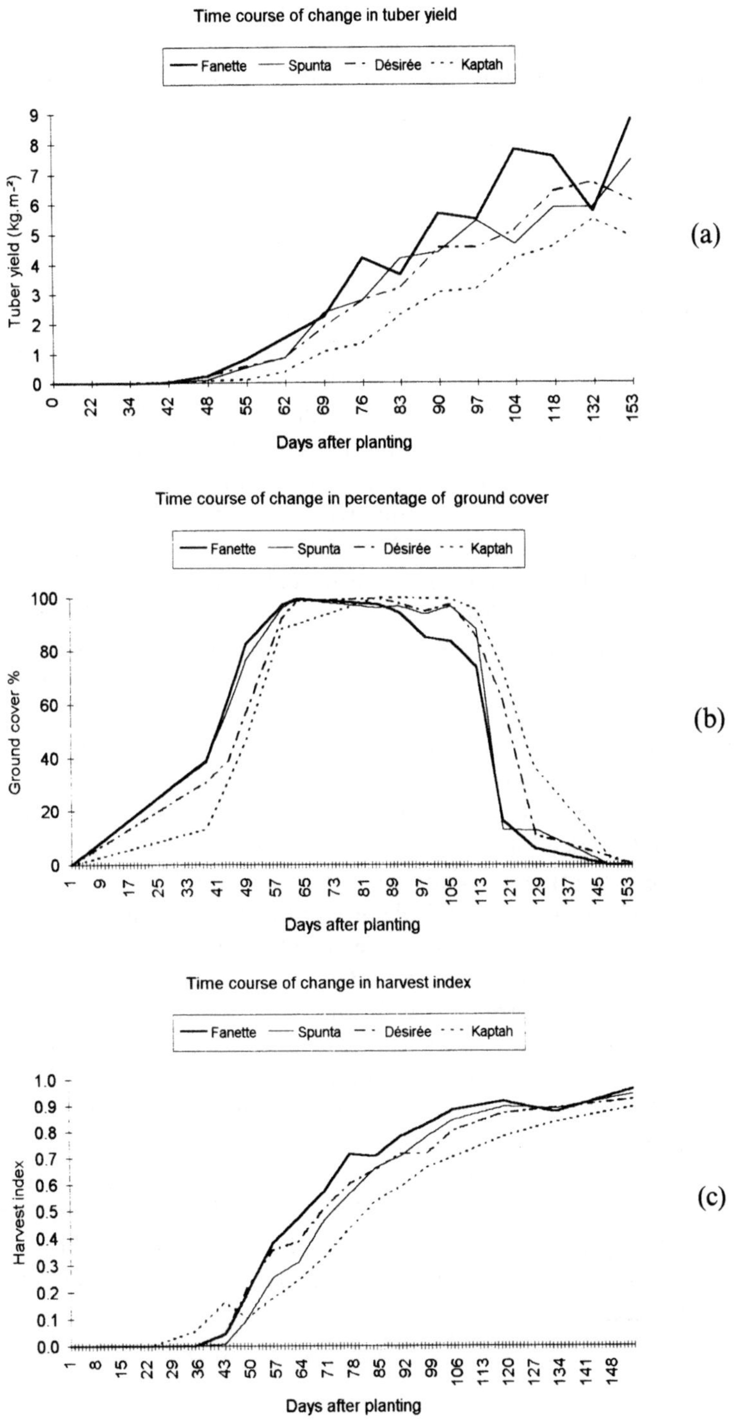

Figure 1. General trends in tuber yield (a), ground cover (b) and harvest index (c), as observed in Ploudaniel (1993) for 4 of the 8 potato cultivars tested.

No significant differences existed for these traits, except for Dm, the values of which ranged from 0.256 for Kaptah, a widely used starch cultivar, to 0.170 for Fanette, a cultivar which is well adaptated to early cropping (Figure 1).

This absence of cultivar differences for yield and yield components is not only in contradiction with the conclusions of the afore-mentioned authors, but it is also in contradiction with our usual observations. In this experiment the yields of the earlier cultivars such as Fanette and Spunta were normal to high while the yields were relatively low for the medium to late cultivars like Désirée and Kaptah. This may be due to strong winds which occured in September and shortened the growing period of the medium-late and late cultivars.

From this single experiment we cannot draw any conclusion about which of the yield components it would be sensible to select for in our conditions to improve yield. This may be due to the year effect and the experiment will be repeated in 1994. However the results suggest that yield components are probably as sensitive to environmental fluctuations as yield itself.

For selection, improved cumulative light absorbtion appears to be a promising trait when breeding for higher tuber yields. This should nevertheless not lead to selection for lateness but it should be wise also to select genotypes showing a quick establishment of their foliage especially when spring temperatures are rather low. As stated by van der Zaag and Burton (1978) it is not sensible to lengthen the growing period towards the end of it and it is unlikely that selection for plant architecture (i.e. improvement of extinction coefficient K) will be soon relevant. Selection for harvest index is complicated by the fact that there is a somewhat contradictory trend between early genotypes with generally higher HI and lower ΣPARa and the late ones with lower HI and higher ΣPARa. Selection for HI therefore should be carried out within maturity groups. The magnitude of genotypic differences in RUE is not clear (Jefferies 1993) and perhaps limited in C3 plants like the potato. This trait needs further investigation and more precise assessment which should be done in the framework of physiological studies and of genetic resources evaluation. Genetic differences in Dm are large, very well known by the breeders and easy to assess. Increasing yield by selection for this trait is however only feasible within the limits of the specific requirements of industry, processing and the fresh market.

The models we discussed underline the importance of improved ΣPARa and HI for tuber yield, often negatively correlated characteristics. The common practice of breeding medium early and medium cultivars can be seen as as a sensible compromise between optimal ΣPARa and HI. Nevertheless, simulation modelling for specific climatic conditions may indicate whether it is possible to further optimize yield by either ΣPARa or HI for genotypes targetted to the regions in question.

Drought tolerance

Differences exist in the varietal response of potato cultivars to water stress (Harris 1978; Levy 1983) but selection for drought tolerance is difficult because several plant attributes are involved in drought tolerance mechanisms and because water stress itself varies in time and in intensity. Crop growth models may be a convenient approach for investigating such a complex situation.

Spitters and Schapendonk (1990) studied the influence of various plant attributes on the drought tolerance of genotypes by means of a model in which the seasonal course of crop dry matter accumulation and soil moisture availability were simulated. It was applied to a medium-early and to a medium-late cultivar grown in temperate climatic conditions. In the model the main effects of water shortage on growth and tuber yield were a reduction in total dry matter increment per unit of intercepted light, a reduction in degree and duration of light interception due to less leaf area growth when the stress occurred early in the growing period, accelerated leaf senescence when the stress occurred later in the growing period, and a slight increase in the proportion of assimilates allocated to tuber growth when the stress occurred during tuber bulking. The simulated effects of transient drought were stronger in early cultivars by increasing their natural tendencies to produce less leaf area and to have a shorter growing period than later cultivars. Early cultivars escaped late water stress effects which affected late cultivars more drastically.

The influence of other plant attributes, i.e. water use efficiency, critical suction, leaf area, wilting point and rooting pattern, was evaluated in the case of a sustained water-stress period. The model indicated that selection for deeper rooting would contribute to drought tolerance. However, the evaluation of this characteristic is very laborious, even with rhizoscopy. Genetic variation exists for root systems but it is likely that this is related to earliness since late genotypes develop more extensive root systems (MacKerron and Peng 1989).

In the model water use efficiency – the dry matter production per unit of transpired water – and yield under stress would be increased if the rate of transpiration were lowered without affecting the potential rate of growth. Both traits are, in fact, modified in the same direction so that selection for low water consumption should lead to smaller relative yield reduction in dry conditions but also to smaller absolute yields under optimal water supply. Increasing water use efficiency at constant transpiration rate would give the highest yields under sufficient water supply or moderate stress. Combining an increased water use efficiency with a lower transpiration rate would result in higher yields under severe drought, according to the model. Genetic variation for water use efficiency has been demonstrated and it can be identified using carbon isotope discrimination (Vos and Groenwold 1989). However such selection procedures based on that technique are debateable, since the relation between increased water use efficiency and drought tolerance is not proven (Hoogendoorn and Groot 1993). Critical soil moisture tension is associated with time of stomata closure under water stress. Photosynthesis requires the opening of stomata while drought stress results in stomata closure. Therefore, according to the model, early stomatal closure associated with a lesser critical tension would save water for later use, at the expense of current growth, whereas retarded stomatal closure would allow the maintenance of growth rates at the expense of the soil moisture reserves, and at the expense of later growth. Cultivar differences in time of stomatal closure have been demonstrated (Levy 1983) when plants were submitted to water stress and high temperatures. The trait is difficult to measure and whether one needs to select for higher or lower critical suction obviously depends very much of the type of drought to be expected in the target areas of the new cultivars. As differences between genotypes in time and degree of wilting of plants have also been demonstrated (Beekman and Bouma 1986), the wilting point was delayed in the simulation until greater soil moisture tension, which resulted in an increase of yield under stress.

The model studies showed that increasing the rate of leaf area expansion with unchanged dry matter distribution was of benefit to yield under optimal conditions, but under drought led to enhanced transpiration, more rapid depletion of soil moisture and therefore to a reduced average light use efficiency, a lower harvest index, and no increase in yield. Foliage activity may also be altered by leaf yellowing due to water stress and by the rate of recovery of plants after water stress. Cultivar differences have been shown for both traits (Beekman and Bouma 1986; Schapendonk et al. 1989). In the model, a reduction of the effect of drought on leaf senescence diminished the yield reduction and this may be a promising characteristic for breeding for drought tolerance.

Jefferies (1993) used a simulation model to assess the possible strategies for drought tolerance in potato. Part of his conclusions were relevant to breeding. He proposed that strategies for drought tolerance would be classified into three groups: a) improving available water supply either by increasing the rooting depth or by improving the water use efficiency, b) improving leaf growth by increasing the potential rate of extension of individual leaves or by improving the relation between fractional leaf expansion rate and the fraction of plant available water remaining in the soil profile (F), c) improving photosynthetic performance by improving the relation between the coefficient for the conversion of intercepted radiation into dry matter (C) and the soil moisture status or by improving the slope of the relation between C and F. He concluded that it was only under severe drought cycles that improving rooting depth did slightly improve yields. The effects of improving C diminished when the severity of drought increased. The best strategy was to improve the relation between leaf expansion rate and soil moisture status as it was efficient under the whole range of drought patterns simulated.

Although it is difficult to compare the conclusions of Spitters and Schapendonk and of Jefferies because of differences in methodology and in the choice of the plant attributes they investigated, both papers suggest that sustained leaf growth under stress is important.

Up until now no single trait has been identified conferring drought tolerance. Therefore, breeders have to carry out field trials corresponding to diverse drought patterns. Models have suggested the importance of earliness, and sustained leaf growth rate under drought. In contrast, water use efficiency does not seem to be an important trait.

Resistance to *Phytophthora infestans*

Probably the most devastating disease of the potato crop is late blight, caused by the fungus *Phytophthora infestans*, which without control can result in total crop loss in temperate climates. Leaves are infected by conidia or zoospores, which after successful penetration give rise to fast growing sporulating lesions which kill the leaves quickly. The spores of the next generation infect other leaves of the same or neighbouring plants. The disease can also kill stems and destroy tubers.

From around 1930 cultivars were introduced with resistance against the disease from Mexican wild *Solanum* species, e.g. *S. demissum* (Colon et al. in 1995a). This resistance is determined by single resistance genes, many of which gave full protection. However, it rapidly became clear that the fungus, when challenged with such resistance genes, very quickly shifted to virulence, which rendered the single dominant resistance genes from *S. demissum* useless for breeding programmes (Ross 1986). It has been claimed since then that polygenic partial resistance will be of much greater value (Vanderplanck 1971). However, such resistance is difficult to handle in breeding programmes, because it is much more environment-sensitive and because different genes may control different steps in the plant-pathogen interaction, which are likely to differ in their final contribution to the disease development.

Simulation studies were carried out by Bruhn and Fry (1981), who studied the effects of environment, fungicide and host, and by van Oijen (1991; 1992a,b), who modeled the interaction between plant and pathogen in an attempt to identify the major characteristics of potato cultivars that affect yield loss due to late blight.

Neither research group took into account genetic variability within the pathogen population, and treated the potato genotypes studied as either susceptible or partially resistant to the complete pathogen population. Although different terminology is used, both studies have modeled similar components of the infection process. For the remainder of this discussion, the terminology of van Oijen (1991) is used: infection efficiency (IE), latent period (LP), lesion growth rate (LG), sporulation intensity (SI), infectious period (IP).

Bruhn and Fry (1981) included susceptible and moderately resistant cultivars in their models. The level of resistance was not high enough to obviate protective spraying, but did increase the optimal interval between sprays with 50% for the moderately resistant cultivar Kennebec in comparison with that for the susceptible cultivar Monona. The results of the simulations showed that even relatively minor differences in susceptibility can have significant effects on the needs for protective spraying, but Bruhn and Fry did not explain which phases in the plant-pathogen interaction are slowed down in the moderately resistant genotypes.

Van Oijen (1991, Chapter 2) showed that the effect of *P. infestans* on tuber yield is likely to be only the result of the reduction in green leaf area, and not of an additional effect on the photosynthesis of (as yet) uninfected, green leaf tissue. This is in agreement with many other foliar diseases on different crops (Waggoner and Berger 1987), although some exceptions have been reported, such as leaf blast in rice (Bastiaans 1991). Van Oijen (1991, Chapter 4) reviewed the literature for the observed genetic variation for the various components of the plant-pathogen interaction, and identified some 40 publications on the subject. In general it appears that the variation for components is larger in wild *Solanum* species than within *S. tuberosum*. Van Oijen used the data on the genetic variation for the components for a sensitivity analysis of the effect of the most resistant value on the disease progress rate (van Oijen 1992b). He identified LG and IE as the components offering the best perspectives for resistance breeding, while the genetic variation for latent period (LP) and infectious period (IP) is not large enough to justify extensive efforts at selection. These conclusions have been confirmed for LG by Colon et al. (1995a), who showed that the differences in partial resistance between R-gene free cultivars as observed in the field were closely correlated to genetic differences in LG between these cultivars.

The above mentioned simulation studies have shown breeders that even low levels of partial resistance may be worth selecting for because, if combined with a proper disease development forecasting system, it may reduce the number of sprays needed to prevent economic damage significantly. It appears to be most effective to select for reduced LG and/or IE, since these are two important components of the infection process, and there appears to be enough genetic variation within the potato genepool to achieve high levels of protection. Both LG and IE are fairly easy to determine in the field or in the glasshouse, and therefore it might be possible to incorporate these selection criteria in a breeding programme. Furthermore, genetical studies of segregating populations derived from crosses between wild species and the cultivated potato *Solanum tuberosum* have indicated a high level of correlation between these two characteristics (Colon et al. 1995b), suggesting that by selecting for reduced LG an indirect selection for reduced IE will be carried out, or *vice versa*, making life easier for the breeder. Whether this is due to tight linkage between genes, or pleiotropic effects of the same genes, remains to be investigated.

Resistance to the potato cyst nematodes

Other important pests of the potato crop, in particular in narrow rotations, are the potato cyst nematodes (*Globodera rostochiensis* and *G. pallida*). Throughout the world, but in particular in Europe and South America, resistance to these nematodes is a breeding objective, in order to prevent yield reduction and/or to be able to reduce soil fumigation. Breeding for resistance is made complicated because high levels of resistance are only found in wild *Solanum* species, which means that an extensive and time-consuming backcrossing programme is needed to introduce resistance into agronomically acceptable germplasm. Furthermore, a *Globodera* population in the soil is often a mixture of genotypes and/or species, which complicates selection procedures to achieve a high level of durable resistance. The nematodes, however, differ from the *Phytophthora infestans* fungus, in that there is only a limited multiplication per potato crop, only one cycle each year which is equal to a multiplication rate of aproximately 25 (Bakker et al. 1993), compared to 10^{15} for the fungus in a susceptible potato crop in the Netherlands (Colon, pers. comm.). Furthermore, the major resistance genes found against the nematodes appear only to be active against avirulent females, but not against avirulent males (Janssen et al. 1992).

Simulation studies were carried out by Spitters and Ward (1988) and Schouten (1993; 1994a,b) to evaluate strategies for breeding and cultivar implementation, in which various types of resistance were evaluated for durability. In contrast to the simulation studies described above for resistance breeding to *P. infestans*, the models discussed here are population models rather than crop growth models, in which the genetics of virulence and avirulence within the nematode population are evaluated when challenged with different types of resistant cultivars.

Spitters and Ward (1988) described a basic model for population development of the nematode, in combination with the effects on nematode density and on yield reduction, using fixed rates for multiplication and damage due to nematode invasion (intolerance). Schouten (1993; 1994a,b) made use of very similar models, but has focussed on virulence frequencies within the nematode population. Both Spitters and Ward and Schouten started their models with some virulent genotypes already being present in the population at very low frequencies, and the simulated breakdown of the resistance is, therefore, the result of a changing genetic composition of the nematode population due to selection and recombination, but not due to mutation. It is generally thought that in the major potato growing areas of Europe, where potatoes are grown in a three year rotation or longer, the multiplication rate of the nematodes is too low to generate enough mutated individuals to break resistance. This is, for example, illustrated by the fact that in the UK no new pathotypes of *G. rostochiensis* have been found with virulence to the H1 gene, even although cultivars with this resistance gene are widely grown (Bakker et al. 1993).

Spitters and Ward (1988) demonstrated that in the case of the cyst nematodes, complete resistance based on a single gene would offer a good protection against damage and virulence increase during the first potato cycles, but that it would break down afterwards. Non-specific partial resistance at low levels offers very little protection, while very high levels of partial resistance could offer long lasting protection. The higher the resistance level is, the more likely it is, however, that the resistance is based on major genes and may be pathotype specific. When different pathotypes are present in the nematode population alternating cultivars with different resistance genes gives a less durable protection than growing a cultivar that combines both genes. Schouten (1994b) demonstrated clearly that the durability of resistance genes is likely to be much more dependent on the intitial frequency of virulent genotypes in the nematode population than on the number or the partiality of the resistance genes involved. He also described how, due to the fact that the selection pressure from resistance genes such as H1 is towards virulent females and avirulent males, the frequency of virulent individuals in the nematode population reaches an equilibrium, which would manifest itself as a durable partial resistance of the host. However, the level of this durable resistance is too low to be of agronomic interest (Schouten 1993; 1994a). For breeding purposes it can be deduced that resistance genes against which little or no virulence is present in the nematode population, are most likely to give a long-lasting protection. However, since some cysts are usually found even on resistant plants, it needs to be established whether such females are virulent, or whether the resistance of the potato genotype is of a high but nevertheless partial nature.

Table 1. Nematode progeny test to separate partial from complete resistance. A situation is described when similar levels of multiplication on the resistant plant genotype of 0.20 are observed after a first test (generation), which are either due to complete monogenic resistance in the plant and virulence in the nematode population, or a high level of partial resistance in the plant and no virulence present in the nematode population. The nematodes 'harvested' are subsequently used for the test in the next generation.
Assumptions as in Spitters and Ward (1988): 67% of eggs hatch, 50% of larvae penetrate the roots, 50% develop into females and 150 eggs/newly formed cyst. A/a – symbol for avirulence genes of the nematode, PP – symbol for complete avirulence of the genotype of the nematode. The A/a system is effective against females but not against male nematodes

Description of the progeny tests	Description of the plant-pathogen interaction	
	Complete resistance & 99.2% avirulent (A.) + 0.8% virulent nematodes (aa)	99.2% partial resistance & 100% avirulent (PP) + no virulent nematodes
1st Generation		
(on resistant genotype)	aa × Aa => 50%Aa & 50%aa	PP × PP => 100%PP
multiplication rate:	0.20	0.20
2nd Generation		
(on susceptible genotype)	Aa × Aa => 75%A. & 25%aa	PP × PP]> 100%PP
	Aa × aa => 50%Aa & 50%aa	
	aa × aa => 100%aa	
multiplication rate:	25 (aa nematodes > 25%)	25 (no virulent nematodes)
3rd generation		
(on resistant genotype)		
multiplication rate:	> 12.5	0.20

This requires progeny testing of the nematodes by rearing the eggs from the few cysts formed in one generation on a susceptible cultivar, followed by a next generation on the resistant potato genotype (Table 1). In the first case the level of virulence will be markedly increased (to at least 25% of the population), in the latter case the virulence of the population will be similar to that in the initial experiments. Such high levels of partial resistance will be preferable. Breeding for durable resistance to the nematodes therefore requires both studies of the genetics of the resistance in the plants, as well as virulence frequency studies of the nematode populations. This type of research however, is likely to be too complicated and costly to be carried out by individual breeding companies.

The results of the simulation studies indicate further that with careful cultivar implementation, by alternating cultivars with different resistance genes, the durability of both genes may be prolonged, although not as effective as when a double resistant cultivar would be grown. However it is easier to breed a cultivar with a single resistance gene than with two or more genes. Molecular marker technologies, implementing RFLPs, RAPDs, AFLPs etc, may, in the future, offer scope for cost-effective, indirect selection for combinations of resistance genes.

Discussion

The case studies presented have shown that simulation modelling can contribute to the efficiency of potato breeding programmes, because modelling analyses complex characteristics, indicates the most promising components for selection, can forecast plant growth under various conditions, including biotic and abiotic stress, and helps, therefore, to formulate breeding strategies.

For breeding purposes new model studies would be welcomed for:

- Adaptation to various combinations of temperature and daylength:
  If by knowing a few characteristics, such as daylength sensitivity, of a putative cultivar it would be possible to forecast its potential for certain climates, the number of on-the-spot trials with this newly selected genotype could be more efficient. Correspondingly, for each new environment the number of genotypes to be screened could be reduced.
- Correspondance between single plants and plots for yield:
  At the early stages of a breeding programme observations are made on single plants only. It is difficult to relate these observations to later ones in replicated trials, due to increased experimental error and intra-progeny competion in the early generations. Modelling the relationships between generations may help to make the observations in the early generations more reliable as indicators of subsequent performance.
- Combination of resistance against and tolerance for *Globodera* spp.:
  Potato genotypes do not only differ in resistance for potato cyst nematodes, but also for tolerance (Arntzen et al. 1994). Spitters and Ward (1988) have already developed a model including both resistance and damage, and the genetic differences observed for tolerance could be added to this model. Since, however, one of the components of tolerance appears to be slower

hatching (Arntzen et al. 1993), an interaction between tolerance and resistance could be expected, which would need to be modeled as well.

- Resistance to other nematodes:
  In contrast to the potato cyst nematodes, other nematodes, e.g. the root knot nematodes, are hermaphrodites, and have more than one cycle a year. Therefore, other resistance strategies than those applicable to the potato cyst nematode may be required.
- Predicting the efficiency and durability of engineered resistance:
  Transformation and regeneration are expensive techniques, while also public acceptance of transgenic plants is not yet very high. If enough information is available about the mechanism of the engineered resistance, model studies may help to compare the effectiveness and durability with that of resistance which could be obtained by more classical breeding methods.
- Optimizing population size and selection efficiency in breeding programmes:
  In the course of a breeding programme many undesirable genotypes are kept on for years, while probably also promising genotypes are discarded. By modelling population size, selection efficiency, number of replicates and variances within segrating populations and within single clones, it may be possible to find the optimal numbers for the breeders to minimize the number of mistaken decisions either to discard or to select certain genotypes.

Developing and validating models takes time and investment, requires large amounts of data and may not be cost-effective if it is only done for breeding purposes. However, most of the modelling work discussed and suggested above is also relevant for other purposes, such as advisory systems, plant protection strategies and fundamental research. Therefore, modelling can be considered to be of general interest, and should be a collaborative effort of both governmental and industrial research and development organizations.

Acknowledgements

The authors thank Mr. M. Bozec for is valuable assistance, Mrs. L.T. Colon, Mr. R. Bonhomme, Mr. M. van Oijen and Mr. H.J. Schouten for their helpful comments and discussions.

References

Anonymous (1993) Bulletin des Variétés de Pomme de Terre. GEVES, Palaniandy, Paris, France.

Arntzen F K, Visser J H M, Hoogendoorn J (1993) Hatching of *Globodera pallida* juveniles by diffusate of potato genotypes, differing in tolerance to *G. pallida*. Annals of Applied Biology 123:83–91.

Arntzen F K, Visser J H M, Wouters D, Hoogendoorn J (1994) Inheritance of tolerance of *Globodera pallida* and the relationship between tolerance of and resistance to *G. pallida* in potato genotypes. Potato Research 37:65–76.

Bakker J, Folkertsma R T, Rouppe van der Voort J N A M, De Boer J M, Gommers F J (1993) Changing concepts and molecular approaches in the management of virulence genes in potato cyst nematodes. Annual Review of Phytopathology 31:169–190.

Bastiaans L (1991) Ratio between virtual and visual lesion size as a measure to describe reduction in leaf photosynthesis of rice due to leaf blast. Phytopathology 80:611–615.

Beekman A G B, Bouma W F (1986) A possible screening technique for drought tolerance in potato. Pages 55–65 in Beekman A G B, Louwes K M, Dellaert L M W, Neele A E F (Eds.) Potato Research of Tomorrow. Pudoc, Wageningen, The Netherlands.

Bruhn J A, Fry W E (1981) Analysis of potato late blight epidemiology by simulation modelling. Phytopathology 71:612–616.

Burstall L, Harris P M (1983) The estimation of percentage light interception from leaf area index and percentage ground cover in potatoes. Journal of Agricultural Science 100:241–244.

Caligari P D S (1992) Breeding new varieties. Pages 334–372 in Harris P M (Ed.) The Potato Crop. The Scientific Basis for Improvement. Chapman & Hall, London, UK.

Colon L T, Turkensteen L J, Prummel W, Budding D J, Hoogendoorn J (1995a) Durable resistance to late blight (*Phytophthora infestans*) in old potato varieties. European Journal of Plant Pathology (in press).

Colon L T, Budding D J, Keizer L C M, Pieters M M J (1995b) Components of resistance to late blight (*Phytophthora infestans*) in wild tuber-bearing South American *Solanum* species. European Journal of Plant Pathology (in press).

Damay N, Le Gouis J L (1993) Radiation use efficiency of sugar beet in Northern France. European Journal of Agronomy 2:179–184.

Dwelle R B, Kleinkopf G E, Steinhorst R K, Pavek J J, Hurley P J (1981) The influence of physiological processes on tuber yield of potato clones (*Solanum tuberosum* L.): Stomatal diffusive resistance, gross photosynthetic rate, leaf canopy, tissue nutrient levels, and tuber enzyme activities. Potato Research 24:33–47.

Harris P M (1978) Water. Pages 244–277 in Harris P M (Ed.) The Potato Crop. Chapman and Hall, London, UK.

Hoogendoorn J, Groot P J (1993) Genetic differences for drought tolerance. Pages 5–6 in Rousselle-Bourgeois F (Ed.) Abstracts of the 12th Triennial Conference of the EAPR. INRA, France.

Janssen R, Bakker J, Gommers F J (1992) The effect of heterozygosity for virulence on the development of the potato cyst nematode, *Globodera rostochiensis*. Fundamental & Applied Nematology 15:463–466.

Jefferies R A (1993) Use of a simulation model to assess possible strategies of drought tolerance in potato (*Solanum tuberosum* L.). Agricultural Systems 41:93–104.

Khurana S C, McLaren J S (1982) The influence of leaf area, light interception and seasonal potato growth and yield. Potato Research 25:329–342.

Levy D (1983) Varietal differences in the response of potatoes to repeated short periods of water stress in hot climates. A. Turgor maintenance and stomatal behaviour. Potato Research 26:303–313.

MacKerron D K L, Peng Z Y (1989) Genotypic comparisons of potato root growth and yield in response to drought. Pages 199–206 in Aspects of Applied Biology 22, Roots and the Soil Environment. Association of Applied Biologists, Wellesbourne, UK.

Rabbinge R, De Wit C T (1989) Systems, models & simulation. Pages 3–15 in Rabbinge R, Ward S A, Van Laar H H (Eds.) Simulation and systems management in crop protection. Pudoc, Wageningen, The Netherlands.

Ross H (1986) Potato Breeding – Problems and Perspectives. Paul Parey Verlag, Hamburg, Germany.

Schapendonk A C M, Spitters C J T, Groot P J (1989) Effects of water stress on photosynthesis and chlorophyll fluorescense of five potato cultivars. Potato Research 32:17–32.

Schouten H J (1993) Models of incomplete selection for virulence of potato cyst nematodes caused by sex determination that depends on host resistance. Netherlands Journal of Plant Pathology 99, Supplement 3:191–200.

Schouten H J (1994a) Preservation of avirulence genes of potato cyst nematodes through environmental sex determination. A model involving monogenic complete resistance. Phytopathology 84:771–773.

Schouten H J (1994b) Don't shut the doors on avirulent nematodes. Prophyta 46 (2):26–27.

Spitters C J T (1987) An analysis of variation in yield among potato cultivars in terms of light absorption, light utilization and dry matter partitioning. Acta Horticultura 214:71–84.

Spitters C J T, Ward S A (1988) Evaluation of breeding strategies for resistance to potato cyst nematodes using a population dynamic model. Euphytica S:87–98.

Spitters C J T, Schapendonk A H C M (1990) Evaluation of breeding strategies for drought tolerance in potato by means of crop growth simulation. Plant and Soil 123:193–203.

Spitters C J T, Neele A E F, Schapendonk A H C M (1989) Crop physiological components of tuber yield and their use in potato breeding. Pages 51–60 in Parental Line Breeding and Selection in Potato Breeding. Proceedings of the Joint Conference EAPR Breeding Section and Eucarpia Potato Section, Wageningen, The Netherlands. Pudoc, Wageningen, The Netherlands.

Vanderplanck J E (1971) Stability of resistance to *Phytophthora infestans* in cultivars without R-genes. Potato Research 14:236–270.

Van der Zaag D E, Burton W G (1978) Potential yield of the potato crop and its limitations. Pages 7–23 in Survey Papers 7th Triennal Conference of the European Association of Potato Research, Warsaw, Poland.

Van der Zaag D E, Doornbos J H (1987) An attempt to explain differences in yielding ability of potato cultivars based on differences in cumulative light interception, utilization efficiency of foliage and harvest index. Potato Research 30:551–568.

Van Oijen M (1991) Identification of the Major Characteristics of Potato Cultivars which Affect Yield Loss Caused by Late Blight. Ph.D. Thesis, Wageningen Agricultural University, Wageningen, The Netherlands.

Van Oijen M (1992a) Evaluation of breeding strategies for resistance and tolerance to late blight in potato by means of simulation modelling. Netherlands Journal of Plant Pathology 98:3–11.

Van Oijen M (1992b) Selection and use of a mathematical model to evaluate components of resistance to *Phytopthora infestans* in potato. Netherlands Journal of Plant Pathology 98:192–202.

Varlet-Grancher C, Bonhomme R (1979) Application aux couverts végétaux des lois de rayonnement en milieu diffusant. II. Interception de l'énergie solaire par une culture. Annales Agronomiques 30:1–26.

Varlet-Grancher C, Chartier M, Gosse G, Bonhomme R (1982) Rayonnement utile pour la photosynthèse des végétaux en conditions naturelles: caractérisation et variations. Acta Oecologica Plantarum 16:189–202.

Varlet-Grancher C, Gosse G, Chartier M, Sinoquet H, Bonhomme R, Allirand J-M (1989) Mise au point: rayonnement solaire absorbé ou intercepté par un couvert végétal. Agronomie 9:419–439.

Vos J, Groenwold J (1989) Genetic differences in water-use efficiency, stomatal conductance and carbon isotope fractionation in potato. Potato Research 32:113–121.

Waggoner P E, Berger R D (1987) Defoliation, disease, and growth. Phytopathology 77:389–398.

23. Agro-ecological zonation for potato production

H. VAN KEULEN and W. STOL
DLO-Research Institute for Agrobiology and Soil Fertility (AB-DLO)
P.O. Box 14, 6700 AA Wageningen, The Netherlands

Abstract. To target research to production problems in those areas of Latin America, Africa and Asia where cultivation of potato is most promising, CIP and collaborating NARC's need to know what tuber yields can be expected under irrigated and rainfed conditions in different agro-ecological zones. In this study we developed an agro-ecological zonation for potato production with a crop simulation model using local environmental data. Potential growth duration, potential yield and water-limited yield of a short-to-medium duration potato variety were estimated, as determined by environmental conditions. Potential tuber dry matter production ranges from 2 to 25 t ha^{-1}, that under rainfed conditions from 0 to 20 t ha^{-1}. These results, tested against field data and exposed to expert knowledge, appeared to be acceptable and consistent.

The results, though still of a preliminary nature, in their degree of detail and in the number of interactions considered, are useful in supporting the setting of priorities in research on crop production and crop protection, and on sustainable use of resources and inputs.

Introduction

In order to estimate the scope for increases in potato production at global scale, as a basis for research planning, an assessment is required of the possible geographical extent of the potato crop, and its yield potential under both irrigated and rainfed conditions. Extensive yield trials have provided a good insight to these yield potentials, particularly for traditional potato growing areas (Horton 1988). For non-traditional zones, such information is scarce. Yet, the crop is rapidly expanding into new zones (CIP 1991; Van der Zaag and Horton 1983). To target research to production problems in those areas of Latin America, Africa and Asia where the prospects for potato production are most promising, knowledge is required of the crop's potential under irrigated and rainfed conditions in various agro-ecological zones. To reduce the long, laborious and expensive experimentation and increase the efficiency of scarce resources, especially in many developing countries, a study was conducted, to explore the possibilities for potato production, on the basis of an agro-ecological characterization, applying a crop growth model, using local environmental data (Stol et al. 1991).

More specifically, the objectives of this study, initiated at the request of the

A.J. Haverkort and D.K.L. MacKerron (eds.), Potato Ecology and Modelling of Crops under Conditions Limiting Growth, 357–372.

International Potato Center (CIP), were: to establish the spatial distribution of potential growth duration, potential yield and water-limited yield of potato crops as accurately as possible. Data quality control and validation of the model was part of the exercise.

In this paper, the method used is briefly described, and some of the results are presented.

Data sources

Weather

The weather data used in this study were obtained from the Müller database, a global set of long-term monthly average values of weather variables from 978 meteorological stations (Müller 1987; 1982). Generally these covered the period 1931–1960, although for some stations available time series were shorter. The following characteristics were used: average temperature, maximum temperature, minimum temperature, precipitation, number of days with more than 0.1 mm precipitation, sunshine duration and average global radiation. To estimate daily total global radiation from sunshine duration, the Ångstrøm formula (Black et al. 1954), requiring two regression parameters, was used. The values of these parameters were estimated using Köppen's climate classification (Bartholomew et al. 1988), in combination with values reported by FAO (Frère and Popov 1979).

Before application, data were checked for consistency and quality (Stol et al. 1991).

Identification of potential growing seasons and application of the crop growth model requires daily values of weather variables. These were derived from the mean monthly values by assigning these values to day numbers at the middle of the months and subsequent linear interpolation. Daily rainfall, however, was determined via a random distribution of the long-term average monthly rainfall over the long-term average monthly number of rainy days.

Soils

The FAO/UNESCO soil map of the world (FAO/UNESCO 1974–1981), scale 1:5,000,000, is the basic document for the classification of the world's soil resources. As a digitized version of this map was not available, a world soil data base derived from the FAO/UNESCO soil map was used (Zobler 1986). Each record covers a cell of 1° latitude by 1° longitude comprising more than 50% of land. When a cell is occupied by more than one map unit, it has been characterized by the unit covering the largest area. For the dominant soil unit, slope class, phase and texture class (referring to the upper 30 cm of the profile) are given. In this study only the texture class was used to represent a cell.

Three main classes have been distinguished: coarse, medium and fine, which leads to seven classes of mineral soils, and peat is added as a separate class. A characterization of the eight texture classes, in terms of clay, silt and sand content and the area of each class as proportion of the total map area is given in Table 1 (Zobler 1986).

Table 1. Characterization and proportion of map area of texture classes (Zobler 1986)

	coarse	coarse – medium	medium	medium – fine	fine	coarse – medium – fine	coarse – fine	peat
% clay	9	20	30	48	67	35	38	–
% silt	8	20	33	25	17	19	12	–
% sand	83	60	37	27	17	46	50	–
% map area	24	7	42	6	12	< 1	< 1	1

The water holding capacity of the soils (Table 2) was derived from their texture class, on the basis of an estimate of the water holding capacity of soils in the European Communities (Reinds et al. 1991), also based on the classification according to the FAO/UNESCO system.

Table 2. Volumetric water content (cm/m) at field capacity and wilting point, and water holding capacity (cm/m)

	coarse	coarse – medium	medium	medium – fine	fine	coarse – medium – fine	coarse – fine	peat
field capacity	13	24	32	46	54	24	24	52
wilting point	4	10	10	30	44	10	10	13
water holding capacity	9	14	22	16	10	14	14	39

Spatial distribution of weather and soils

The basis for the spatial component of the study was a terrestrial grid with a resolution of 0.5° longitude and latitude, obtained from IIASA (Leemans and Cramer 1991). Each grid cell was allocated to the nearest available weather station, by means of tesselation resulting in a map of climate zones, represented by the centrally located weather station. For each climate zone, the number and lengths of suitable cropping periods were calculated from the temperature data (Map 1).

For the calculation of potential and water-limited yields with the crop growth

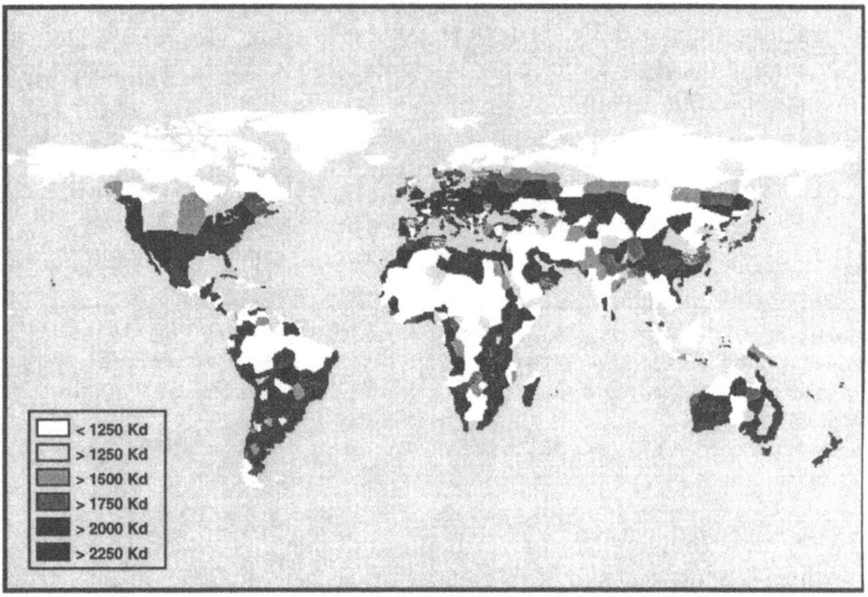

Map 1. Agro-ecological zonation for potato production, indicating lengths of growing seasons in Kd. Grey dots indicate climate zones with more than one potential growing period.

model, a more complete set of monthly climate data is needed: in addition to temperature also sunshine duration or global radiation, precipitation and number of rainy days are required. This combination of data was available for a smaller number of weather stations than the temperature data. On the basis of these stations a new climate zonation was established, forming the basis for further model calculations and presentation of results.

For the calculation of water-limited yields, climate data had to be combined with soil data, available for a 1° longitude by 1° latitude grid. By means of an overlay, each terrestrial 0.5 × 0.5 grid cell within each climate zone was allocated to the corresponding soil grid cell. In this way, the relative area of each soil type within a climate zone was determined. For each soil type in a climate zone, water-limited yield was calculated separately, and subsequently the area-weighted average yield within a zone was determined and indicated on the map.

The calculation procedure

Identification of potential growing seasons

The climate zones suitable for potato growth and the number and lengths of potential growing seasons within each zone were derived from the temperature data, on the basis of the following criteria:

- daily minimum temperature above 278 K,

- daily maximum temperature below 301 K,
- minimum thermal time requirement of 1250 Kd (base temp. 275 K),
- maximum thermal time requirement of 2000 Kd (base temp. 275 K).

The minimum requirement of 1250 Kd matches the thermal time of a short growing season for the production of seed-potatoes in temperate regions or that of an early maturing variety (60-days CIP-genotype) in tropical regions.

If temperature requirements are never met or only during a period that is too short to meet the minimum thermal time requirement, the climate zone is classified not suitable for potato production. If temperature conditions are suitable during part of the year, the first day that is suitable is identified, and a planning procedure for 365 days is started. Planting date for the first growing season is set at the first suitable day, if daily minimum temperature was the constraint, or two weeks earlier, if maximum temperature was the constraint. The latter measure is intended to accelerate breaking of dormancy in physiologically young seed potatoes. The growing season is terminated, when the accumulated thermal time is at least equal to the minimum, and the minimum or maximum day temperature reaches a value outside the defined limits. If both maximum and minimum temperatures stay within their defined limits then the growing season is terminated once the maximum value for thermal time has accumulated. A subsequent growing season may be identified if temperatures return to within the defined limits, either following a period with unsuitable conditions, or directly following the first growing season. Up to three seasons will be considered within one year. As temperatures are the only criteria, the selected cropping calendar(s) is (are) not necessarily synchronized with rainfall. The potential durations of the growing seasons as calculated with the procedure are shown on Map 1. Climate zones allowing more than one growing season are indicated with black dots.

Potential production

The crop growth model used is a modified version of LINTUL (Light INTerception and UtiLization), a relatively simple model for the calculation of daily crop growth that requires relatively few parameters (Spitters 1987), which makes it suitable for a global study where available information is limited.

Potential production is mainly dependent on the PAR (Photosynthetically Active Radiation, i.e. between 400 and 700 nm wavelength) intercepted by the foliage (Monteith and Elston 1983), which is a function of incident light intensity and the fraction of light interception. The latter is described by a logistic function with the thermal time ($t = \Sigma(T_{av} - 275)$ from emergence) as the driving variable:

$$F_{LINT} = \frac{N{*}F_0{*}e^{R_0{*}t}}{N{*}F_0{*}e^{R_0{*}t}+1-N{*}F_0} \qquad (1)$$

with N plant density, F_0 initial light interception and R_0 initial relative growth rate. Senescence of the foliage is assumed to reduce light interception linearly from 1 at 1200 Kd, to 0, over a period of 600 Kd. The simulated interception of PAR is calculated from the lower bounds of the functions for growth and senescence.

The model has been parametrized for a standard crop with a density of 4 plants m^{-2}, while F_0 and R_0 have been estimated for Dutch conditions, and then adjusted for growth in warmer climates. The senescence parameters have been set for a mid-late variety in the temperate zone.

To obtain daily dry matter production in kg m^{-2}, intercepted PAR is multiplied by the light use efficiency (LUE):

$$W_{tot} = LUE * F_{LINT} * PR \qquad (2)$$

LUE has been set at a constant value of 2.9 g MJ^{-1} on the basis of experimental results worldwide, although in literature, values exceeding 4 g MJ^{-1} have been reported (Haverkort and Harris 1987; Khurana and McLaren 1982) under very favourable growth conditions.

Finally, total tuber production is calculated by multiplying total dry matter production by a harvest index (HI):

$$W_{TU} = W_{TOT} * HI \qquad (3)$$

The harvest index is derived from the average temperature during tuber growth (t_{av}) (Haverkort 1990; Bodlaender 1960). Up to 288 K, the harvest index is set at a constant value of 0.8, and decreases at higher temperatures to reach 0 at 301 K.

Water-limited production

A simple soil water balance model for a freely draining soil profile was used (Spitters and Schapendonk 1990). The maximum rooting zone is a single homogeneous layer of 0.6 m thickness. The rate of change in soil moisture in this layer follows from the balance of infiltration (I_n), percolation (P), soil surface evaporation (E) and transpiration (T):

$$\delta S = I_n - P - E - T \qquad (4)$$

All precipitation in excess of field capacity percolates from the rooting zone. Soil surface evaporation rate is calculated according to:

$$E = E_r * (1 - e^{-0.7k \times LAI}) * F_{sm2} \qquad (5)$$

E_r is the reference evapotranspiration calculated according to Makkink (1957) from radiation and temperature. F_{sm2} accounts for the drying of the top 2 cm of soil (Van Keulen and Seligman 1987), for which soil moisture content is calculated separately. Potential crop transpiration rate is dependent on LAI, assuming LAI equal to 4 for the reference transpiration:

$$T_P = E_r * F_{cr} * \frac{(1 - e^{-0.7k*LAI})}{(1 - e^{-0.7k*4})} \qquad (6)$$

Actual transpiration falls short of the potential, under limited soil moisture availability. The ratio between actual and potential transpiration is assumed to decrease linearly with soil moisture content from unity at the critical soil moisture content (Driessen 1986) to zero at wilting point:

$$\frac{T_A}{T_P} = \frac{SM - SM_{wp}}{SM_{cr} - SM_{wp}} \quad \text{with } 0 \leq \frac{T_A}{T_P} \leq 1 \qquad (7)$$

The value of the critical soil moisture content increases with increasing reference evapotranspiration.

Water stress affects two major crop growth processes: growth and leaf senescence. The level of drought stress, calculated each day, is expressed as:

$$S_d = (1 - \frac{T_A}{T_P}) - 0.2 \qquad \text{with } S_d \geq 0 \qquad (8)$$

The reduction by 0.2 accounts for the tolerance of the leaves to low degrees of stress. The multiplication factor for growth rate due to drought stress decreases linearly with increasing S_d from 1 at 0, via 0.5 at 0.5 to 0 at 0.75. The effect on leaf senescence is dependent on cumulative drought stress (i.e. the integrated value of S_d), in such a way that, due to accelerated leaf senescence, the crop canopy does not expand any further beyond a cumulative drought stress value of 10. At still higher values, light interception decreases irreversibly. Effects of water-logging were not taken into account.

Initialization

For simulation of water-limited production, the soil water balance has been initialized 80 days prior to planting, with soil moisture content set at field capacity. During these 80 days, the soil may dry out or remain wet, depending on local climatic conditions. After planting, for both potential and water-limited production, a thermal time of 200 Kd (with a base temperature of 275 K) is needed to complete emergence and start crop growth. The minimum period needed for emergence, is set at two weeks.

Results

Model evaluation

Potential yields depend on weather conditions and varietal properties, which, in this study, are considered to be fixed, ceiling yields. However, higher potential yields might be attained using varieties with an extended period of foliage production. Extending the duration of an active green foliage could be obtained through varieties with increased frost resistance and adaptation to low temperatures on one hand, or through increased adaptation to high temperatures on the other.

Alternatively, higher yields could be achieved through genotypes having a higher light use efficiency or a dry matter distribution pattern favouring tuber production over the production of foliage.

Water-limited yields may in some situations be improved by a better developed root system and improved drought resistance.

Experimental yields

'Experimental yields' are defined here as those obtained in experiments carried out by research institutes for genotypes that make best use of the available growing season (daylength adapted) and with an optimal supply of inputs such as irrigation and fertilizers, and adequate weed, disease and pest control measures. Such yields are often not economically justified, in view of the current prices of inputs and produce. Those experimental yields are close to potential yields, with differences of only a few tonnes per hectare. Such differences are partially due to weather variability, as the input levels are generally based on long-term weather expectations. The actual weather in any given experimental season may deviate from the long-term average, resulting in sub-optimal levels of inputs (nitrogen supply for instance).

Experiments occasionally may be aimed at validating a model for potential production (cf. Neeteson et al. 1987). The gap between experimental and potential yields may be more difficult to close in adverse climate conditions for potato growth than in more suitable conditions, as the quantitative effects of many factors leading to yield reduction are not fully understood.

Attainable yields

'Attainable yields' are defined in this study as the yields that the best farmers can attain when applying all available techniques economically. Attainable yields are often considerably lower than experimental yields, because the expected rates of return for inputs are so low that their application is too risky.

Figure 1 shows the relation between calculated potential yields for nine selected sites and their attainable yields. The latter were obtained in the global CIP-Canadian experiments carried out in the mid-eighties (Tai and Young, pers. comm.). National Agricultural Research Centres around the world received a set of genotypes and were invited to grow them under the best farmer's conditions. The proportion of the potential yield that was attained varied from 0.4 to 0.8, depending on the degree of adaptation of the best genotype to the environmental conditions and on unpredictable hazards, such as excessive rainfall causing an epidemic of late blight or an unexpected dry spell where the crop depended on rainfall only. Moreover, the calculation of potential yield was based on long-term average weather data and not on the actual weather data of the season in which the crops were grown. Generally, attainable yields are about 50% of the potential yields, irrespective of growing conditions. The best farmers in the Netherlands attain about 60 t ha^{-1} (fresh weight), whereas 120 t ha^{-1} is calculated as the potential yield. Similarly, a Tunisian farmer obtains around 40 t ha^{-1} where 80 t ha^{-1} is the potential.

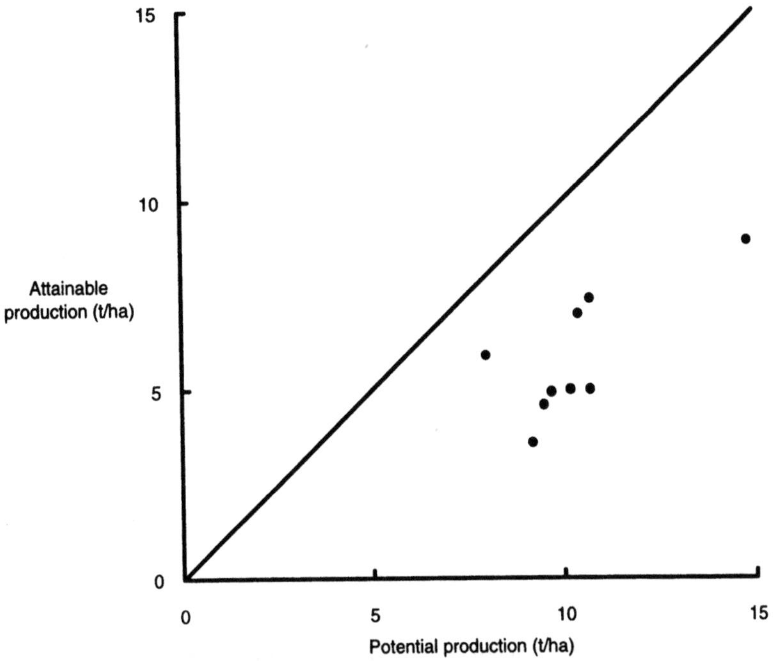

Figure 1. Attainable versus potential production.

Actual yields

'Actual yields' are defined in this study as those obtained in a region under the prevailing weather conditions under the current level of inputs with the cultivars used. In temperate climates, actual yields are often closer to attainable yields than in tropical conditions, where less inputs and sometimes less-adapted cultivars are used. Actual yields in Africa presently are about 8 t ha^{-1} and in Europe about 35 t ha^{-1} (Horton 1988).

Calculation of potential yields allows comparison with the other yield levels: experimental, attainable and actual. The ratios between these various yield levels and potential yield provide indications for the scope for yield increases and the amounts of inputs needed to realize these. The lower the ratio, the more effective better cultural practices, adequate disease and pest control measures, the use of healthy seed and the introduction of more adapted genetic material, are likely to be.

Potential yield

Potential yield of the potato crop, i.e. the maximum tuber production that can be attained for a specific variety, given planting and harvest date (Haverkort 1986), is dependent on solar radiation and temperature only. All other conditions such as water and nutrient supply and weed, pest and disease control, are assumed to be optimal. The calculated potential tuber dry matter production in the highest yielding season for the various agro-ecological zones is shown in Map 2.

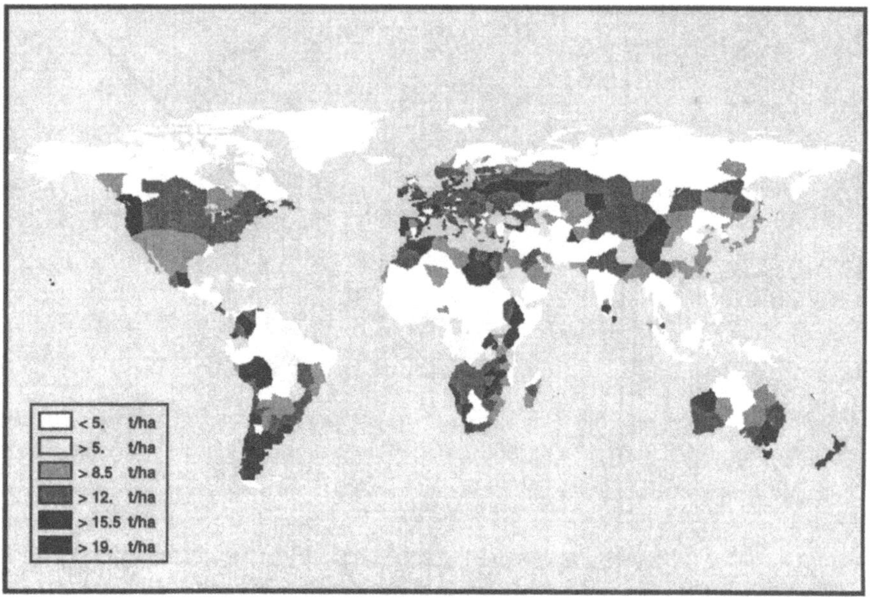

Map 2. Potential production of potato per agroclimatic zone.

Potential production in the highest yielding season ranges from 2.2 to about 25 t ha^{-1} (dry matter), with an average close to 13 t ha^{-1} (Figure 2). Assuming a dry matter concentration of 23% in the tubers, global mean potential fresh tuber yield is then estimated at about 56 t ha^{-1} per season. Yield potentials are highest in the temperate zones at latitudes between 40° and 50°, both in the northern and southern hemisphere. At higher latitudes, yield potentials are lower, because low temperatures in spring and autumn restrict the length of the growing season. Closer to the equator, high summer temperatures reduce the length of the growing season and lead to low harvest indices. Yield potentials are also high in the tropical highlands with suitable temperatures year-round.

On an annual basis potential yields in Mediterranean climates are higher than those in the temperate zones because at least two (spring and autumn) crops can be grown.

Water-limited production

Water-limited yields are dependent on precipitation and soil type, in addition to temperature and radiation. Average water-limited tuber dry matter production for each climate zone (calculated as the area-weighted mean for different soil types) is shown in Map 3. Water-limited production in the highest yielding season ranges from 0 to about 20 t ha^{-1} (dry matter) with an average close to 8 t ha^{-1} (about 35 t ha^{-1} fresh weight). The values are highest in climate zones in Canada (East), Uruguay, Ecuador, Chile, Ireland, India (South), Sri-Lanka

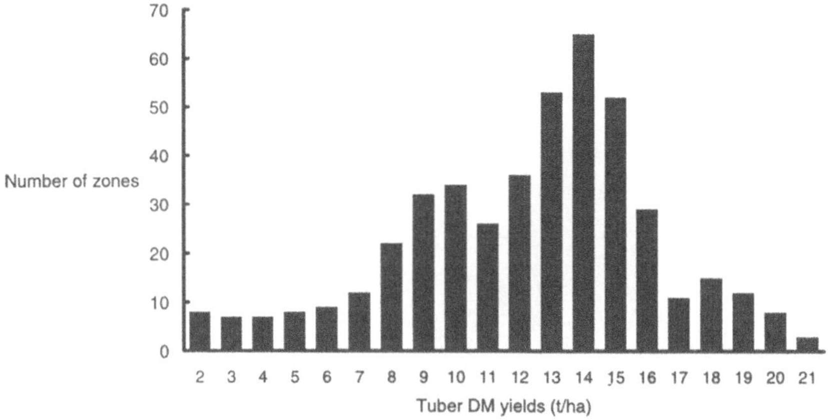

Figure 2. Frequency distribution of potential yields (dry weight) in the highest yielding season.

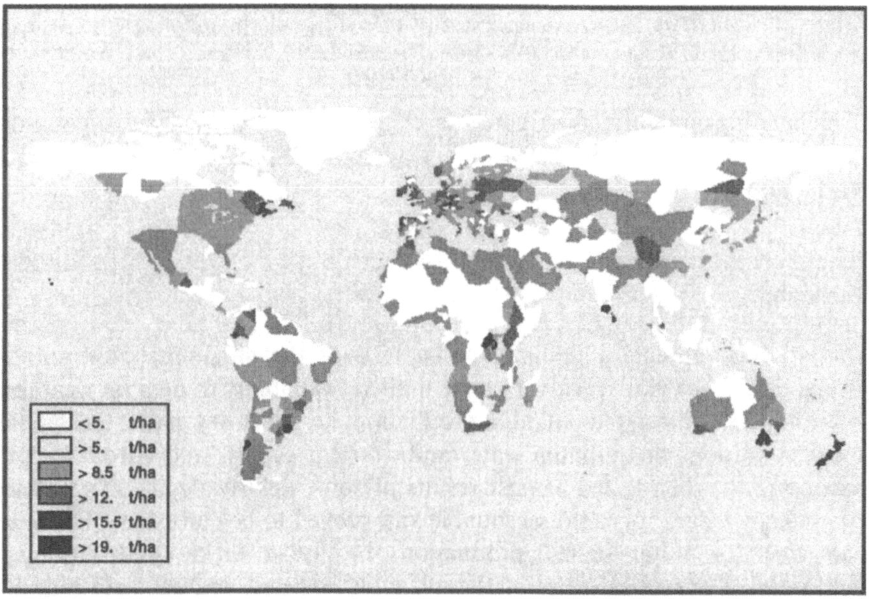

Map 3. Water-limited production of potato per agroclimatic zone.

and New-Zealand. In Latin America, yields are severely reduced by low precipitation in Colombia, Peru, Bolivia, Chile (South) and Argentina (Mid and South). In Africa, yields are reduced north of the tropic of Cancer, and in South Africa between 20° S and 30° S. In Asia, yields are severely reduced due to low precipitation in China (NW), Kazachstan (W) and Mongolia, in Australia in Western-Australia and part of New-South Wales. In climate zones with extended areas of soils with a low water holding capacity (textures coarse and fine), yields are more severely reduced.

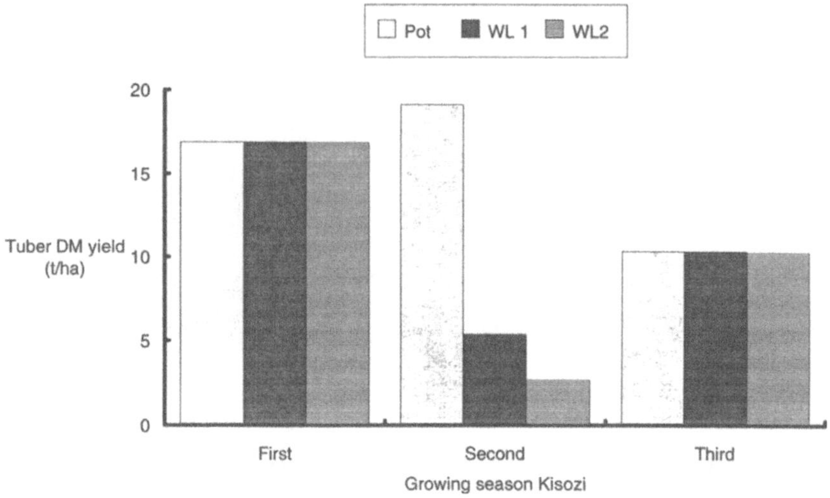

Figure 3. Potential (Pot.) and water-limited yield (WL1: soil-type with highest yield, WL2: soil-type with lowest yield), in a triple cropping in Kisozi (Burundi).

Depending on rainfall distribution over the year, the highest yielding season for potential production may not coincide with the best season for water-limited production.

Discussion

This study represents a preliminary analysis of the possibilities for potato production at the global scale. Despite limited availability of data on weather, soils and crop, calculated potential production, i.e. tuber dry matter attainable under conditions of optimum water and nutrient supply, and adequate pest management, when tested against results of some field trials and exposed to expert knowledge, appeared reasonable and proved to be consistent. This was also true for water-limited production, i.e. tuber yields under rainfed conditions. The spatial scales of the available soil and weather data blurred much of the local topography, so that cautious interpretation of the results is necessary especially in hilly and mountainous areas.

The length of the growing season was best described by assuming that crop growth is restricted to periods where daily minimum temperatures exceed 278 K and maximum temperatures do not exceed 301 K. Increasing the lower limit would exclude too many cool areas such as regions above 60° N or high mountainous areas. Lowering the maximum temperature would have resulted in exclusion of relatively warm sites and/or warm parts of the growing season, where potato is grown currently. Comparison of the results of the present study with the survey map of areas where the crop is currently grown (CIP 1991; Rhoades 1987; FAO 1981; 1980; 1978), indicates: (i) that the latter are all located within areas identified by the model as suitable, which inspires confidence in the discriminating characteristics of the model; and (ii) that the current extent of the crop is much more limited than the suitable areas identified in this study. This may be partly due to the fact that moisture availability has not been considered as a constraint, but even if the areas with water-limited yields below 5 t ha^{-1} are excluded, the suitable area is still much larger than that currently in use.

Comparison of the results of this study with those of a recent agro-ecological characterization study for the EC (Van Lanen et al. 1992), shows good agreement for potential production. Water-limited yields in the Mediterranean area are higher in our study, because of a preference for winter-crops, instead of crops planted in spring.

A number of options are possible to refine the current study:

- availability of additional long-term weather data bases from agricultural research stations could increase the location-specificity of the study.
- refining the weather data base on the basis of a digital elevation map would allow construction of interpolated weather surfaces, which might improve delineation of the suitable areas.
- a more accurate assessment of the area suitable for potato production (Map 1), including a distinction between irrigated production and rainfed production would be possible if more detailed information were available on the distribution of precipitation, irrigation possibilities and soil properties.

Production constraints such as nitrogen deficiency or late blight pressure could be included in the study to gain insight to production levels below potential or water-limited yield.

By including different crops in a rotation, optimal lengths of the growing seasons for each crop may be assessed and potential potato production for the proper length of the growing season may be calculated. Present production systems based on a combination of crops, such as potato and maize (Peru), potato and wheat (Turkey) or potato and rice (Madagascar) could be evaluated for their potential to feed increasing populations. Aiming for longer or shorter growing seasons of one crop or the other will give information on which system may produce the largest amount of food annually.

Refinement of the characterization may further increase the relevance of the simulation results for strategic decisions for research planners and policy makers. Additional information on the nutritional value of the various commodities may be an asset to arrive at a more balanced diet.

Further analysis, especially on the effects of adaptation of the crop to higher and lower temperatures may be useful for breeding purposes. Decreasing the lower limit of 278 K will only be realistic if increased frost resistance were to be incorporated in the cultivars grown in areas with a cold start or end to the growing season. Simulation can show what the global impact of frost resistance will be and which areas will benefit most from such a trait. A combination of adaptation to low and high temperatures would be beneficial in continental climates with frost in winter and rapidly increasing temperatures in spring.

Agro-ecological characterization may also be useful in estimating the influence of global climate change. If temperatures rise by 2 K, what would be the influence on the distribution of the potato crop? In the tropics the crop will move to higher elevation and in the temperate zones to higher latitudes. In the latter case, cultivars adapted to longer photoperiods would be needed.

References

Bartholomew J C, Christie J H, Ewington A, Geelan P J M, Lewisobe H A G, Middleton P, Winkleman H (Eds.) (1988) The Times Atlas of the World. Times Books Limited, London, UK.

Black J N, Bonython C W, Prescott G A (1954) Solar radiation and the duration of sunshine. Quarterly Journal of the Royal Meteorological Society 80:231–235.

Bodlaender K B A (1960) De Invloed van de Temperatuur op de Ontwikkeling van de Aardappel. Mededeling 112, Instituut voor Biologisch en Scheikundig Onderzoek van Landbouwgewassen, Wageningen, The Netherlands.

CIP (1991) Meeting the Challenge. The International Potato Center's Strategy for the 1990s and Beyond. International Potato Center, Apartado 5969, Lima, Peru.

Driessen P M (1986) The water balance of the soil. Pages 76–116 in Van Keulen H, Wolf J (Eds.) Modelling of Agricultural Production: Weather, Soils and Crops, Simulation Monographs. Pudoc, Wageningen, The Netherlands.

FAO (1978) Report on the Agro-Ecological Zones Project. Vol. 1. Methodology and Results for Africa, FAO World Soil Resources Report Nr. 48/1.

FAO (1980) Report on the Agro-Ecological Zones Project. Vol. 4. Results for South-East Asia, FAO World Soil Resources Report Nr. 48/4.

FAO (1981) Report on the Agro-Ecological Zones Project. Vol. 3. Methodology and Results for South and Central America, FAO World Soil Resources Report Nr. 48/3.

FAO/UNESCO (1974–1981) Soil Map of the World 1:5,000,000. Vol. I–X. FAO, Rome, Italy.

Frère M, Popov G F (1979) Agrometeorological Crop Monitoring and Forecasting. FAO Plant Production and Protection Paper, No. 17. FAO, Rome, Italy.

Haverkort A J (1986) Yield levels of potato crops in central Africa. Agricultural Systems 21:227–235.

Haverkort A J (1990) Ecology of potato cropping systems in relation to latitude and altitude. Agricultural Systems 32:251–272.

Haverkort A J, Harris P M (1987) A model for potato growth and yield under tropical highland conditions. Agricultural and Forest Meteorology 39:271–282.

Horton D (1988) Underground Crops. Long-Term Trends in Production of Roots and Tubers. Winrock International, Morrilton, USA.

Khurana S C, McLaren J S (1982) The influence of leaf area, light interception and season on potato growth and yield. Potato Research 25:329–342.

Leemans R, Cramer W P (1991) The IIASA Database for Mean Monthly Values of Temperature, Precipitation and Cloudiness of a Global Terrestrial Grid. RR-91-18, International Institute of Applied Systems Analysis, Laxenburg, Austria.

Makkink G F (1957) Testing the Penman formula by means of lysimeters. Journal of the Institute of Water Engineers 11:277–288.

Monteith J L, Elston J (1983) Performance and productivity of foliage in the field. Pages 499–518 in Dale J, Milthorpe F L (Eds.) The Growth and Functioning of Leaves. Cambridge University Press, Cambridge, UK.

Müller M J (1987) Handbuch ausgewählter Klimastationen der Erde. 5. Heft: Forschungsstelle Bodenerosion Mertesdorf (Ruwertal), Universität Trier, Germany.

Müller M J (1982) Selected Climate Data for a Global Set of Standard Stations for Vegetation Science. Dr. W. Junk Publishers, The Hague, The Netherlands.

Neeteson J J, Greenwood D J, Draycott A (1987) A model to predict yield and optimum nitrogen fertiliser application rate for potatoes. Proceedings Fertilizer Society 262:3–31.

Reinds G J, De Koning G H J, Bulens J D (1991) Crop Production Potential of the Rural Areas within the European Communities. Vol. V: Soils, Climate and Administrative Regions. Technical Working Document, Netherlands Scientific Council for Government Policy, The Hague, The Netherlands.

Rhoades R E (1987) Potato Production Zones of Developing Countries. Computerized Maps. International Potato Center, Lima, Peru.

Spitters C J T (1987) An analysis of variation in yield among potato cultivars in terms of light absorption, light utilization and dry matter partitioning. Acta Horticulturae 214:71–84.

Spitters C J T, Schapendonk A H C M (1990) Evaluation of breeding strategies for drought tolerance in potato by means of crop growth simulation. Plant and Soil 123:193–203.

Stol W, De Koning G H J, Kooman P L, Haverkort A J, Van Keulen H, Penning de Vries F W T (1991) Agro-Ecological Characterization for Potato Production. Report 155, CABO-DLO, P.O. Box 14, 6700 AA Wageningen, The Netherlands.

Van der Zaag D E, Horton D (1983) Potato production in world perspective with special reference to the tropics and sub-tropics. Potato Research 26:323–362.

Van Keulen H, Seligman N G (1987) Simulation of Water Use, Nitrogen Nutrition and Growth of a Spring Wheat Crop. Simulation Monographs, Pudoc, Wageningen, The Netherlands.

Van Lanen H A J, Van Diepen C A, Reinds G J, De Koning G H J, Bulens J D, Bregt A K (1992) Physical land evaluation methods and GIS to explore the crop growth potential and its effects within the European Communities. Agricultural Systems 39:308–329.

Zobler L (1986) A World Soil Profile for Global Climate Modelling. NASA Technical Memorandum 87802.

24. Future focus of physiology and modelling of potatoes

C.D. VAN LOON

Research Station for Arable Farming and Field Production of Vegetables P.O. Box 430, 8200 AK Lelystad, The Netherlands

Abstract. The need for research on the physiological aspects of potato in support of modelling is approached from the point of view of the potato industry. Against the background of increasing public concern about the environment and demand for healthy food of excellent quality, major buyers of potatoes are exerting increasing influence on production methods (Total Quality Management-TQM).

It is concluded that models have the potential to be useful tools enabling the farmer to deal with TQM. However, insufficient is known about important characteristics of tuber quality such as dry matter concentration, blackspot,fry colour and nitrate concentration to be able to include these aspects in detailed crop models. The same is true for the population dynamics of some of the main diseases and pests and for the behaviour of nitrogen in the soil.

Some of the prerequisites for successfully introducing models into farming practice are discussed.

Introduction

The development of crop modelling in potato has parallelled that of other crops: first, potential production was calculated under ideal conditions for growth; this was followed by yield simulation under various yield-limiting factors and other applications. A comparison between the programmes of the 1st and 2nd Potato Modelling Conferences underlines this observation. In 1990 mainly general potato models were presented, but at this Conference there are models describing in detail the effect of drought, temperature and day-length, CO_2 and nutrients on yield. Recently, more attention has been focused on modelling the effects of diseases and pests on potato yield. Some of these models are intended for farmers to use.

In this contribution the 'Future focus of the physiology and modelling of potato' will be approached from the point of view of the potato industry including breeders, producers (farmers) and users (consumers, processing industry) of potatoes. The questions that arise are: Which areas of physiology need further exploration to solve the problems of the potato industry? And, in which cases would modelling provide a useful tool for farmers, breeders,researchers and policymakers? These aspects will be discussed against

A.J. Haverkort and D.K.L. MacKerron (eds.), Potato Ecology and Modelling of Crops under Conditions Limiting Growth, 373–380.

the background of the current constraints under which arable farmers operate in most European countries.

Environment and healthy food

Public concern about the environment and demand for healthy food are growing in many countries. This will have a major impact on the production of agricultural products including potatoes. For example, arable farmers in the Netherlands are legally required to reduce their annual use of pesticides by the year 2000 to 50% of the mean amount used annually in 1986–1988 (Van Loon et al. 1994). Meanwhile, consumers are increasingly demanding certain external and internal quality characteristics in potatoes. Potatoes should not only have an attractive appearance and be clean and undamaged but they should not be contaminated by toxic compounds such as residues of pesticides and nitrate. Since they are concerned about the environment, consumers are also interested in the quality of the production process, and wish to have environmentally-friendly production methods.

Against this background supermarket chains (and the processing industry) are aiming for quality products that are produced in an environmentally-friendly way. In order to achieve these aims, supermarket chains and processing companies are forcing cooperatives and merchants to exert more influence on production methods. They tell the farmer how much fertilizer he may use, which pesticides are permitted and the maximum amounts of these that can be used.This development is called 'supervised production' or 'total quality management' (TQM). Concomitantly, the farmer's income is under pressure and, therefore, farmers are being forced to produce at a lower cost. In summary, the farmer is faced with the problem of how to produce potatoes with a number of desired quality characteristics in an environmentally-friendly way and as profitably as possible. Models would appear to have great potential for indentifying the best combination of input factors and management measures allowing a good yield, the desired product quality, an environmentally-friendly production system and a low cost.

Potato physiology

To be able to build useful models we must have sufficient information at our disposal; for example, a sound theoretical basis for the physiological processes governing the main quality characteristics of potato and for the population dynamics of the crop's main diseases and pests. However, this is certainly not the case for quality characteristics such as dry matter concentration, blackspot, fry colour of french fries and potato chips (crisps), after cooking blackening and the NO_3 concentration in tubers. Our knowledge is particularly deficient in the effects of the environment on the establishment of quality characteristics, and the effect of the storage period on quality.

Tuber size

Tuber size distribution is an aspect of quality in both seed and ware potatoes. Seed growers would find it particularly useful to be able to predict the optimum harvest time for economic yield and tuber size distribution. Several attempts have been made to model the development of tuber size distribution on the basis of the results of an early harvest (Sands and Regel 1983; MacKerron and Marshall 1990). But as far as I know this has not yet resulted in models that are used in practice by farmers. It is even more difficult to make such a model on the basis of plant or stem density. Here, too, knowledge is lacking about the effects of environmental conditions on tuber initiation and the growth of individual tubers.

Nitrogen

Nitrogen is a growth factor with major implications for yield, product quality and the environment. For example, an excess of nitrogen adversely affects several quality characteristics of potatoes including dry matter concentration, fry colour, enzymatic and non-enzymatic blackening of cooked and pre-fried potatoes, NO_3 concentrration, tuber malformation and hollow heart (Van Loon 1989). Optimum N supply to the crop is, therefore, extremely important for product quality. However, such an optimal supply is difficult to achieve, particularly on soils rich in organic matter. Soil processes carried out by micro-organisms, including mineralization, immobilization and denitrification, combined with the effect of leaching result in large differences in N availability for the crop. These soil processes are influenced by temperature, moisture content and aeration of the soil. In the Netherlands differences in N availability greater than 150 kg ha^{-1} have been found between fields during the growing season (Neeteson et al. 1989). This emphasises the urgent need for a reliable model describing the behaviour of N in soil under various conditions. Some results of work already carried out on this subject have been presented during this conference, but the problem is so complex that much work still remains to be done.

Diseases and pests

There is increasing interest in using models as tools in disease and pest control. Some examples have been presented during this conference. Models can be helpful in determining the need for chemical or biological control of diseases and pests as damage is not always sufficient to merit control. Models should include quantitative estimates of reductions in yield and quality in relation to levels of infestation (MacKerron et al. 1990). These, in combination with the cost of control allow damage thresholds to be set.

There is also an urgent need for models that reliably forecast the main potato diseases such as late blight, early blight and *Rhizoctonia solani*. A number of forecasting models are already available for late blight. They include the 'Negativ-Prognosis' (Schrödter and Ulrich 1967), which predicts the start of an epidemic, and models defining risk periods for infection e.g. Blitecast (MacKenzie 1981), Milsol (Duvauchelle 1991) and Prophy (Bus et al. 1993). However, the last-mentioned group of models are often insufficiently reliable to assure both a good protection of the crop and no more fungicide applications than are strictly necessary.

Rhizoctonia. This widespread soilborne disease can be controlled effectively by fungicides. Soil treatment, often applied in the Netherlands prior to the planting of a seed crop, requires large amounts of fungicide and is expensive. The need for fungicides depends on very many factors, such as cultivar, level of infection of seed and soil, vitality of the fungus on the seed, cropping frequency of potatoes, soil type, seed preparation etc. Lamers and Campmans (1993) have supplied a basis for a *Rhizoctonia* model that could be useful to farmers and that is currently being evaluated.

Potato cyst nematode. For regions where the potato cyst nematode, especially *Globodera pallida*, is a problem, notably in the Netherlands, there is an urgent need for a simulation model indicating population dynamics under various conditions e.g. initial population, soil type, cultivars, cropping frequency etc. Progress has been made in this field, as has been shown by Been and Seinhorst (this volume) and by the prototype of a decision support system (TERRA) demonstrated by Jansen and Nijboer.

Aphids, Colorado beetle, tuber moth. In some countries farmers need models to forecast the population dynamics of aphids, Colorado beetle and potato tuber moth (*Phthorimaea operculella*). Nemecek (1993) has shown an example of such a model for aphids.

Other diseases for which a forecasting model would be valuable in certain geographical regions are early blight (*Alternaria solani*) and brown rot (*Pseudomonas solanacearum*).

Models to be used in breeding

Breeders can be of great help to farmers by developing appropriate cultivars that enable them to meet quality and environmental requirements. However, so many properties have to be combined in a modern cultivar that the task is daunting.

Recently, Haverkort et al. (1994) showed that modelling can improve the efficiency of breeding. They presented models that can be used in developing selection criteria for drought tolerance and resistance to late blight and potato cyst nematode. The selection criteria for drought tolerance include ability to maintain LAI under dry conditions, and the soil water potential at which leaf expansion is constrained. The selection criteria for increased resistance to *Phytophthora infestans* are a low infection efficiency and a slow rate of lesion growth. But there are still many selection criteria for resistance and tolerance to biotic and abiotic stresses that need to be developed and for which modelling could be a useful technique.

Total Quality Management (ware potatoes)

Total Quality Management forces the farmer to make well balanced decisions about input and management strategy and models appear to be very valuable tools for this. Rather than use one very comprehensive model it seems more likely that farmers or advisers will use a number of models.

Planning the production process

A basic model for planning a sound basis for the production process, taking into account the following groups of boundary conditions:

- quality specifications for a number of characteristics;
- environmental requirements;
- cost price;
- whether a certain field is infested with certain diseases and pests;
- maturation time of the crop.

The model should indicate the best cultivar(s) to be grown, the preparation of seed, pre-planting fertilization and the strategy for disease and pest control.

One of the drawbacks of optimization of for example, fertilization is the spatial variation of soil and soil fertility in a field. Global Positioning Systems (GPS) could be of great help in overcoming this problem (Long et al. 1991). GPS assist precision farming, since they enable farmers to locate deviant spots in the field (e.g. in terms of soil fertility). The farmer is then able to adjust the amount of fertilizer applied in accordance with this spatial variation, and thus can fertilize the entire field optimally. This is not only interesting from the point of view of cost effectiveness but it also benefits tuber quality and the environment.

Crop management during the growing season

The most important topics requiring attention during the growing season are N fertilization, irrigation and late blight control.

For the first two aspects, useful information could be generated by models simulating:

- plant growth, absorption of N by the crop and the amount of N available to the crop in the soil; such a model could indicate whether the crop needs additional N;

- evapotranspiration based on crop development and meteorological conditions; a comparison with the amount of plant-available water in the soil would show the need for irrigation.

For late blight control a forecasting model should define periods with a high risk of infection and indicate the best choice of fungicide.

Post-harvest behaviour of the product

Based on the expected post-harvest behaviour of a certain batch of potatoes, the desired quality level at the end of the storage period and the length of the storage period, a model could indicate the required storage conditions.

An alternative model could indicate, the most appropriate destination of the crop, and also the storage temperature and the maximum length of the storage period, on the basis of cultivar, the initial quality of the product at harvest and the growing conditions.

Within TQM or otherwise, models could also provide a prognosis of supply and demand. This is vitally important for merchants positioned in between growers and the major buyers such as supermarket chains and the processing industry.

From the foregoing it can be concluded that there is a potential market for models as planning and management tools in the potato industry. But is it realistic to assume that these kinds of models can be developed in such a way that they can indeed aid the farmer? And will farmers accept them?

To date, farmers' acceptance of crop management models has not been overwhelming (Seligman 1990). This might be because the present generation of farmers did not grow up with the computer. But is this the only cause? There appear to be three conditions for the success of such models:

1. Models should be robust: the results should be reliable under all kinds of (field) conditions.
   An important question in this respect is, will science be able sufficiently to simulate the often-very-complicated mechanisms of processes and – even more difficult – the effect of the environment on these mechanisms. Much work remains to be done here.
2. The model should match the wishes of the users and be user-friendly. Models that have to be used several times during a growing season appear to be more successful than models that can be used only once or twice. The users must be involved in the process of development of the model at an early stage.
3. The model should generate additional revenue for the farmer by reducing production costs and/or by enhancing the value of the product.
   A thorough evaluation of each model before it is marketed should make clear that the user is indeed receiving value for money.

These conditions should be given careful consideration before much effort is put into models intended for farmers. In a small country such as The Netherlands, with only a limited number of potato farmers, the small market for potato models is a problem. However, international co-operation, although not easy, could remedy this. Clearly, the development of successful models for the potato industry is a challenge for future agricultural research.

References

Bus C B, Schepers H T A M, Ridder J K (1993) Beheersing van *Phytophthora infestans* bij gebruik van minder gewasbeschermingsmiddelen. Pages 82–95 in Van Loon C D (Ed.) Themadag Aardappelen. PAGV, Lelystad, The Netherlands.

Duvauchelle S (1991) Lutte contre le mildiou de la pomme de terre en France: modèles de prévision pour les avertissements agricoles et stratégie d'utilisation des spécialités contenant des matières actives systémiques face à la résistance. EPPO Bulletin 21:49–53.

Haverkort A J, Kooman P L, Van Oijen M A (1994) Simulatiemodellen als Hulpmiddel bij Veredeling en Rassenkeuze van de Aardappel. Rapport 12, AB-DLO, Wageningen/Haren, The Netherlands.

Lamers J G, Campmans H E (1993) Een nieuw adviessysteem voor de geïntegreerde bestrijding van *Rhizoctonia solani* in aardappelen. IKC-informatie Akkerbouw en Vollegrondsgroente 4 (2):51–60.

Long D S, DeGloria S D, Galbraith J M (1991) Use of the global positioning system in soil survey. Journal of Soil and Water Conservation 46 (4):293–297.

MacKenzie D R (1981) Scheduling fungicide applications for potato late blight with Blitecast. Plant Disease 65:394–399.

MacKerron D K L, Greenwood D J, Marshall B, Rabbinge R, Schöber B (1990) Forecasting systems for the potato crop. Pages 85–106 in Proceedings 11th Triennial Conference EAPR, Edinburgh, UK.

MacKerron D K L, Marshall B (1990) Applications of a model of tuber size distribution. Pages 129–130 in Abstracts of Conference Papers and Posters, 11th Triennial Conference EAPR, Edinburgh, UK.

Neeteson J J, Greenwood D J, Draycott A (1989) A dynamic model to predict yield and optimum nitrogen fertilizer application rate. Pages 82–109 in Neeteson J J (Ed.) Assessment of Fertilizer Nitrogen Requirement of Potatoes and Sugar Beet. Ph.D. Thesis, Wageningen, The Netherlands.

Nemecek T (1993) The Role of Aphid Behaviour in the Epidemiology of Potato Virus Y: A Simulation Study. Ph.D. Thesis, Swiss Federal Institute of Technology, Switzerland.

Sands P J, Regel P A (1983) A model of the development and bulking of potatoes. V. A simple model for predicting graded yields. Field Crops Research 6:25–40.

Schrödter H, Ulrich J (1967) Eine mathematisch-statistische Lösung des Problems der Prognose von Epidemien mit Hilfe meteorologischer Parameter, dargestallt am Beispiel der Krautfäule (*Phytophthora infestans*). Agricultural Meteorology 4:119–135.

Seligman N G (1990) The crop model record: promise or poor show? Pages 249–280 in Rabbinge J et al. (Eds.) Theoretical Production Ecology Reflections and Prospects. Pudoc, Wageningen, The Netherlands.

Van Loon C D (1989) Enkele kwaliteitsaspecten van consumptie-aardappelen in relatie tot bemesting. Aardappelwereld 9:18–20.

Van Loon C D, Bedin P, Reust W (1994) Vers une production de pomme de terre respectueuse de l'environnement. Pages 36–56 in Proceedings 12th Triennial Conference of the EAPR, Paris, France.

Current Issues in Production Ecology

1. P.A. Leffelaar (ed.): *On Systems Analysis and Simulation of Ecological Processes.* With Examples in CSMP and FORTRAN. 1993
ISBN 0-7923-2434-X; Pb 0-7923-2435-8.
2. J. Goudriaan and H.H. van Laar: *Modelling Potential Crop Growth Processes.* Textbook with Exercises. 1994 ISBN 0-7923-3219-9; Pb 0-7923-3220-2
3. A.J. Haverkort and D.K.L. MacKerron (eds.): *Potato Ecology and Modelling of Crops under Conditions Limiting Growth.* Proceedings of a Conference (Wageningen, The Netherlands, May 1994). 1995 ISBN 0-7923-3412-4

KLUWER ACADEMIC PUBLISHERS – DORDRECHT / BOSTON / LONDON

GPSR Compliance

The European Union's (EU) General Product Safety Regulation (GPSR) is a set of rules that requires consumer products to be safe and our obligations to ensure this.

If you have any concerns about our products, you can contact us on ProductSafety@springernature.com

In case Publisher is established outside the EU, the EU authorized representative is:

Springer Nature Customer Service Center GmbH
Europaplatz 3
69115 Heidelberg, Germany

Batch number: 10755329

Printed by Printforce, the Netherlands